ROUTLEDGE HANDBOOK OF TEA TOURISM

The *Routledge Handbook of Tea Tourism* provides comprehensive and cutting-edge insights into global tea tourism. With contributions from leading scholars and experts across 19 countries, it demonstrates the interdisciplinary nature and breadth of topics associated with global tea tourism.

Tea is deeply connected to tourism through both travel and consumption. For host communities it provides an opportunity for diversification from the production and/or serving of tea while sharing cultural traditions and improving livelihoods. The *Handbook* is organized into five parts, with an introduction and epilogue, and the first part begins with an overview of historical and contemporary perspectives on the foundations of tea tourism. It digs into the roots of such tourism in China, the relationship of wild tea to indigenous tourism in Vietnam, heritage railways to tea tourism, and tea tourism in Africa. The second part examines sustainable tea tourism, with examples from Thailand, Turkey, Sri Lanka and India. The third part explores the management and marketing of tea tourism, highlighting tools and techniques for development and the impact of social media on the tea tourism experience. It draws on examples of tea tourism experience in diverse settings, such as the English tea room, a pearl milk tourism factory in Taiwan and a hot spring tea destination in Japan. The fourth part provides perspectives on innovation and practice in tea tourism, such as gastronomical tea tourism in Turkey, Japan and Thailand; tea cafés and community diversification in Japan; the role of GIAHS designation in tea tourism; and tea tour guiding in Iran. Finally, the fifth part provides insights on resilience in tea tourism, examining topics such as human–wildlife conflicts and the impact of the COVID-19 pandemic on the sector in both Asia and Europe.

This *Handbook* provides a valuable resource for students and researchers, presenting a rich collection of theoretical and empirical insights, an agenda for future directions in the field and end-of-chapter discussion questions. It also serves as a useful tool for key stakeholders, aiming to increase interaction between academia and industry, encouraging the development of sustainable responsible tea tourism that benefits local communities on a global basis.

Lee Jolliffe is visiting professor at Ulster University, UK. She has written extensively on heritage tourism topics including tea tourism through her 2007 edited book, *Tea and Tourism: Tourists, Transitions and Transformations*. In researching tea and tourism she has visited tea gardens and estates in many countries, completing a Japanese Tea Master Course in Japan and the World Tea Tours Darjeeling Immersion program in India.

M.S.M. Aslam is professor in tourism management, Department of Tourism Management at Sabaragamuwa University of Sri Lanka, and Editor in Chief, *Asian Journal of Management Studies*.

He has carried out research on tea and tourism from different perspectives published individually and jointly in international journals and conference proceedings. He initiated and is working with Sabaragamuwa University of Sri Lanka to establish the International Tea Tourism Institute (ITTI). He works with public and private organizations to develop tea tourism in Sri Lanka.

Amnaj Khaokhrueamuang is an associate professor in tourism at the School of Management and Information, University of Shizuoka, Japan. His research interest focuses on rural tourism-related issues associated with community development, agricultural extension, culture and heritage. Tea tourism is one of his research focuses, particularly in the international exchange of tea-related business between Japan and Thailand, which expects to provide the lessons learned to global tea industry communities.

Li-Hsin Chen is an assistant professor, International Masters' Program of Tourism and Hospitality, National Kaohsiung University of Hospitality and Tourism. She is Associate Editor of the *Journal of Responsible Tourism Management* and also serves as editorial board member for other journals in tourism and hospitality. Her research interests include coffee and tea tourism, bicycle tourism, dual attitudes model, indirect measurement, experiencescapes, service design and multisensory marketing.

ROUTLEDGE HANDBOOK OF TEA TOURISM

Edited by Lee Jolliffe, M.S.M. Aslam,
Amnaj Khaokhrueamuang and Li-Hsin Chen

Routledge
Taylor & Francis Group

LONDON AND NEW YORK

Designed cover image: © Getty Images

First published 2023
by Routledge
4 Park Square, Milton Park, Abingdon, Oxon OX14 4RN

and by Routledge
605 Third Avenue, New York, NY 10158

Routledge is an imprint of the Taylor & Francis Group, an informa business

British Library Cataloguing-in-Publication Data
A catalogue record for this book is available from the British Library

ISBN: 9781032053233 (hbk)
ISBN: 9781032053240 (pbk)
ISBN: 9781003197041 (ebk)

DOI: 10.4324/9781003197041

Typeset in Bembo
by Apex CoVantage, LLC

CONTENTS

Contents

CASE STUDIES

FIGURES

TABLES

CONTRIBUTORS

R.S.S.W. Arachchi graduated from Sabaragamuwa University of Sri Lanka before receiving his master's degree from Colombo University, Sri Lanka, and his PhD from the Management and Science University in Malaysia. His major teaching and research areas are ecotourism, community-based tourism, sustainable tourism development and homestay tourism in Asian countries.

M.S.M. Aslam is professor in tourism management, at the Department of Tourism Management, Sabaragamuwa University of Sri Lanka, Sri Lanka. He has published on tourism and sustainability, including tea tourism and advised on many related tourism development projects in Sri Lanka.

Hartwig Bohne is professor of international hotel management, lecturing in Germany and Austria, external examiner for Hotel Management of the TU Dublin, Ireland, and a passionate author and researcher in European tea culture and consumption traditions.

Wayne Buente is a professor at the University of Hawaii at Mānoa, School of Communications. His research examines the interplay between society and information and communication technologies. His recent work on the social media platform Instagram examined the use of betel nut.

Kadir Çetin is an assistant professor at Burdur Mehmet Akif Ersoy University, Turkey. He has an academic background in gastronomic studies. He has published academic articles on the history and culture of gastronomy.

Li-Hsin Chen is an assistant professor at the National Kaohsiung University of Hospitality and Tourism, Taiwan. Her research interests include tourism experience and experienscape.

Piyaporn Chueamchaitrakun is an assistant professor in food science and technology at the School of Agro-Industry, and the Head of the Tea and Coffee Institute at Mae Fah Luang University, Thailand. She researches tea processing and innovation, sensory evaluation and food product development.

Chandan Datta is a senior researcher at Vivekananda College for Women under University of Calcutta of West Bengal in India. He submitted his PhD thesis at University of Calcutta. His current research interests are in human–wildlife conflicts, environmental crime, tea tourism and wildlife tourism in the Dooars of India.

Belinda Davenport is the owner of Davenports Tea Room, a quintessential English tea room in the UK. She has an academic background in travel and tourism, an MA in tourism management and a keen interest in heritage, tea and what influences people. Published papers include "Audience Development Survey for Manchester Science and Industry Museum: Cutting Edge Research" and "What Influences Young People to Study Travel and Tourism in Cheshire".

G.V.H. Dinusha is a lecturer in the Department of Tourism Management of Sabaragamuwa University of Sri Lanka. He obtained his BSc in 2016 and is currently reading for an MBA at the same university. His teaching and research areas are concentrated in intercultural communication for tourism, sociology of tourism, tourism economics, greening hospitality and travel agency operations.

Saurabh Kumar Dixit is associate professor and Head, Department of Tourism and Hotel Management, North-Eastern Hill University, Shillong, India. His research interests include consumer behaviour, gastronomic tourism and experiential management and marketing in hospitality and tourism contexts. He has published 15 books, including *The Routledge Handbook of Consumer Behavior in Hospitality and Tourism* (2017), *The Routledge Handbook of Gastronomic Tourism* (2019), *The Routledge Handbook of Tourism Experience Management and Marketing* (2020), and *The Emerald Handbook of Luxury Management for Tourism and Hospitality* (2021).

Hilary du Cros is an honorary research associate of the University of New Brunswick, Canada, and concurrently an adjunct professor at Western Sydney University. She has over 35 years' experience focusing on the Asia Pacific region regarding heritage, arts and tourism management studies.

Baghva Erathna is a lecturer in the Sri Lanka Technological Campus and following a postgraduate diploma in tourism economics and hotel management in the Department of Economics, University of Colombo, Sri Lanka.

Emre Erbaş is associate professor at Burdur Mehmet Akif Ersoy University, Turkey. With an academic background in tourism management studies, he has written extensively on tourism management, including a focus on managerial capabilities.

Imali N. Fernando is a professor in management at the Faculty of Management, Uva Wellassa University of Sri Lanka, and former dean of the faculty. She is mainly interested in marketing, hospitality and tourism marketing research disciplines, and her teaching area is in management.

Krishantha Ganeshan is a front office executive at 98 Acres Resort and Spa. He has engaged with research work on tourism management, sustainable tourism and destination management.

P. Gayathri is an assistant lecturer of tourism and cultural resources management at University of Kelaniya, Sri Lanka. She has completed a BA special degree in tourism and cultural resources management at University of Kelaniya and enrolled in the post-graduate diploma in tourism economic and hotel management in University of Colombo, Sri Lanka. She is interested in tourism management, destination management and Chinese language.

Sousuke Goto majored in tourism at the University of Shizuoka, Japan. His graduate research focused on developing wellness tourism routes for senior markets. He also researched creating tea tourism routes by linking tea cafés with community resources.

Yanting Gu is a teaching assistant at the Yunnan University of Finance and Economics. Her research interests are in the area of tourism education and psychology in tourism.

Unathi Sonwabile Henama is senior lecturer at the Department of Tourism and Events Management at the Central University of Technology, Free State in South Africa. He has a PhD from Mid-Sweden University and a master's degree from the University of the Free State. He has written numerous articles. As the leading tourism commentator in South Africa, his views are highly sought after by TV, print and online news outlets.

H.M.J.P. Herath is a lecturer at the University of Vocational Technology, Sri Lanka. She obtained her bachelor's in hospitality, tourism, and events management at Uva Wellassa University. Her research interests are specifically in tourism, sustainability, community empowerment and destination development.

Kyoko Ishigami is the World Green Tea Association's assistant director. Since 2020, she has served as the secretariat of the Council for the Promotion of the World Agricultural Heritage "Traditional Tea-grass Integrated System".

Kunihiko Iwasaki is professor in marketing and Director of the Center for Regional Management Studies, University of Shizuoka, Japan. He writes marketing books related to regional development through tourism and agriculture. His textbook on branding tourism destinations won an award from the Japan Institute of Tourism Research.

Lee Jolliffe is visiting professor at Ulster University, U.K. She has written on heritage tourism, including tea tourism through *Tea and Tourism: Tourists, Transitions and Transformations* (2007). Researching tea and tourism she has visited tea gardens and estates in many countries.

Daisuke Kanama is a professor at the Institute of Transdisciplinary Sciences, Kanazawa University, Japan. His research focuses on innovation management including start-up ecosystems, university-industry collaboration, intellectual properties, motives, and incentives of research and development professionals.

Kannapat Kankaew is assistant professor and Deputy Dean of Research and Academic Services at the College of Hospitality Industry Management, Suan Sunandha Rajabhat University, Thailand. His research interests include human resource management, learning and development, hospitality and tourism, air transport and business management. He serves as an editorial board member, reviewer and associate editor in various journals, as well as a book editor.

Annette Kappert is a highly experienced educator who defines herself as a neo-generalist through her ability to create networks between international hospitality education, educational leadership and strategic management. Her research interests include business pedagogy, diversity and inclusive learning environments, decolonial feminism, occupational identity and autoethnography.

A.C.I.D. Karunarathne is a senior lecturer at Uva Wellassa University, Sri Lanka, with an academic background in tourism and hospitality and event management. She extends her expertise in sustainable tourism development to hospitality and tourism marketing, event management, and women and marginalized community groups in tourism.

Amnaj Khaokhrueamuang is an associate professor in tourism at the School of Management and Information, University of Shizuoka, Japan. His research interest focuses on rural tourism-related issues associated with community development, agricultural extension, culture and heritage.

Koichi Kimura is completing his bachelor's degree in management and information at the University of Shizuoka, Japan. Innovation in tea tourism products is one of his interests. He studies branding tourism destinations through glamping development in the tea-growing area of Kawanehon town, Shizuoka.

Balvinder Kler is senior lecturer in tourism at Universiti Malaysia Sabah. Her research explores people-place relationships to uncover meanings and attachments for destination communities experiencing tourism development. She has published on niche tourism markets.

K.M.B.S.Y. Kulasekara is an english instructor in the University of Peradeniya. Currently, he is reading a postgraduate diploma in tourism economics and hotel management in the Department of Economics, University of Colombo, Sri Lanka.

Jianming Li is an assistant research fellow at the University of Yunnan Minzu University. His current research interests include China Southeast Minority Traditional Political System with significant publications in the field.

Lebogang Matholwane Mathole is a lecturer in tourism management, business management and marketing at the Tshwane University of Technology, South Africa. She holds a master's degree specializing in digital marketing. She is currently pursuing interdisciplinary doctoral research focusing on entrepreneurship, consumer psychology and local economic development. She has presented papers at several conferences and is growing as a researcher by publishing in journals and book chapters.

Lehlohonolo Gibson Mokoena is lecturer, Department of Tourism and Event Management at Central University of Technology, Free State, South Africa. He holds a master's degree in tourism and hospitality focusing on medical tourism in South Africa. He is currently working on his PhD studies with North-West University focusing on destination management of small towns. Research interests include poverty alleviation through tourism development, destination management, services marketing and cultural tourism. He has presented papers at local and international conferences.

Kohei Nagaoka is completing his bachelor's degree in management and information from the University of Shizuoka, Japan. His graduate research concerns marketing tea tourism to Kawanehon town in Shizuoka for young travellers. He is also interested in global tea culture.

Kazuyoshi Nakakoji is the director of the Chamber of the Tea Association of Shizuoka Prefecture and serves as editor-in-chief of the *TEA*, Japan's monthly tea industry magazine. He also publishes books on varieties of tea and pests, which tea companies in Japan widely use.

Evarisa M. Nengnong is a research scholar, Department of Tourism and Hotel Management, North-Eastern Hill University, Shillong, India. She completed a master's degree in tourism and travel management and a BSc in hotel management and administration. Apart from five years of work experience in the hotel industry, media, events, sales, and marketing, she has also been a guest lecturer at North-Eastern Hill University and ICFAI University. Her research focus is on food tourism and tea tourism.

Cuong Duc Hoa Nguyen is the Head, Tourism Department, Hanoi University, Vietnam. Over the past 26 years, he has established a reputation as a tourism professional in Vietnam and Asia. He is the author of several publications, including for Routledge and Vietnam Tourism Review.

Mai Chi Nguyen is pursuing a master's degree in Italy and currently working at the Pacific Asia Travel Association. Deeply interested in tourism studies, Mai has contributed to a few tourism publications for Kyoto University of Foreign Studies and the Ministry of Science and Technology of Vietnam.

Joan Pan is a PhD candidate in the Department of Geography and Environment at University of Hawaii at Mānoa, in Honolulu. Her research interests include the relationships between tea, nature and well-being, and the use of social media by businesses and consumers.

Brian Park is founder of the International Teaics Education Center in Jeju, South Korea. With an academic background in teaics, agronomy, chemistry, statistics, food hygiene and law, he developed the structure of teaics and the one-word term "teaics" for tea studies. He also wrote a dictionary of teaics.

Athitaya Pathan is a lecturer at Mae Fah Luang University, Thailand, and Head of the Tourism, Hospitality, and Events Research Group. As a researcher, she has been working on tourism and event studies. Her research philosophy is to widen perspectives in both practice and the academic disciplines of tourism and events horizons. Her previous studies were about community-based meetings, incentives, conventions and events tourism (MICE), community-based tea tourism, festival economy and creative industry.

P.G.S.S. Pattiyagedara is a lecturer at Uva Wellassa University, Sri Lanka. She has completed the BBM in hospitality, tourism and event management and is interested in tourism and community development activities. Her main research specialization areas are sustainable tourism, responsible tourism and destination management and marketing.

Madiseng Messiah Phori is Lecturer, Department of Tourism Management, Tshwane University of Technology, South Africa. He holds a master of management sciences, specializing in hospitality and tourism from Durban University of Technology, Kwa-Zulu Natal, South Africa. He is currently completing doctoral studies in tourism from Tshwane University of Technology. His research focus includes community-based tourism, sustainable tourism development, heritage and cultural tourism, rural tourism, and safety and security in tourism.

Jose Soares de Albergaria Ferreira Pinto is a PhD candidate (University of Macau) and a visiting lecturer (University of Saint Joseph). His background is in marketing (MBA) and food science and nutrition (MSc). He worked for 18 years in the hospitality industry and currently studies resilience, sustainability, foresight and sense-making methods.

J.P.R.C. Ranasinghe is a professor at Uva Wellassa University, Sri Lanka, with an academic background in tourism and hospitality. He has a strong interest in sustainable tourism development and management and is the Pioneering Chairman of Uva Provincial Tourism Promotion Board in Sri Lanka.

Sujama Roy is Assistant Professor at the ICFAI University Sikkim, India. With an academic background in literature and tourism studies, she has done extensive research on rural tourism in North Bengal in India. Her research interests include tourism in North East India and in India's Borderlands.

Masako Saito is a coordinator of Urban-Rural Exchange. She moved to Umegashima, Japan, and got to know the difficult situation of tea farmers. In order to increase the farmer's income, she created the KAKURECHA tea brand in 2009, started tea tourism and developed fermented tea (after Lapet in Myanmar), which won the Grand Prix in Undiscovered Gems in Shizuoka.

U.G.O. Sammani is a lecturer at Uva Wellassa University of Sri Lanka. Sammani is specialized in hospitality, tourism and events management and has contributed many scholarly works including journal papers and conference papers. She is currently working on projects related to sustainable tourism development.

J.A.R.C. Sandaruwani is a lecturer, Department of Tourism Management of Sabaragamuwa University of Sri Lanka. She obtained her BSc. and MBA specialized in tourism management from the same university in 2014 and 2019. Her research interests span sustainable tourism, cultural and heritage tourism, and e-tourism.

Gary Sigley is a professor in human geography in the Faculty of Geographic Sciences at Beijing Normal University. He has published widely on the topics of tea, tourism and heritage in contemporary China. His book *China's Route Heritage* was published in 2020.

D.A.C. Suranga Silva is professor at University of Colombo, Sri Lanka. With an academic background in tourism economics, he has written extensively on related topics, including a focus on sustainable tourism.

Bussaba Sitikarn is an assistant professor at Mae Fah Luang University, Thailand. With an academic background on sustainable tourism planning and development, she focuses on community-based tourism (CBT), especially in ethnic communities in northern Thailand. Most of her research projects focus on using CBT as a tool for poverty deduction, enhancing locals' well-being, and conservation of environment and socio-culture of community destinations.

Risa Takano is a PhD candidate at the Graduate School of Human and Socio-Environmental Studies, Kanazawa University, Japan. Her research has focused on the value creation process and new market development, especially authentic experience and the customer journey.

Akari Takeguchi is completing her bachelor's degree in management and information at the University of Shizuoka. She is interested in researching tea cafés in the tea gardens, although her graduate dissertation focuses on educating the Sustainabile Development Goal's movement and sustainable tourism for Japanese tourists.

Lysbeth Vink is a lecturer at the Hotelschool The Hague. She received her education in hotel management. Ms. Vink has a research interest in tea and the cultural heritage of tea. She has written papers on the cultural heritage of tea in the Netherlands and the art of tea in the hospitality industry.

Teeshakya Weerakotuwa is working as a team leader of customer relationship management at Hilton Colombo. She is reading a postgraduate diploma in tourism economics and hotel management in the Department of Economics, University of Colombo, Sri Lanka.

Paulin Wong is Senior Lecturer, Faculty of Business and Management, Quest International University, Malaysia. She obtained her PhD from Universiti Malaysia Sabah. Her research focuses on understanding place meanings among host communities in tourism destinations.

Nikki Wu is now travelling around Taiwan through work exchange. She graduated from the International Master's Program of Tourism and Hospitality, National Kaohsiung University of Hospitality and Tourism, Taiwan. Her research interests include glamping and immersion.

Kunbing Xiao is an associate professor at Southwest Minzu University, China. Her research interests include anthropology of food, cultural heritage studies and particularly the cultural history of tea in China, on which she has published three monographs.

Haruna Yagi is completing her bachelor's degree in management and information at the University of Shizuoka, Japan. Her research interests are in tea tourism experiences and regional development. She created the tourism development model for revitalizing rural communities through the music festival.

Naoko Yamada is an associate professor at the Institute of Human and Social Sciences, Kanazawa University. Her research interests are centred around visitor experiences in free-choice learning settings, focusing on applied communication of natural and cultural resources with visitors for heritage conservation, and tour guiding for sustainable tourism.

Libo Yan is an associate professor at Macau University of Science and Technology, Macau, China. His current research interests include tourism history, tourist experience and destination marketing. He has a series of publications in tourism history.

Gulsun Yildirim is Associate Professor in Gastronomy and Culinary Arts at Recep Tayyip Erdogan University, Turkey With an academic background in rural tourism studies, she has written on sustainable and ecotourism topics, including tea tourism through her 2021 book chapter on "Traditional Turkish Tea Culture in Gastronomy Tourism".

Mutsumi Yokota has a background in tourism management with a bachelor's degree in management and information at the University of Shizuoka, Japan. Her research interest in tea tourism is in strengthening tea-producing areas. She researched tourism contents to promote local communities.

Min (Lucy) Zhang is an assistant professor at the School of Hospitality and Tourism Management, Yunnan University of Finance and Economics, Kunming, China. Her research interests include border tourism governance, community-based tourism, cultural heritage and sociocultural sustainability in tourism destinations. She has consulted for the United Nations Development Programme on various ethnic tourism projects in border regions of China and the surrounding countries.

Hamira Zamani-Farahani has a PhD in tourism management. She is the head of the Astiaj Tourism Consultancy and Research Centre in Tehran, Iran. Her publications include journal papers, book chapters and five books. She has a wide interest in tourism research, including special interest tourism.

ACKNOWLEDGEMENTS

The *Routledge Handbook of Tea Tourism* is the result of the dedication and efforts of many individuals both within the tourism academy and the tea, tourism and hospitality industries. The editors would especially like to acknowledge the contributions to the volume.

Piyaporn Chueamchaitrakun, assistant professor and director of the Tea and Coffee Institute at Mae Fah Luang University, Thailand, brought together individuals with an interest in tea tourism at the Tea and Coffee International Symposium 2019 at Chiang Rai, Thailand. At that symposium the Working Group on International Tea Tourism (WG-ITT) was established with academics, destination planners and industry members interested in that topic, and the idea for this *Handbook* evolved from the work of that group. The editors would like to thank the members of the WG-ITT for their ongoing support and contributions towards the completion of this volume.

Concurrently with these efforts, academics including our editor Li-Hsin Chen edited and published a special issue on coffee and tea tourism in the *International Journal of Culture, Tourism and Hospitality Research* (renamed as *Consumer Behaviour in Tourism and Hospitality*) published as issue 3 in 2021, that influenced the development of the *Handbook*.

The home universities of the editors provided various supports for research leading up to this volume, including Ulster University, Northern Ireland, UK; Sabaragamuwa University of Sri Lanka, Sri Lanka; University of Shizuoka, Japan; and National Kaohsiung University of Hospitality and Tourism, Taiwan. The University of New Brunswick, Canada, also provided support for Lee Jolliffe's earlier research on tea tourism. In Japan, the Toyota Foundation was influential in funding the project Revitalizing Tea Industry Community through Gastronomical Tea Tourism, thus facilitating partnership between tea communities in Japan and Thailand. This project was spearheaded by our editor Amnaj Khaokhrueamuang at the University of Shizuoka, Japan. In Sri Lanka, our editor M.S.M. Aslam is laying the foundations for an International Tea Tourism Institute based at Sabaragamuwa University of Sri Lanka.

The editors would like to acknowledge the tea workers, garden owners, tea farmers, tea factory owners, homestay operators, hoteliers, tea communities, tea schools, tea tour leaders, tea stores and tea rooms that contribute to the resource for destination-based tea tourism experiences. We are grateful for author contributions to the *Handbook* from around the globe contributed by both academics and practitioners.

On behalf of the contributing authors, we thank those that provided permission for the use of images and other materials that contribute to illustrating the chapters, including Ian Gibbs, An

Zhuolin, Republic of Amsterdam Radio, Jetwing Symphony PLC, Suresh Sathyanathan, Belinda Jannette Davenport (Davenports of Cheshire) and Ismal Martin Kong (Sabah Tea Resort).

Our editor at Routledge, Faye Leerink, was encouraging regarding our contribution and patient with all of our questions as we assembled the manuscript. We also appreciate the use of the open access eReviewer system created by the University of South Florida M3 Center (https://m3center. org/) and AcademiaCentral (www.academiacentral.org/).

Finally, thanks to all of our family, colleagues and friends who put up with our many late afternoon and evening meetings in Asia to accommodate our editor in Canada and provided support and encouragement in many other ways.

Lee Jolliffe, Canada
M.S.M. Aslam, Sri Lanka
Amnaj Khaokhrueamuang, Japan
Li-Hsin Chen, Taiwan
April 2022

INTRODUCTION

Lee Jolliffe and M.S.M. Aslam

> Looking deeply into your tea, you see that you are drinking fragrant plants that are the
> gift of Mother Earth. You see the labor of the tea pickers; you see the luscious tea fields
> and plantations in Sri Lanka, China, and Vietnam.
>
> *(Hanh 2013, 91)*

The preceding quote, by author Thich Nhat Hanh (2013), shows the complexity of tea. It is an agricultural product of exotic lands, and brewing the leaf into a beverage provides a connection to nature,
to the tea workers and to the tea landscapes of production areas. These are all experiences that are
inherent in tea tourism whether through preparing the beverage at home from souvenir tea purchased
during travel, consuming it at a local café or in a tea course, or visiting with tea workers in the tea fields
and factories where it was produced and experiencing local tea culture. This volume delves into the
historic and evolving relationship between tea and tourism through tea communities and their visitors
around the world. It presents detailed studies of the various aspects in the development of tea tourism.

Tea

Tea is a plant (*Camellia sinensis*) with the most dominant varieties being *C. sinensis* var. *sinensis* and
C. sinensis var. *assamica*. Tea prepared from the processed leaf is often cited as the most consumed
beverage in the world, after water (Jolliffe 2007). Tea as a beverage may be prepared from the leaf
of *C. sinensis* or as an infusion from herbs and botanicals, referred to as a tisane. Tea is grown mainly
in Asia, Africa and South America. The main tea-producing countries today are China, India, Sri
Lanka and Kenya, according to the Food and Agriculture Organization (FAO 2021b). For these
producing regions tea is one of the most important cash crops, having a significant role in rural
development, food security and poverty reduction in exporting and developing countries. For millions of smallholder producers in these latter countries it is a principal livelihood source. In terms
of heritage and of relevance to tourism, China, Korea and Japan have four tea cultivation sites as of
2021 designated as Globally Important Agricultural Heritage Systems (GIAHS) by FAO (2021b),
and an additional site from China is on the potential list for designation.

Beyond its economic importance, tea is culturally and spiritually significant in many societies
(Pratt 1982). Tea culture is defined by the way that people prepare and interact with tea, and in
many cases elaborate ceremonies accompany the consumption of tea. The offer of and serving of

DOI: 10.4324/9781003197041-1

Table 0.1 Tea history collection plan

	Categories
1	A history of *Camellia sinensis* dating back to the very beginning to include the cataloguing and digitalization of the Tea Collection at the Royal Botanic Gardens at Kew, London
2	A historical collection of important books on tea and tea companies
3	A historical collection of items associated with the business of tea in the UK; information on tea brokers, manufacturers and the London Tea Auctions with four plaques from the London Tea Auction Room at Plantation House
4	A collection of vintage tea tins and tea advertisements
5	Historical items from current UK tea manufacturers

Source: Ian Gibbs, personal correspondence

tea is a sign of hospitality in many cultures. Recognizing the importance of tea, the United Nations declared the first International Tea Day on 21 May 2021, focusing on the cultural heritage, health benefits and economic importance of tea while highlighting sustainable production and ensuring benefits for people, cultures and the environment (FAO 2021a).

Tea has a long history that underlies the development of any tea-related products for tourism. This is conveyed through both tangible and intangible traditions, a trend noted by the recognition by UNESCO of Frisian tea culture as intangible cultural heritage (Bohne 2021). Museums around the world preserve and interpret tea cultures, acting as a resource for the relevant development of tea tourism at their locations and in their regions. For example, in 2018 Shizuoka Prefecture in Japan opened a Tea Museum as part of the prefecture's objective to be known as the tea capital of Japan. In the UK, a region with a rich history of tea trading, consumption and related traditions, the Tea History Collection has opened at Banbury. Developed by an entrepreneur with a background in the tea trade, this centre documents the history of tea and offers educational activities. The planned collection will provide a resource for the development of tea-related offerings in the UK and beyond (Table 0.1). The collection development process is supported by the London Tea History Association. At the time of writing, viewing of the collection is by prior appointment only.

Tea and travel

Tea has an ancient history and has long been associated with travel. Tea as a product has travelled in caravans from Yunnan to Tibet on the Ancient Tea Horse Road and on clipper ships from China to England (Pratt 1982). In the spread of tea from country to country, monks, explorers and traders travelled with the tea plant and aided in its dissemination, as with those who travelled to China and brought the tea plants back to Darjeeling (Koehler 2016). Officials in the east and early travellers in the west also travelled with their own tea.

Intentional travel to experience tea as a primary tourist motivation may have its origins in more recent times, although in colonial times in Asia officials and missionaries travelled to cooler resort areas in the mountains, often where tea was grown. Consequently, they probably experienced tea there, both through the tea landscapes and culture and the consumption of the beverage, encouraged by afternoon tea traditions, as in the colonial areas of the British realm (Skinner 2019). The remains of colonial reigns in tea regions provide a heritage resource for tea tourism, as with the colonial lodgings of Sri Lanka (Aslam and Jolliffe 2015). In Darjeeling, where many of the colonial tea manager bungalows (referred to locally as burra bungalows, a culture developed by the 19th-century British planters) have been repurposed, both tea and tourism are recognized as the leading industries of the region (Bose 2018). However, the emphasis on the colonial narrative may detract from the potential of such regions for other forms of tourism based more on the natural environment, such as in India's Dooars region where tea tourism is proposed alongside forest tourism (Datta 2018).

Tea and tourism

Tea is connected to tourism through both travel and consumption, and is ever-present in hospitality situations (Jolliffe 2006). Encountering these settings around the world, tea is often provided as a welcoming ritual in home and commercial hospitality settings. Where tea has a strong connection to tourism is through the direct experience of tea in hospitality and lodging settings; at production areas; through visits to or stays at tea estates or gardens, such as with tea heritage tourism in Sri Lanka (Jolliffe and Aslam 2009); and visits to tea-themed restaurants and cafes both in producing and consuming locations. Here, too, enhanced experiences are offered through tea events, tours and courses that provide more encompassing and intensive involvement, as reflected by food and beverage experiences (Knollenberg *et al.* 2021).

In recent times, when travel has been limited by the COVID-19 pandemic, direct experiences with tea have been offered online, often accompanied by tastings with the tea being sent to participants ahead of time. This was the case with the Japanese Tea Marathon, sponsored by the Global Japanese Tea Association and the Japan Tea Central Council PIIA, held in parallel to the delayed Tokyo 2020 event in 2021. For the duration of the Olympics, participants were able to experience tea from 15 regions of Japan, meeting the tea producers in their fields and factories through Zoom and consuming the tea they brewed in their own homes (GJTA 2021).

Researchers have linked tea and related tourism to the image of destinations (Fernando *et al.* 2016). Recent work has shown the relationship of the tea preferences of tourists to destination development (Gupta *et al.* 2020). Many destinations are identifying the potential to capitalize on their tea heritages for tourism, such as in areas of the Netherlands and Germany with the Frisian tea culture that has been recognized by UNESCO as intangible cultural heritage (Bohne 2021).

Tea tourism

Tea tourism primarily includes visits to and experience at tea gardens, plantations, factories, schools, cafés, lodgings and events at a destination (Jolliffe 2007). Understanding such activity requires a knowledge of tea consumer behaviour and tourist involvement with and attachment to tea while at a destination and at home at consuming locations (Chen *et al.* 2021). As a niche, tea tourism includes tea festivals and events, tea tourism experience, tea tourism marketing and management, and responsible and community-based tea tourism (Jolliffe 2007; Jolliffe and Nakashima 2020).

As a development tool, tea tourism is noted to have benefits for host communities and related tea industries in terms of improving livelihoods and diversifying business. Researchers have noted the importance of local stakeholder views in the development of tea tourism (Cheng *et al.* 2012), as well as the integration of tea and tourism into sustainable livelihood approaches (Su *et al.* 2019). Tea tourism is consequently of relevance to the UN World Tourism Organization's (UNWTO's) World Tourism Day 2020 theme of Tourism and Rural Development and the memorandum of understanding to work on this theme signed by the UNWTO and the FAO (UNWTO 2020).

Relationship to other forms of tourism

Tea tourism has been identified as a form of cultural heritage tourism with a relationship to many other forms of special interest tourism. Types of niche tourism closely aligned with tea-related tourism include green tourism, nature tourism, ecotourism and agritourism (Dixit 2019). Tea tourism forms a part of culinary, food and beverage tourism, especially in hospitality settings where hotels are embracing tea culture to differentiate and enhance their offerings (Bohne and Jolliffe 2021). At tea-producing locations, tea tourism is closely related to concepts of responsible and rural tourism (Jolliffe and Nakashima 2020), as well as green tourism (Yeap *et al.* 2021). As tea consumption and settings can benefit personal well-being, tea tourism is in addition related to health and wellness tourism (Su and Zhang 2020). As it develops alongside other types of tourism, the editors of this volume observe tea tourism moving towards being more of a form of inclusive tourism, benefiting tea producers and workers alongside with local communities and providing responsible experiences for visitors.

Towards tea tourism theories

As noted above, scholarship on tea tourism has evolved from the identification of the niche and its characteristics advocated by Jolliffe (2007). This has included more detailed analyses of sustainability, marketing and management (Fernando *et al.* 2016). Recent work has shown the relationship of the tea preferences of tourists to destination development (Gupta *et al.* 2020). Research also represents innovation in tea tourism, for example the efforts of a tea region in Japan to attract Thai tourists through the marketing of their bottled tea in Thailand (Khaokhrueamuang *et al.* 2021). The product life cycle theory can be applied to the study of tea tourism (Jolliffe 2022) to assist in understanding the management and marketing of this form of tourism. However, Chen *et al.* (2021) in their systematic review of tea tourism literature identified that most previous studies of tea tourism have been fragmented, often using the lens of tea tourism to identify aspects of such tourism from the perspective of both stakeholders and consumers. A further stage in tea tourism studies is the proposition of models and methods of studying this form of tourism (Mondal and Samaddar 2021). At this stage, tea tourism studies, while still being dominated by descriptive works, are thus moving towards more systematic and theoretical approaches.

Organization of the Handbook

The 32 chapters in this volume are divided into five thematic parts in order to provide insights into the various aspects of tea tourism (Figure 0.1). The collection of themes represented by these parts

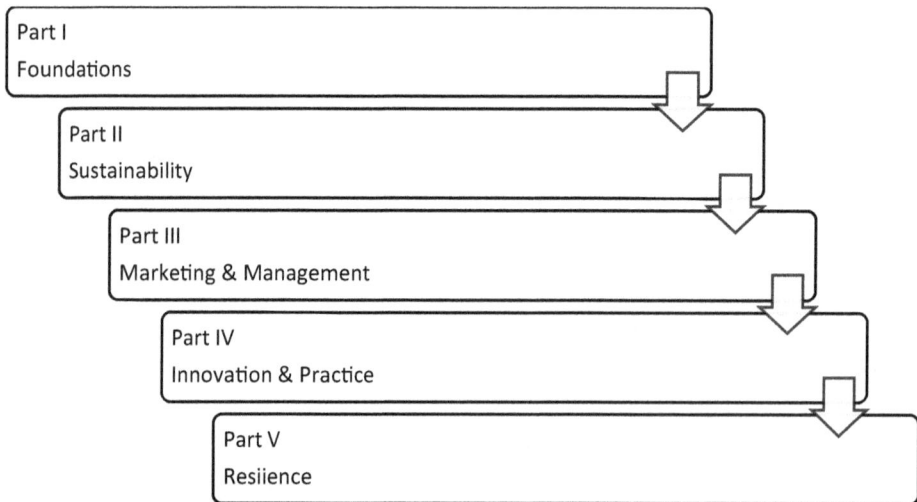

Figure 0.1 Scheme for organisation of the volume
Source: Authors

aims to extend the body of knowledge. Including topics of interest to both academic and industry sectors, the volume is designed to increase interaction and communication between the two, thus encouraging the development of sustainable and responsible tea tourism that benefits local communities on a global basis.

The *Handbook* is interdisciplinary, with authors from 19 countries representing various disciplines including tourism, hospitality, geography, management, marketing and tea science, and the practices of operating tea rooms, conducting tea tours and marketing tea tourism. These diverse perspectives provide new insights into and appreciation of the various aspects of global tea tourism. A critical aspect is presented as authors identify issues surrounding tea tourism development including shared tea heritages, decolonization, employment issues and working sustainably to improve livelihoods in tea-producing communities and enhance the quality of life and experience in tea-consuming places.

Part I: Foundations of tea tourism

Tea tourism to a certain extent is founded on the history of tea and our knowledge about it. Understanding the origins of this form of tourism provides perspective and informs current and future development. In Part I, knowledge about tea history, traditions and heritage is combined with an understanding of different forms of tea tourism highlighting how tea heritage is recognized, interpreted and reinvented for contemporary tourism. Beginning with a historical perspective in Chapter 1, authors Libo Yan and Kunbing Xiao investigate the ancient origins of tourism related to tea in China, showing the relationship of tea and tourism and its connection to travel in imperial China. This is the first study to explore this link, adding to understanding of the past of tea tourism in ancient societies. Author Gary Sigley in Chapter 2 provides a case study of a village located on China's Ancient Tea Horse Road that has been recreated for touristic purposes. This chapter examines tea tourism within the concept of route tourism and local identity in the age of mass mobility, identifying a distinction between the old and the new Ancient Tea Horse Road. In another contribution from China for Chapter 3, authors Jianming Li, Min (Lucy) Zhang and Yanting Gu use the theory of embodiment to examine the connection between tea and spiritual tourism through Panchen Tuo tea produced in Yunnan Province. They found that the experience of a local tour guide turned tea entrepreneur influenced the interplay of tea with spirituality through personal experiences and social operations, providing an example for cooperation and integration of the tea and tourism industries there.

Moving on to Vietnam in Chapter 4, Cuong Duc Hoa Nguyen and Mai Chi Nguyen define and discuss the importance of wild tea and indigenous tourism. Using the case study of tea tourism in Moc Chau District, Vietnam, they observe that the local tourism industry has not explored the full potential of the local Snow Shan tea to offer distinctive experiences and tourism services. Identifying the significance of heritage railways to tea tourism, Amnaj Khaokhrueamuang, Akari Takeguchi, Kohei Nagaoka and Koichi Kimura in Chapter 5 profile the case of Senzu, a tea-growing community in the Kawanehon town of Shizuoka, Japan, where the terminal station of a heritage steam locomotive service is located. The chapter advocates optimizing the value of rail transportation as a community asset and tourist attraction for creating tea tourism offerings.

In Chapter 6, Lysbeth Vink, Annette Kappert and Hartwig Bohne look at the cross-border (Netherlands and Germany) cultural heritage of Friesland tea that has been recognized as Intangible Cultural Heritage by UNESCO. This case considers the development and potential for tea culture and heritage to contribute to the brand value of the tourism industry here. Hilary du Cros in Chapter 7 examines the recognition of the cultural heritage of tea from an international perspective and tracks the progress of the development of tea tourism. This chapter will assist readers in understanding the cross-cultural and international perspectives on the cultural heritage of tea that are part of an evolving process reflected in tea tourism. In Chapter 8, Brian Park discusses the theory of teaics, a logical, objective and structured framework of knowledge relating to tea, which is presented as a useful tool for the development of tea tourism and related products. In Chapter 9, authors Madiseng Messiah Phori, Lebogang Matholwane Mathole, Unathi Sonwabile Henama and Lehlohonolo Gibson Mokoena provide perspectives of tea tourism in the Global South, examining how it has emerged as a form of special interest tourism. Case studies of a tea-producing region in Kenya and a tea tour route in South Africa provide evidence of the potential for tea tourism development in this region.

Part II: Sustainability in tea tourism

Part II of the *Handbook* focuses on sustainability in tea tourism from the perspectives of various global tea communities in terms of embracing their tea culture, interpreting it for tourism and developing new forms of hospitality that could benefit local communities. This is highlighted through the development of community-based tea tourism in Thailand, the recognition of community tea culture in Turkey, the use of tea in hospitality in Sri Lanka, the implementation and development of the homestay concept in small tea gardens in India, and the use of authentic line room accommodation on tea plantations in Sri Lanka for tourism. Also in Sri Lanka, employment issues within the tea tourism industry are examined from a sustainable development perspective. In all of these situations to a certain extent, the implementation of tea tourism can contribute to the achievement of the UN's Sustainable Development Goals (SDGs), particularly in the areas of promoting sustainable agriculture, improving livelihood and increasing gender equality. Tea tourism can contribute by complementing agriculture, diversifying agriculture and encouraging inclusive growth by diversifying employment opportunities, especially in terms of the development of homestays and other alternate tourism forms. The chapters in this section encompass a preliminary exploration into aspects of sustainability in tea tourism that is worthy of more dedicated study.

In Chapter 10, authors Bussaba Sitikarn, Kannapat Kankaew and Athitaya Pathan document the implementation and management of community-based tea tourism (CBTT) in Thailand, applying the principles of community-based tourism to tea tourism. The study illustrated that the key integrated management of creative CBTT contributes to value prepositions that include emotional, experiential, functional, educational, economic, environmental and social aspects. In sum, implementation of CBTT could lead to value added in tea and local cultural products and value symbiosis amongst the host and the guest in the long term.

In Chapter 11, authors Kadir Çetin and Emre Erbaş investigate the case of the tea culture of Turkey as a resource for the development of tea-related tourism. This chapter illustrates the socioecological practices of tea in Turkey as cues for sustainable tea tourism. This can contribute to understanding the structure of tea tourism and its sustainability in a socially constructed world. Findings elucidate to what extent cultural practices in a tea community can be repurposed, as a sustainable basis for tea tourism could be applicable to other countries with strong and identifiable tea cultures.

Turning to the case of tea culture in relation to hospitality operations in Sri Lanka, in Chapter 12 authors J.A.R.C. Sandaruwani, G.V.H. Dinusha and R.S.S.W. Arachchi profile how Sri Lankan commercial hospitality takes tea culture to the next level, mainly focusing on tea-related hospitality practices of the hotels located in the main tea-growing district, Nuwara Eliya. This contributes to overcoming the limitations in literature, expanding upon scientific communication on the role of tea culture in the hospitality industry.

The next two chapters examine different forms and developments of the homestay concept in tea gardens and estates in India and Sri Lanka. In Meghalaya, India, authors Evarisa M. Nengnong and Saurabh Kumar Dixit in Chapter 13 use the case studies of two tea gardens to show how the homestay concept has been implemented as a diversification from tea production. The authors interpret results from community landscapes where tea is grown along with small homestay projects.

This reveals that homestay has the prospect of improving local livelihoods, boosting rural economies, and addressing socioeconomic challenges in tea landscapes. The developed homestays in small tea estates of Meghalaya, India, elicit and interpret that tea tourism offers potential for tourists to explore experience and knowledge associated with tea culture and consumption. In Chapter 14, on Sri Lanka, the author team of J.A.R.C. Sandaruwani, G.V.H. Dinusha and R.S.S.W. Arachchi investigates repurposing line rooms for experiential and creative tourism practices in comparison with other formally established homestays in light of authenticity and inclusivity. Interpretations of both tea homestays products studied in Chapter 14 unveil multiple truths of the line room and its cultural experiences through delineating both authenticity and staged authenticity in conjunction with tea tourism visitors as emerging segment in tea landscapes. This demarcates a staged authenticity that is also subjective to the tea tourism visitor segment.

Finally, once again in Sri Lanka, P. Gayathri, K.M.B.S.Y. Kulasekara, D.A.C. Suranga Silva, Krishantha Ganeshan, Baghva Erathna, K.M.B.S.Y. Kulasekara and Teeshakya Weerakotuwa in Chapter 15 investigate various issues related to overall employment in the tea tourism sector. They highlight challenges faced by employees in the Sri Lankan tea tourism industry. These include issues related to training and development, the welfare system, the competitive market and employee satisfaction. Special attention is given to the impact of COVID-19 and the problematic situation of employees in the tea tourism industry after the pandemic. This analysis brings forth lessons for sustainable tea tourism development through a symbiotic approach in tea-producing landscapes across the globe.

Part III: Management and marketing of tea tourism

Tea plays an essential role in the marketing mix of both tea-producing and consumption destinations. Although previous studies have emphasized the importance of management and marketing for developing attractive tea tourism businesses, few have explored and explained how the tea tourism experience is influenced by every detail and component in a tea tourism destination. Furthermore, as the mainstream of management and marketing literature focuses on the "growth" and "competitiveness" of tea tourism business, this part of the *Handbook* indicates a successful destination should be "mindfully" managed to enhance the well-being of staff, local residents and visitors. Only when the marketing and management strategies can balance the needs and satisfaction of both hosts and guests can the tea tourism business thrive in the current experience economy and develop a sustainable future. Part III starts with a study examining the service quality of an English tea room. In Chapter 16, Belinda Davenport identifies the role of an innovative marketing strategy using customers' own photographic equipment to capture the memory of the tea experience and to increase the word-of-mouth for the business. In Chapter 17, Nikki Wu and Li-Hsin Chen adapt the orchestra model to investigate the tourism experience in a Taiwanese pearl milk tea tourism factory. They found that offering interactive activities and integrating five main components of the tourism experience can generate more pleasurable memories and satisfaction for visitors. Recognizing the importance of social media in tourism marketing, Joan Pan and Wayne Buente explore how Instagram facilitates storytelling and marketing experiences of two tea brands throughout, during and after those experiences. They conclude in Chapter 18 that storytelling is vital and that social media influencers provide a significant contribution toward improving engagement for the brands.

Moving from local tea business to a broader context, in Chapter 19, Balvinder Kler and Paulin Wong introduce an interdisciplinary conceptual framework to understand tea tourism as a niche market. They suggest that nurturing the elements of genius loci inherent to place creates rewarding, enriching, adventuresome and informative experiences at tea tourism destinations. In Chapter 20, J.P.R.C. Ranasinghe, A.C.I.D. Karunarathne, U.G.O. Sammani, H.M.J.P. Herath and P.G.S.S. Pattiyagedara put forward a model to explore the impact of tea tourism–driven activities, services and product offerings on the attractiveness of a destination. The positive outcome of the model indicates the integration of resources within a destination is essential for increasing the perceived values of tourists. Meanwhile, in Chapter 21, Masako Saito introduces an action research case demonstrating how tea tourism revitalizes a declining tea-producing community in Umegashima, Shizuoka, Japan. Similarly, Imali N. Fernando in Chapter 22 found tea tourism may be a hidden salvation to connect the tea industry and the local community. The Porter Diamond model is adapted to explore tea tourism in Central and Uva Provinces, Sri Lanka. Finally, this part concludes with a study of tea-producing regions in Shizuoka, Japan, in Chapter 23. Kunihiko Iwasaki and Amnaj Khaokhrueamuang explore the possibility for marketing a tea-producing region through green tea tourism.

Part IV: Innovation and practice in tea tourism

The global tea industry is witnessing several innovations. Today, tea leaves are used for producing consumer products such as cosmetics, dyed fabrics and foods. Tea offers valuable experiences in tourism and hospitality such as tastings, brewing courses, picking and tea factory visits. These products and services add significant value to the tea industry. Part IV of the volume comprises six chapters that highlight innovative forms of tourism linked to the world of tea.

In Chapter 24, Gulsun Yildirim examines the emergence, development and problems of tea gastronomy tours in Turkey as potential tourism products. The author notes that travel agencies face difficulties organizing this kind of tour even though it has the potential to attract visitors other than tea experts. Regarding gastronomy, tea tourism can strengthen tea-producing communities, especially when innovative ideas lead to new products. In Chapter 25, for example, Amnaj Khaokhrueamuang, Piyaporn Chueamchaitrakun and Kazuyoshi Nakakoji apply the concepts of tea culture commodification (Khaokhrueamuang and Chueamchaitrakun 2019) and tourism area life cycle (Butler 2006) to gastronomical tea tourism projects in a declined tourism destination in Japan and a new tea tourism village in Thailand. Thanks to the exchange programme between the two countries, innovative ideas are generated, including tea cuisines and tea tourism routes.

In Chapter 26, Amnaj Khaokhrueamuang, Haruna Yagi, Mutsumi Yokota and Sousuke Goto discuss creative tea tourism routes in Japan's Shizuoka urban area that link tea cafés and different community resources. As an inventive tea tourism product, tea tourism routes diversify the tea café community to serve four functions for both residents and tourists: communicating, relaxing, tasting tea and learning about and exchanging tea culture. In Chapter 27, Kyoko Ishigami and Amnaj Khaokhrueamuang examine the Chagusaba, the Japanese practice of nurturing Shizuoka's Traditional Tea-Grass Integrated System, which is designated as a GIAHS. The authors' analysis of documents regarding the promotion of this agricultural system reveals that various tourism forms have been developed to add value to the local tea gardens. They include green tourism, ecotourism, heritage tourism, gastronomic tourism and sports tourism.

To document innovative forms of value addition in the tea industry, the authors of Chapter 28, Risa Takano, Naoko Yamada and Daisuke Kanama, explore the process of value creation in the Japanese black tea market and tourism. They discuss four practices that play a key role in providing customers with authentic experiences: curation of experiences for people who share the same interests, community building, convening real and rare conversations, and evaluation through contests. These practices are based on authenticity, which is a crucial factor in building the relationship between product innovation and tourism. Finally, in Chapter 29, Hamira Zamani-Farahani investigates tea tourism in the Middle East. The author reviews the role of tea tour guiding and its training and considers the characteristics of tea tourism and tea tour guides in Iran based on her previous research and the analysis of published materials. This case study contributes to filling a gap in the literature on guiding for tea tourism, both in Iran and other global tea regions.

Part V: Resilience in tea tourism

As evidenced by the previous chapters of this volume, the emergence and development of tea tourism today faces many challenges and issues including colonial and postcolonial legacies, cultural knowledge and interpretation, authenticity and the relationship to both local communities and visitors. Challenges are posed by the environment in tea-producing regions in terms of natural disasters, human-wildlife interactions and climate change. Most recently, the tea tourism industry has been faced with the challenges related to COVID-19, causing related enterprises to pivot in terms of their offerings and to address the challenges of reduced visitation and work forces. In these contexts, resilience is required for tea tourism to develop and prosper.

The three chapters in Part V on resilience in tea tourism focus on tea tourism in a postcolonial context in a tea region in India; human-wildlife interactions and tea tourism in a West Bengal region of India; and tea tourism development and recovery in the Azores, Portugal. In Chapter 30, author Sujama Roy examines the development of tea tourism (or lack of it) in the tea region of the Dooars in West Bengal, India. This is a case of resilience in terms of the tea estate workers and an instance where tea tourism, if implemented with sustainable goals, might assist in improving the livelihoods of workers who often have limited opportunities. In Chapter 31, Chandan Datta provides an in-depth case study analysis on the same area of the Dooars, covering the challenges of human-wildlife interactions in and near tea gardens and how they affect tea tourism – a persistent issue for tourism in producing locations. Finally, in Chapter 32 Jose Soares de Albergaria Ferreira Pinto examines the historic tea industry in the Azores, Portugal. This chapter explains how resilience is encrypted in the culture of local people and how tea contributes to tourism under the most challenging times, such as during the COVID-19 pandemic. The role of tea tourism is noted in co-creating and enhancing experiences that satisfy and prompt tourists to recommend and return to the Azores.

Conclusion

Edited by a team of international academics from Canada, Sri Lanka, Japan and Taiwan, the *Handbook* is intended to extend the body of knowledge on tea tourism. It provides practical guidance and cases related to marketing, management, development and governance of tea tourism in a responsible and sustainable manner. It represents a concerted effort to gather and highlight the existing information on tea tourism, adding new theoretical and practical insights on the niche.

The volume has been designed to be useful to a broad segment of readers ranging from academics (both teachers and students at undergraduate and graduate levels) to practitioners (including but not limited to tea producers, companies and tea tourism–related enterprises, as well as destination planners and managers). The *Handbook* could serve as a catalogue for researchers, learners, teachers, employers, employees and others concerned with tea and related tourism. The chapters can stand alone, each including a narrative or case study, discussion questions and references, and can be used for readings, workshops and assignments in college or university courses relating to cultural heritage tourism, hospitality and niche tourism. These chapters can be used for practical training, for example of tea enterprise operators, tea tour guides and tourism planners. The volume creates a foundation for future work on and application of the tea tourism concept whilst setting out a research and action agenda for the next five to ten years. The volume will serve as a reference work and guide to the current issues facing this tourism niche, including achieving the SDGs, tea tourist behaviour and experience, management, marketing, development, governance, innovation and crisis management.

The editorial team is most appreciative of the chapter authors for sharing their up-to-date research and knowledge of tea tourism through this volume. It is especially notable that the authors completed their work during the global COVID-19 pandemic, which presented limitations to field work and created personal challenges due to lockdowns, online teaching and meetings, and the infection of friends, family, co-workers and even the authors themselves.

References

Aslam, M. and Jolliffe, L., 2015. Repurposing colonial tea heritage through historic lodging. *Journal of Heritage Tourism*, 10, 111–128.

Bohne, H., 2021. Uniqueness of tea traditions and impacts on tourism: the East Frisian tea culture. *International Journal of Culture, Tourism and Hospitality Research*, 15(3), 371–383.

Bohne, H. and Jolliffe, L., 2021. Embracing tea culture in hotel experiences. *Journal of Gastronomy and Tourism*, 6(1–2), 13–24.

Bose, S.P., 2018. Growth and development of tea industry in Darjeeling: colonial period. *International Journal of Research*, 7(2), 15–17.

Butler, R.W., 2006. *The tourism area life cycle, vol. 1: applications and modifications.* Clevedon: Channel View Publications.

Chen, S.-H., Huang, J. and Tham, A., 2021. A systematic literature review of coffee and tea tourism. *International Journal of Culture, Tourism and Hospitality Research*, 15(3), 290–311.

Cheng, S., Hu, J., Fox, D. and Zhang, Y., 2012. Tea tourism development in Xinyang, China: stakeholders' view. *Tourism Management Perspectives*, 2, 28–34.

Datta, C., 2018. Future prospective of tea-tourism along with existing forest-tourism in Duars, West Bengal, India. *Asian Review of Social Sciences*, 7, 33–36.

Dixit, S.K., 2019. Gastronomic tourism: a theoretical construct. *In:* Dixit, S.K., ed. *The Routledge handbook of gastronomic tourism.* London: Routledge, 13–23.

Fernando, P.I.N., Rajapaksha, R.M.P.D.K. and Kumari, K.W.S.N., 2016. Tea tourism as a marketing tool: a strategy to develop the image of Sri Lanka as an attractive tourism destination. *Kelaniya Journal of Management*, 5, 64–79.

Food and Agriculture Organization of the United Nations (FAO), 2021a. International tea day. Available from: www.fao.org/international-tea-day/en/ [Accessed 21 October 2021].

Food and Agriculture Organization of the United Nations (FAO), 2021b. Tea markets and trade. Available from: www.fao.org/markets-and-trade/commodities/tea/en/ [Accessed 21 October 2021].

Global Japanese Tea Association, 2021. Japanese tea marathon. Available from: https://gjtea.org/japanese-tea-marathon/ [Accessed 21 October 2021].

Gupta, V., Sajnani, M., Dixit, S.K. and Khanna, K., 2020. Foreign tourist's tea preferences and relevance to destination attraction in India. *Tourism Recreation Research*, 1–15.

Hanh, N., 2013. *Love letter to the earth.* Berkeley, CA: Parallax Press.

Jolliffe, L., 2006. Tea and hospitality: more than a cuppa. *International Journal of Contemporary Hospitality Management*, 18(2), 164–168.

Jolliffe, L., ed., 2007. *Tea and tourism: tourists, traditions and transformations.* Clevedon: Channel View Publications.

Jolliffe, L., 2022. Tea tourism. *In:* Buhalis, D., ed. *Encyclopedia of tourism management and marketing.* Cheltenham: Edward Elgar Publishing, 1–3.

Jolliffe, L. and Aslam, M.S., 2009. Tea heritage tourism: evidence from Sri Lanka. *Journal of Heritage Tourism*, 4, 331–344.

Jolliffe, L. and Nakashima, M., 2020. Responsible rural tourism in Japan's tea villages. *In:* Nair, V., Hamzah, A. and Musa, G. eds. *Responsible rural tourism in Asia.* Bristol: Channel View Publications, 61–74.

Khaokhrueamuang, A. and Chueamchaitrakun, P., 2019. Tea cultural commodification in sustainable tourism: perspectives from Thai and Japanese farmer exchange. *Special Issue of Rajabhat Chiang Mai Research Journal*, 104–117.

Khaokhrueamuang, A., Chueamchaitrakun, P., Kachendecha, W., Tamari, Y. and Nakakoji, K., 2021. Functioning tourism interpretation on consumer products at the tourist generating region through tea tourism. *International Journal of Culture, Tourism and Hospitality Research*, 15(3), 340–354.

Knollenberg, W., Duffy, L.N., Kline, C. and Kim, G., 2021. Creating competitive advantage for food tourism destinations through food and beverage experiences. *Tourism Planning and Development*, 18, 379–397.

Koehler, J., 2016. *Darjeeling: A history of the world's greatest tea.* London: Bloomsbury Publishing.

Mondal, S. and Samaddar, K., 2021. Exploring the current issues, challenges and opportunities in tea tourism: a morphological analysis. *International Journal of Culture, Tourism and Hospitality Research*, 15(3), 312–327.

Pratt, J.N., 1982. *The tea lover's treasury.* New York: 101 Productions.

Skinner, J., 2019. *Afternoon tea: a history.* Lanham, MD: Rowman and Littlefield.

Su, M.M., Wall, G. and Wang, Y., 2019. Integrating tea and tourism: a sustainable livelihoods approach. *Journal of Heritage Tourism*, 27, 1591–1608.

Su, X. and Zhang, H., 2020. Tea drinking and the tastescapes of wellbeing in tourism. *Tourism Geographies*, 1–21.

United Nations World Tourism Organization (UNWTO), 2020. Tourism and rural development. Available from: www.unwto.org/news/tourism-and-rural-development [Accessed 21 October 2021].

Yeap, J.A., Ooi, S.K., Ara, H. and Said, M.F., 2021. Have coffee/tea, will travel: assessing the inclination towards sustainable coffee and tea tourism among the green generations. *International Journal of Culture, Tourism and Hospitality Research*, 15(3), 384–398.

PART I

Foundations of tea tourism

1

ANCIENT ORIGINS OF TEA TOURISM

Libo Yan and Kunbing Xiao

Introduction

As the most consumed beverage in the world, tea has had a rich connection with travel since the mediaeval world. Jolliffe (2007) disclosed the travel-tea link and suggested that the rich connection between the two can be investigated through the study of the travellers who were interested in the history, tradition, and consumption of tea. This study aims to explore the link between tea and tourism in imperial China. It collects historical records as source material (translation by authors) to investigate the historical connection between tea and travel. The study focuses on leisure travel; business travel associated with the tea trade is beyond the research scope. Case studies were adopted to showcase the link between tea and leisure travel in imperial China.

Literature review

Elite travel from the Tang to the Qing dynasties (618–1912)

Educated men formed a social elite in imperial China since this social group constituted the candidate pool of government officials. They were called literati or scholar-officials. The civil service examination system was already well established in the Tang dynasty (618–907), and this system of recruitment of officials largely increased the scale and frequency of literati travel (Zhang 2011). Since the seventh century, starting from a young age, men of the literati class extensively travelled for examination learning and participation in the provincial and imperial examinations (Feng 2008). Once an educated man successfully passed the imperial examination, he was appointed to a government position. The Tang dynasty established an empire-wide bureaucracy with its members regularly rotated from post to post. This bureaucracy system made the literati-officials spend much of their adult life travelling, moving between the provinces and the capital (Ridgway 2005).

Infrastructure significantly improved from the Tang to the Song dynasties (Ridgway 2005). The road system was greatly expanded to facilitate the transmission of official personnel and documents across the country. A hospitality system, with inns and postal stations every 25 miles or so along the roads, was established to facilitate official travels.

The elite travels were characterised by government assistance and sponsorship, as well as the activities of the travelling officials (Zhang 2011). The government assisted the officials on the road, providing lodging and catering services, soldier-porters, and transport means (e.g., horses, carriages,

DOI: 10.4324/9781003197041-3

and boats). Travelling associated with court assignments was fully funded by the government and the itineraries were predetermined by the relevant court assignments. However, the literati-officials enjoyed the flexibility of travel arrangement in temporal terms. They were allowed to spend longer to arrive at their new posts than expected. They tended to spend extra time on their journeys to visit sites of historical and cultural significance, as well as to meet with local officials, acquaintances, or persons they admired. In other words, sightseeing and building social networks were major activities of the travelling literati-officials. Such activities were based on the financial and material support of the government. The state sponsorship and assistance "freed the travellers from many mundane worries when they headed off and allowed them abundant time and freedom to engage in activities of their own choice" (Zhang 2011, 7–8).

Travelling for on-site brewing fine tea with mountain spring water

Chinese literati-officials' travelling in pursuit of mountain spring water to brew the finest tea was motivated by remarks by Lu Yu (733–804; the Sage of Tea and the author of the highly regarded monograph *The Classic of Tea*), on a hierarchy of water sources for brewing tea. Based on his extensive experience of tea tasting, Lu Yu concluded that different water sources led to different qualities of tea tastes. His remark reads: "For water, mountain water is the best, river water the second, and water from well is the least suitable" (Jade Tea 2020). However, the mountain water would become undrinkable if stored stagnantly for a long time, and thus the most desirable water for brewing tea should be freshly drawn from clear flowing springs in the wilderness (Wang 2005).

Lu Yu's classic remark initiated a tea culture tradition passed from generation to generation, and the principle of selecting the correct water remained the first critical step of tea making and tea tasting throughout mediaeval China (Jade Tea 2020). The water-tea link motivated many literati to travel to scenic mountains to access the quality spring water needed to brew tea for the desired taste. Chang's (2004) study of the travel life of literati in the Ming dynasty has noted the tea-water-travel link, quoting the works of three literati-officials. The present study adopts the case of the Huishan spring (the second-best spring in China) for an inductive analysis of the water-tea-travel link.

The region south of the Yangtze River was one of the major tea-planting regions, hosting many famous temples and well-known springs. The best teas and springs in this region attracted literati from different parts of imperial China (Wu 1993). The strong interest of the Chinese literati class in tea connoisseurship was cultivated in their daily life. Tea drinking had been a leisure activity amongst the literati class since the Tang dynasty. Literati, monks, and Taoists often hosted tea parties for socialisation purposes (Pu 2015). The tea-gathering setting evolved from urban tea houses in the Song dynasty to mountain temples in the Ming dynasty (Wu 1993). Tea-gathering permeated the travel life of literati-officials.

Travelling for researching and producing desirable tea

Celebrated teas have a historical link to famous mountains. The natural environment, the geographical position, the altitude, the spring water, the precipitation, and the micro-environment typical of famous mountains are excellent conditions for the growth of tea plants. Many specialty teas are directly connected to famous mountains and their names underline this connection, such as *Huangshan Maofeng* (Maofeng tea from Mt. Huangshan), *Wuyi Dahongpao* (Big-red-robe tea from Mt. Wuyi), *Junshan Yinzhen* (Silver-needle tea from Mt. Junshan), *Mengding Ganlu* (Sweet Dew tea from Mt. Mengding), and *E'mei Xueya* (Snow-buds from Mt. E'mei).

Literati-officials in imperial China were strongly interested in visiting famous mountains and rivers (Wu 2003) and meeting tea-making experts such as Buddhists and Taoists who dwelled in the mountain temples. Such meetings were driven by the travelling literati's intention to produce desirable tea. The cultural history of Wuyi tea showcased that, since the Tang dynasty, literati-officials from all over China frequently visited Mt. Wuyi and sojourned in the region (Xiao 2013). More than just praising the remarkable landscape of Wuyi, literati-officials' travelogues reflect their great interest in researching and producing the desirable tea. Their engagement in producing desirable tea normally resulted from their interaction with Buddhists and Taoists in the mountains. Yuan Zhen (779–831) wrote that tea was a drink that "poets admire and monks adore" (see Liu and Wu 2020, 84). This poetic line implies that tea was a social medium for the interaction between the secular (poets or literati) and the sacred (monks).

Case study context

The Huishan spring is located at the foot of Mt. Huishan (Wuxi, China). The spring is named after the mountain. Spring water is filtered through rock strata, and thus it acquires a clean taste. The tea experts, Lu Yu and his contemporaries Liu Bochu (755–815) and Zhang Youxin (fl. 813) evaluated the Huishan spring as the second-best spring in China for brewing tea. Apart from the water quality praised by these early experts, the location of the Huishan spring also accounts for its popularity amongst the literati class. The spring was located in the centre of the Jiangnan region (the south of the lower reaches of the Yangtze River) where most of the fine-tea consumers resided (Jiang 2019).

Mt. Wuyi, in the southwest of China, is not only known for its picturesque landscape, but it is also revered by tea connoisseurs as the birthplace of both Oolong tea and black tea. The spectacular landscape and the specialty tea jointly contributed to the attractiveness of Mt. Wuyi. Since the Tang dynasty, many literati-officials visited this mountain and have written about their visitations. Ding Wei (966–1037), Fang Zhongyan (989–1052), Cai Xiang (1012–1067), and Su Shi (1037–1101) are the most representative amongst them. They were all renowned scholars and fans of Wuyi tea.

Case study 1.1   The Huishan spring and the related poems and artworks

The Huishan spring became famous in the Tang dynasty (618–907) due to the praise of Lu Yu (the Sage of Tea) and other tea experts. The tea experts agreed that the Huishan spring was the second best in China while disputing which spring was the first. Since that time, the Huishan spring has attracted literati to brew tea on-site. The most relevant literary works and art works depicting travels to the Huishan spring follow.

Li Shen (772–846), a literati-official in the Tang dynasty, wrote: "In front of the Huishan study and under the pine trees and bamboos is a sweet and cool spring. It is unearthly dew in the world. . . . Once tea has this spring water, it will give off all its fragrance of this tea" (Bie quanshi 別泉石〈有序〉 by Li Shen 李紳, in Chen n.d.).

In the Song dynasty, Wang Yuchen (954–1001) wrote a poem to record his official travel to Mt. Huishan: "I travelled to take office with a ship of books and the lute and arrived in the ancient county of Piling adjacent to the river. Having disembarked to the shore, I looked for the Ganlu Temple. When entering the town, I first asked for the location of the Huishan spring" (Wang n.d.).

Su Shi (1037–1101) wrote in a poem that he had visited the Huishan spring with a prestigious teacake. His lines read: "I have climbed almost all the mountains in the south of the Yangtze

River, but still cannot help lingering on this hill. With the Little Dragon Round Teacake from heaven, I came to try the second-best spring in the world" (Su n.d.).

In the Yuan dynasty, Zhao Mengfu (1254–1322), the greatest calligrapher of his period, wrote five Chinese characters for the spring: "The Second Spring Under Heaven". Inscriptions of his calligraphy have been preserved in the spring pavilion. Zhao also wrote the following lines: "The Huishan spring is located in the ancient temple built in the Southern dynasties; I visit the Second Spring because of its reputation" (Zhao n.d.).

In the Ming dynasty, the famous artist Wen Zhengming (1470–1559) visited Huishan in 1518 with his friends and drew a celebrated painting titled "Huishan Tea Party" to depict the gathering (The Palace Museum 2021). Five literati participate in the tea gathering. Two of them sit around the well fence, and the third who just arrives is greeting them. Another two literati are leisurely walking around with a guide. Several boy attendants are busy preparing tea.

Case study 1.2   Travelling for researching and producing desirable tea in Mt. Wuyi

This case study was adapted from Wu (n.d.). Mt. Wuyi is located at the junction of Jiangxi and northwestern Fujian and has been celebrated as a famous scenic site since the Tang dynasty (618–907). The fame of Mt. Wuyi was based on its unique natural landscape, the prevalence of Buddhism and Taoism, the numerous artworks and literary works dedicated to this site, and its renowned tea culture. Researching and producing desirable tea in Mt. Wuyi motivated many literati officials to visit Mt. Wuyi.

Literati-officials' travelogues, notes, and poems unveil their involvement in processing desirable tea in Mt. Wuyi. When Chang Gun (729–783) served as the governor of the Jianzhou (present Jian'ou), teacakes were steamed, roasted, and ground. This specially processed tea was named *yangaocha* (grinding-cream tea). Xu Yin (849–938), a renowned scholar in the late Tang dynasty, was the first to describe the Wuyi tea production method: "In warm spring when the moon is just full, picking tea sprouts and offering them to the god of the earth. . . . I feel the affectionateness from the gift tea and boil it with the north mountain spring" (Xu 2020, 25). His poem not only highly praises Wuyi tea but also illustrates how Wuyi tea was already an appreciated gift amongst the upper class in the Tang dynasty.

The cultural relevance of Wuyi tea reached an unprecedented peak during the Song dynasty due to the literati's wide-spreading enthusiasm for tea. Su Shi (1037–1101) described the literati craze in his poem: "Millet-sized buds sprouted out along the Wuyi creek; Ding and Cai devoted to this tea one after the other" (Su 2017, 77). "Ding" refers to Ding Wei (966–1037), while "Cai" refers to Cai Xiang (1012–1067), two leading figures in the Song dynasty, renowned not only for their scholarly achievements and political influence but also for their role in supervising the production of the elaborate Wuyi teacake (*Tuancha*) for the imperial court.

During the Ming dynasty (1368–1644), literati-officials' travelling to Mt. Wuyi continued to thrive. Literati essays show that literati-officials were involved in the innovation of tea-processing methods in Mt. Wuyi. Wu Shi (around 1642), a scholar and tea expert from Xiuning (in south Anhui), argued that the tea-processing methods of his time were outdated. When travelling in Mt. Wuyi, he had processed fresh tea leaves according to the *Songluo* technique to make a new tea product.

Yin Yingyin (around 1700) improved the tea-processing technique when he was on the post of the magistrate of Chong'an (present Wuyishan) in 1652. Yin attempted to improve the tea-processing techniques and thus invited monks from Mt. Huangshan to help transform the technique to enhance the fragrance and taste of tea leaves (Jiang 2019). According to Zhou Lianggong (1612–1672), a frequent visitor of Mt. Wuyi and a recognised scholar in his time, Yin's experimentation was very successful. Zhou valued as a treasure the few tales of this tea that he received, praising particularly its colour and fragrance. His remark reads: "Even after months, the purple-red colour remained the same" (Zhou 1985, 12).

Discussion

The why of travelling to noted spring sites

Since the Tang dynasty, Chinese scholar-officials and literati have shown a strong interest in the mountain springs such as the Huishan spring. Whenever possible, they would travel to famous springs for brewing tea on-site. When they had no access to the renowned springs, they might arrange delivery of mountain spring water (Jiang 2019), regardless of the distance between the spring site and their dwelling place. This situation is especially applicable to the Huishan spring. However, since the Song dynasty, literati began to question the ranking of the Huishan spring, noting that there were numerous mountain springs in China and it was impossible for the tea experts represented by Lu Yu to taste and compare each of them (see *A Record of the Huishan Spring* 惠山泉記 by Nie Houzai 聶厚載, in Chen n.d.). It was unconvincing to say that the Huishan spring ranks second in terms of water quality for brewing tea, because the ranking of springs was based on the limited experience of the early tea experts. The Song literati Nie Houzai (around 1049) discussed the reputation of famous springs such as the one of Huishan. He wrote:

> It was difficult to differentiate springs according to water taste, and Lu Yu should have differentiated springs according to his mind rather than tongue. Many literati from that time began to prefer the famous springs Lu ranked because of their admiration of Lu as the Sage of Tea, and such literati differentiated springs with their ears rather than tongues.
>
> *(A Record of the Huishan Spring by Nie Houzai, in Chen n.d.)*

Anecdotes show that the literati who lived far from the spring site could not recognise the fake water which was labelled as the Huishan spring.

> Qiu Zhangru toured Wu County to visit me [Yuan Hongdao who was the magistrate of Wu], and returned with 30 jars of water from the Huishan spring. He went home earlier and asked his servants to carry back the water jars. His servants hated to carry the heavy jars and poured the spring water out. When they were very close to the town where Qiu lived, they filled the jars with river water. Qiu invited his friends for a tea party to sample the famous spring water. The guests smelled, tasted, and highly praised the tea water. None of them could recognise the fake water until the fact disclosed as a result of quarrel amongst the servants.
>
> *(A Record of the Hui Spring* 記惠泉 *by Yuan Hongdao* 袁宏道, *in Chen n.d.)*

This anecdote suggests that literati-officials' preference for famous springs was based on the springs' historical connection with those literati of the past and the contemporary society rather than on the quality of water for brewing tea. The calligrapher Zhao wrote that like many other literati, he visited the Huishan spring for its reputation (see Case study 1.1). The same motivation can be seen in other poets' lines through their frequent use of the term "Second Spring" to refer to the Huishan spring. What essentially motivated the literati-officials' travelling to the spring site was its historical and cultural connection with tea connoisseurship. This finding is consistent with the feature of Chinese literati traveller behaviours, centring on the history of the place of interest (Wang 2006).

The fascination of noted springs as the ideal water source for tea brewing communicated two meanings for the literati travellers. First, this strong interest in the mountain springs helped scholar-officials build a cultural identity and differentiated them from the merchant class and the commoners (Jiang 2019). Travelling to the spring sites provided opportunities for the literati class to show their superiority over other social classes who could not understand the cultural values attached to the well-known springs (in the previous quotation, for instance, Qiu's servants failed to understand the intangible value of the water taken from the Huishan spring).

Second, travelling to the famous spring sites also meant opportunities to interact with the literati's scholar-official friends and Buddhist monks (Jiang 2019). Literati-officials were enthusiastic about building social networks, and thus they would visit the local administrators of stopovers; meanwhile, the latter were usually ready to show hospitality to their travelling colleagues (Zhang 2011). Furthermore, most of the famous springs sites were located in scenic mountainous areas and managed by Buddhist monks, who were excellent makers and connoisseurs of fine tea. Interacting with these tea experts added value to scholar-officials' travel to the spring destinations. In short, the scholar-officials' travels to the renowned springs surpassed the material level of consumption and became a symbol of social status. Since the Tang dynasty, the literati-class constructed one layer of their social identity through the cultivation of an aesthetic taste of the mountain springs. This practice closely reminds one of their predecessors in the Jin dynasty (265–420), who initiated the aesthetic appreciation of natural landscapes to differentiate them from other social classes (Yan 2015). Construction of social identity through tourism activities continued in imperial China. Wu's (2003) study of literati travels in the Ming dynasty also reveals this feature of Chinese literati travel.

The why of travelling to famous mountains to produce the desirable tea

Literati-officials' pleasure of having tea with friends could also be attained in towns. However, the full gratification resulting from making tea following the prescription of historical and literary sources could only be achieved in mountains, because fresh tea leaves must be processed immediately after being plucked. In his book *The Classics of Tea*, Lu Yu explained how to prepare tea with the correct tea instruments and how to process it properly. Since that time, tea connoisseurship has become a must-learn skill for Chinese literati. They had to know the right instruments for containing and brewing tea and the right water to release the fragrance of tea. Some of them managed to process tea in person. The desire to comply with Lu Yu's prescriptions about tea motivated their travels to tea-producing mountains across China.

Literati-officials and the religious practitioners belonged to the educated class, and these two groups maintained a subtle interaction in imperial China. Tea functioned as the social media between the secular and the sacred. Besides debating on Buddha's dharma or Taoist tenets, literati travellers enjoyed producing tea with monks or Taoists.

Most tea trees were grown on the mountain lands, and the tea leaves were processed by resident monks (Benn 2015). This was also the case in Mt. Wuyi: "There were certain ranks of Wuyi tea . . . the best was rock tea. All the rock tea was picked and processed by resident monks and Taoists . . . it was not available on market" (Liu cited in Chen 1999, 721). Processing tea with Buddhists and

Taoists who were tea masters met literati-officials' travel motivations of tacit knowledge-seeking and showcased their privileged access to the rare tea resources.

Making the desirable tea in famous mountains enhanced literati-officials' privileged social status and established social distinctions based on their access to the rare teas. Tea played an important role in literati-officials' social gatherings; serving an unusual tea that was unavailable on the market and crafted by the host in a famous mountain was an effective way to impress the guests. The marvellous tea made by literati-officials during their travelling to famous mountains suggests two symbolic meanings. At the material level, the taste of the desirable tea had to be unique, for it was a response to literati-officials' special requirement for a certain fragrance, obtained through a special processing method. This specially crafted tea was very rare compared to standard, mass-produced tea available on the market, and it could not be purchased even at high prices. At the sociocultural level, literati's travels to famous mountains for seeking tea knowledge and processing their own special tea surpassed the material level of consumption to become a symbol of social distinction (Xiao 2017).

Conclusion

This chapter shows that Jolliffe's (2007) definition of tea tourism is applicable to the investigation of the early tea travellers in imperial China. This special interest type of tourism concerned social elites of imperial China since the seventh century. This study reveals that the literati-officials were motivated by a strong interest in the tea-related heritage of the noted spring sites and famous mountains. The research results add to our understanding of tea tourism history, disclosing that the early tea tourism was driven by the development of connoisseurship of fine tea and spring water, as well as by the desire to research and produce tea. In their journeys to take new posts, Chinese literati-officials enjoyed visiting well-known spring sites near their stopovers for brewing tea and gathering with local officials and friends. Literati-officials' strong interest in mountain springs reflected their intentions to build a historical connection to the spring sites. This aesthetic taste helped to build their social identity. Moreover, literati-officials also travelled to the famous mountains to seek knowledge about tea production methods, aiming at participating in or supervising tea processing. Their interaction with Buddhist and Taoist masters of tea not only added a pleasant element to their journeys but also strengthened their social distinction.

A limitation of this study is that other social classes were excluded from the investigation. Chang (2004) has mentioned that lower social classes would follow the literati class in their taking spring water for tea brewing. Furthermore, this study does not discuss how commoners with tea expertise could climb the social ladder. Future studies are thus suggested to investigate the learned behaviours of other social classes on their mobility related to taking the spring water and brewing tea, or to explore the commoners whose social status was changed because of their tea expertise. In line with this point, the social interaction of the literati class with the mundane experts of tea is worthy of scholarly attention.

Discussion questions

1   Apart from travelling to locations of mountain springs, were there any other ways that Chinese literati officials adopted to showcase their preference of using the noted spring water to brew tea?
2   Would other social classes such as the merchant class have the similar interest in travelling to famous spring sites to brew tea? Why or why not?
3   To what extent did the literati officials, monks, and Taoists promote innovation in tea-processing methods in China?
4   Apart from those specialty teas from famous mountains presented in this chapter, can you list other famed teas originating from scenic mountainous areas?

References

Benn, J.A., 2015. *Tea in China: a religious and cultural history*. Honolulu, HI: University of Hawaii Press.

Chang, C., 2004. *The travel life of the Ming dynasty people*. Yilan, Taiwan: Research Team of the Ming Dynasty History (in Chinese).

Chen, M.L., ed., n.d. The Gujin Tushu Jicheng (Imperial encyclopedia) 欽定古今圖書集成 [online]. Available from: https://zh.m.wikisource.org/wiki/欽定古今圖書集成/方輿彙編/山川典/第098卷 [Accessed 22 March 2022].

Feng, L.R., 2008. *Youthful displacement: city, travel and narrative formation in Tang tales*. PhD Thesis, Columbia University.

Jade Tea, 2020. Water, essence of life and tea [online]. Available from: www.jadetea.co.uk/blogs/cupping-notes/water-essence-of-life-and-tea [Accessed 26 March 2021].

Jiang, Y., 2019. *More than just a drink: tea consumption, material culture, and 'sensory turn' in early modern China (1550–1700)*. PhD Thesis, University of Minnesota.

Jolliffe, L., ed. 2007. Connecting tea and tourism. *In*: Jolliffe, L., ed. *Tea and tourism: tourists, traditions and transformations*. Clevedon: Channel View Publications, 3–20.

Liu, L. and Wu, Y., eds., 2020. *Zhonghua chawenhua gailun (An introduction to Chinese tea culture)*. Beijing: Beijing University Press.

Liu, Q., 1999. Wuyicha (Wuyi Tea). *In:* B. Chen, ed. *Zhongguo chawenhua jingdian (China tea classics)*. Beijing: Guangming Daily Press.

Pu, Y., 2015. An analysis of the social functions of the tea-drinking custom in Southwest China during Tang-Song dynasties. *Canadian Social Science*, 11(1), 211–217.

Ridgway, B.B., 2005. *Imagined travel: displacement, landscape, and literati identity in the song lyrics of Su Shi (1037–1101)*. PhD Thesis, University of Michigan.

Su, S., 2017. The lychee lament 荔枝嘆. *In:* Wang, S.Y. and Zheng, X.S., eds. *Overseas collection of lord Suwenzhong* 蘇文忠公海外集. Haikou: Hainan Press, 77.

Su, S., n.d. A visit of the recluse Qian at Huishan東坡全集/惠山謁錢道人烹小龍團登絕頂望太湖 [online]. Available from: https://zh.m.wikisource.org/wiki/惠山謁錢道人烹小龍團登絕頂望太湖 [Accessed 22 March 2022].

The Palace Museum, 2021. Huishan tea party 惠山茶會圖 [online]. Available from: www.dpm.org.cn/collection/paint/228278.html [Accessed 26 March 2021].

Wang, L., 2005. *Tea and Chinese culture*. San Francisco, CA: Long River Press.

Wang, X., 2006. Travel and cultural understanding: comparing Victorian and Chinese literati travel writing. *Tourism Geographies*, 8(3), 213–232.

Wang, Y., n.d. A letter to Piling Liuboshi小畜集钞/寄毗陵劉博士 [online]. Available from: https://zh.m.wikisource.org/wiki/寄毗陵劉博士 [Accessed 22 March 2022].

Wu, C.H., 1993. Tea-drinking lifestyles among the Ming dynasty literate groups. *In*: Xu, X.H., ed. *Proceedings of the first Chinese food culture symposium*. Taipei: Foundation of Chinese Dietary Culture, 279–307.

Wu, J.S., 2003. Travel and consumption culture in late Ming China: a case study of the Jiangnan region. *Bulletin of the Institute of Modern History*, 41, 87–143 (in Chinese).

Wu, S., n.d. Wuyi Zaji Yijuan 武夷雜記一卷 [online]. Available from: http://kanji.zinbun.kyoto-.ac.jp/kanseki?record=data/FANAIKAKU/tagged/5458061.dat&back=3 [Accessed 26 March 2021].

Xiao, K., 2013. Skill and magic in rock tea making – argument on traditional handicrafts in intangible cultural heritage (in Chinese). *Folklore Studies*, 6, 83–90.

Xiao, K., 2017. The taste of tea: Material, embodied knowledge and environmental history in northern Fujian, China. *Journal of Material Culture*, 22(1), 3–18.

Xu, Y., 2020. Shangshu Hui Lamiancha 尚書惠蠟面茶. *In:* Ye, G.S., ed. *Selected readings of ancient Chinese tea literature works*. Shanghai: Fudan University Press, 25.

Yan, L., 2015. What landscape meant for the early medieval Chinese gentry. *Asia Pacific Journal of Tourism Research*, 20(11), 1195–1211.

Zhang, C.E., 2011. *Transformative journeys: travel and culture in Song China*. Honolulu, HI: University of Hawaii Press.

Zhao, M., n.d. A travel poem of Huishan 松雪齋集/留題惠山 [online]. Available from: https://zh.m.wikisource.org/wiki/松雪齋集_(四庫全書本)/全覽 [Accessed 22 March 2022].

Zhou, L., 1985. *Min xiaoji (A short book on Fujian)*. Fuzhou: Fujian People's Press.

2

TEA TOURISM AND ROUTE HERITAGE

Nakeli village on China's Ancient Tea Horse Road

Gary Sigley

Introduction

The tea mountains (*chashan*) of Southwest China refers to the hilly country on both sides of the Lancang River (Mekong) just as it is about to exit Yunnan Province and enter mainland Southeast Asia. Situated in a subtropical latitude with an altitude ranging from several hundred to several thousand metres, the environment is ideal for tea cultivation. Indeed, the ethnic communities in this region are among the most ancient of tea cultivators. Quite possibly, some argue, humanity's engagement with the tea leaf began in these forests where wild specimens of *Camellia sinensis* var. *assamica* can still be found.

Puer tea is the local tea variety. The tea was typically grown as trees in groves with some specimens reaching venerable ages counted in hundreds of years. After harvesting, sorting, withering, rolling and drying (Zhang 2014, 91), steam compression was used to form the tea into round tea cakes. Once the cakes were wrapped in rice paper, they were piled into batches of seven which were then wrapped in bamboo leaves and twine. This 'seven cake tea stack' (*qizibing*) was then ready for transporting via horse and mule caravans across the mountainous terrain of Southwest China to tea markets as far away as Lhasa. Some of the premium tea made its way to the dynastic court in Beijing as 'imperial tribute tea' (*gongcha*).

This network of tea trading routes was named I 'Ancient Tea Horse Road' (*Chamagudao*) in 1992 by six Yunnanese scholars who regarded tea as the essential shared element of ethnic identity in the Southwest (Mu *et al.* 1992). The naming of this heritage route coincided with China's long decade of double-digit growth. As incomes on the eastern seaboard began to rise, as transport infrastructure was improved, and as domestic tourism became ever larger, the Ancient Tea Horse Road (hereinafter also 'tea road') was used by tea manufacturers and tourism providers as a symbol for toponymic branding. The 1990s and 2000s thus not only witnessed the emergence of numerous Chinese studies of the Southwest tea routes; it was also a time when the concept of the tea road left its scholarly confines to become a well-recognised signifier in China for a touristic tea modernity. In this sense the tea road is an expression of how different actors in China in the present relate to and understand the past.

In this connection, this chapter considers how the development of tea tourism – or more precisely 'tea road tourism' – has been experienced at the village level. In particular, I take the tea road staging post (*yizhan*) of Nakeli in Puer (formerly Simao) as a case study to examine how local government and village households took advantage of a natural disaster (an earthquake) to redevelop the village's economic focus from subsistence agriculture to tea tourism. In terms of cultural heritage

DOI: 10.4324/9781003197041-4

discourse, which here includes heritage tourism, the concept of authenticity has been a central focus for determining the standard by which to measure the nature of tourist sites (that is, do they conform to an objective measure of authenticity). The approach adopted here sidesteps the authenticity quagmire. Instead of determining whether the development of tea route heritage tourism in Nakeli conforms to a certain authorised heritage standard (I leave that to other heritage experts), I divide the tea road into two categories. The first is the 'old Ancient Tea Horse Road'. This refers to the body of discourse concerned with the history and preservation of the remaining tangible and intangible tea road heritage. This is primarily the focus of scholars in fields such as history, archaeology, linguistics, folk studies and so on. The second is the 'new Ancient Tea Horse Road', which I use to refer to the various ways in which the tea road as a signifier for an imagined, and often nostalgic, past has been incorporated into the design and construction of tea route tourist sites and motifs. Hence, rather than taking authenticity as a set standard from which to measure deviation, I see it functioning as a concept across a broad spectrum of meaning. Together with other colleagues, I have referred to this approach as a form of 'relational authenticity' (Su *et al.* 2020). In short, Nakeli is taken as an example of the new Ancient Tea Horse Road and a salutary expression of how the past is imagined and experienced in the present.

Literature review

Taking its lead from the success of the Silk Road as a means of capturing the touristic imagination, locations across the world that have connections to tea cultivation, processing, transportation and exchange are exploring how tea can be used as a means of toponymic branding and connecting different heritage sites in a linear visitation chain. Georgia, Russia, Mongolia and Mauritius, for instance, have all begun to develop tea route tourism. This has been stimulated by the emergence of 'cultural routes' as an official category of UNESCO World Heritage. Gaining inscription as a UNESCO cultural route, as is the case of the Great Tea Road (*Wanli chadao*) that starts in Fujian and terminates in St. Petersburg, is a sure way to increase brand recognition and gain plaudits far and wide.

In this context of greater recognition, this chapter draws upon research focusing on tea routes and tea tourism in the People's Republic of China (PRC). Since the early 1990s when the concept of the Ancient Tea Horse Road was coined, research in China on the tea road has expanded dramatically (Ling *et al.* 2018). Many books and articles have been published exploring different aspects of the tea road and the culture of the peoples en route. This has now been expanded to include tea roads in other parts of China (Bei Chamagudao Yanjiuhui 2011). Not surprisingly, a large body of this literature is devoted to the development of tea road tourism (Liu 2009; Ling 2020). By comparison, there are few publications in English. Some have provided historical and ethnographic overviews of tea road history and puer tea culture in the contemporary context (Zhang 2014; Xiao 2019). Others have explored the issue of cultural heritage and touristic modernity (Sigley 2010, 2013, 2020).

The major contribution of the tea road tourism literature has been to further highlight the complexities of balancing the preservation and revival of cultural heritage whilst using commodification and commercialisation to promote economic development. This is a difficult task to accomplish to the satisfaction of all stakeholders – the government, the locals, the developers, the visitors and the cultural heritage experts – and China has tea road tourism examples of success and failure. What the literature reveals, at least those that engage in detailed case studies, is the important relationship between local government and the host community in the Chinese context. The case of Nakeli is no exception. In this regard, one area where research on tea road tourism, and tea tourism in China more generally, could benefit is with more in-depth engagement with comparative contexts in other countries (Shen and Antolín 2019). This would help identify those features which have particular 'Chinese characteristics' and those which are of a more universal value.

Methodology

The research that forms the basis of this chapter employs a discourse analysis and mixed methods approach. This approach entails employing qualitative and quantitative methods to gather a diverse array of viewpoints which are derived from multiple sources. After completion of a thorough reading of the Chinese tea tourism field, and with the support of local experts, the village of Nakeli was chosen as the case study. Relevant government reports, academic studies, news items and tourist commentary on Nakeli covering the period from 2007 to 2021 were collected and analysed. This was supplemented by site visits in 2017, 2018 and 2019. During the site visits, in addition to general observation, interviews were conducted with local officials, villagers and visitors. With this accumulated data at hand, a textual and discourse analysis of the major themes was undertaken. The results of this research as they relate to the distinction between the old Ancient Tea Horse Road and new Ancient Tea Horse Road are presented in this chapter.

Case study context

Nakeli is an administrative village (*xingzheng cunluo*) located in Ning'er Yi and Ha'ni Autonomous County, Puer City, Yunnan Province (see Figure 2.1). As an administrative village, the actual village

Figure 2.1 Ancient Tea Horse Road of Southwest China

Source: Designed by An Zhuolin

of Nakeli, which is the focus here, is the administrative centre for 15 villages and hamlets. Nakeli village, at an altitude of 1280 m and with an average annual temperature of 20°Celsius, is nestled between two large hills which form part of the very mountainous terrain that characterises the region.

According to figures from 2014, Nakeli consists of 69 households with a population of 268 persons. In 2017 the average household income was 10,261 yuan. The main economic activities are tea cultivation and leisure tourism, but previously it was subsistence agriculture, with some households benefiting from the highway traffic as motor-inn service providers. In addition to tea, which is a relatively new crop, most of the villagers still grow their own produce and raise domestic animals. In that regard Nakeli is still a functioning village, but as we shall see, one that has to adapt to new circumstances. The villagers themselves come from different ethnic backgrounds including Dai, Bai and Han, but the two main groups are Yi and Ha'ni ethnicities.

Nakeli is situated midway between Puer (24 km) and Ning'er (22 km). Puer was historically, and remains to this day, the major wholesale distribution centre for puer tea. From Puer the tea made its way by horse and mule caravan either towards the provincial capital of Kunming and then on towards central China or, more significantly in terms of route heritage tourism, towards Lhasa passing through the major market towns of Yunnan – and now predominantly tourist towns – of Dali, Lijiang and Shangrila (formerly Zhongdian) before entering Tibet. Nakeli was one of many staging posts en route in a journey that could take up to six months from Puer to Lhasa. During its heyday as a staging post Nakeli provided caravansary services for muleteers, travellers and their beasts of burden. At peak seasons Nakeli could be host to hundreds of animals and dozens of muleteers.

Case study 2.1   Nakeli village

On one side of present-day Nakeli, running in a north-south direction, is National Highway 213 (completed in 1958 and upgraded in 1997); on the other is the Kunming-Bangkok International Expressway (operational from 2011). Through the middle, and through the village, are the remnants of the old road nowadays known as the Ancient Tea Horse Road. This preponderance of roads highlights Nakeli's strategic location on the route from the areas of tea production to those of tea consumption. In short, the story of Nakeli is the story of roads. Each instance of road construction and development has meant a new chapter in the story of Nakeli. Combined with an understanding of the prevailing socioeconomic background, in this case that of the transition from the 'socialist planned economy' (1949–1990) to that of the 'socialist market economy' (1992–present), we can understand how local governments and residents have adapted to changing circumstances.

With the cessation of caravans in the 1950s, Nakeli descended into relative obscurity. Nakeli's engagement with travellers continued in the form of the motor-inn (*qiche lüguan*). It became just one of many fuel and rest spots for truck drivers on the new highway (which given the mountainous terrain was still a slow and difficult journey). During this era of the socialist planned economy there was no tourism, no private cars and no emphasis on cultural heritage preservation at the local level. The tea caravan trade became a memory that over time faced possible extinction both physically and mentally.

Nakeli's fortunes changed once again with the onset of the policies of 'reform and openness' (*gaige kaifang*) in the early 1980s. In 1983 the collective economy was dismantled and households were allocated their own plots of land. This indicated the beginnings of a revival in entrepreneurial activity and a mindset getting ready to engage with the broader socioeconomic changes that were taking place across China. Highway 213 went through a major upgrade in 1997, and by the early 2000s business in Nakeli was brisk as the Yunnanese economy continued to expand. Yet overall, Nakeli still remained primarily engaged in subsistence agriculture with some cash income from tea cultivation and a few households benefiting from running motor-inns and roadside restaurants. The arrival of the expressway, however, meant that from 2011 Nakeli was no longer an overnight stop, and the motor-inn business came to an abrupt end (Zhang and Gao 2015). New business opportunities remained to be explored. Hence, although the tea road was gaining fame throughout the Southwest and many sites across Yunnan were exploring how to capitalise on this concept in terms of tea production and tea tourism, Nakeli's connection to the tea road remained undeveloped.

However, on 3 June 2007 Nakeli and other nearby settlements were devastated by a 6.4-magnitude earthquake. Most of the heritage buildings associated with the tea caravan trade were damaged beyond repair. The village now faced not only the extinction in heritage memory but also in tangible terms. All that remained was the remnant old road consisting of a cobble stone path 4.377 km in length ranging from 1.5 to 2 m wide, along with the Rongfa Caravansary (*Rongfa madian*) and the Bridge of Wind and Rain (*Fengyuqiao*). However, as has been the case in other tea road towns hit by either earthquakes (Lijiang Earthquake 1997) or fire (Zhongdian Fire 2014), the local government took crisis as opportunity and produced a tourism development plan in which Nakeli was rebuilt for the specific purpose of promoting tea road tourism.

The model that was developed is very typical of such endeavours across China. Three major stakeholders came together in partnership to develop tourism: the local government, a commercial tourism developer, and the village residents. The Nakeli tourism development plan was released in August 2010. Some of the main physical items include an Ancient Tea Horse Road Heritage Museum, a Visitor Reception Centre, artisan workshops and a cultural plaza. Various other attractions, such as the water mill, are reproductions of pre-earthquake buildings whilst others are completely new, such as the village waterfall. A great deal of capital was also used to upgrade and enhance water treatment, public toilets, village beautification, and car and tour bus parking facilities (Qi 2020). In other words Nakeli, with large investments from government and the active participation of residents, began to undergo a major facelift that would dramatically alter the physical appearance and sense of connection to Nakeli's route heritage past and tourist mobility present. In the context of the revival of cultural heritage across China, I take these to be indicative of the expansion of the socialist market economy.

Some of the villagers, who now had homes rebuilt in a unified traditional style (including families moved from vulnerable sites in the surrounding hills), began to engage in home-stay tourism (*nongjiale*) in which visitors were encouraged to stay with resident families and experience up close the lives of the residents (Zhao 2019). This took the form of overnight stays but more typically was reduced to banqueting and short-term visitations of several hours (highlighting the transient nature of automobile tourism). Other residents lacking in capital had the opportunity to rent one of the new stalls at the refurbished village entrance where they could sell trinkets and/or local food items with an ethnic flavour. One important change was the removal of domestic farm animals to specialised pens outside the village (a significant change in village atmospherics). On a typical day, several hundred people will visit Nakeli. At peak times this could exceed 1000. Most of the visitors are domestic Chinese tourists with a large proportion being visitors from the provincial capital of Kunming on road trips to Puer and Xishuangbanna. Kunming has one of the highest per capita private ownership rates in China.

Since the reconstruction Nakeli has been awarded various prizes and commendations from local, provincial and national governments and institutions covering the areas of tourism, leisure, heritage, ecology, ethnic harmony and civilisation. On top of the list, the remnant road and what remains of the village tangible heritage were included in the 2013 State Council incorporation of the Ancient Tea Horse Road into the 'Seventh National Cultural Artefact Protection List' (*Diqipi guojia wenwu baohu danwei mingdan*). In March 2017 the Ancient Tea Horse Heritage Park, which includes Nakeli, was included as one of 30 sites across Yunnan to receive special attention in the province's plans to develop cultural industries (*wenhua chanye*). Hence from the view of the 'authorised heritage discourse' (Smith 2006), Nakeli has been thoroughly absorbed into the mainstream narrative of 'cultural revival' (*wenhua fuxing*). Nakeli is, therefore, an excellent case study for examining the balance between heritage preservation and revival and commercial development.

Analysis

Tourism has radically changed Nakeli. It has physically changed the village through the reconstruction in ways specifically designed to facilitate and promote the visitation experience. It has economically changed the village by shifting the everyday focus from subsistence agriculture to a more direct engagement in the modern economy through the provision of cash crops such as tea and as service providers through tea road tourism. It has socially changed the village through reworking the villagers' conception of themselves in relation to the broader society, through altering levels of household income and creating a gap between richer and poorer families, and through watering down the strength of kinship connections in favour of non-kin business relationships. In short, with Nakeli's integration into the socialist market economy, the village has experienced a range of transformations that are shared with countless other contexts across the length and breadth of the PRC.

What makes Nakeli's experience different is its ongoing connection to the past through tea route heritage. The village's strategic position on the major tea routes connecting Puer to the rest of China and beyond has meant that 'roads' have always played an important role in Nakeli's economy and identity. The roads, however, in themselves do not determine the nature of that relationship. Rather, it falls to the creation of certain conditions to facilitate the emergence of a touristic modernity. Some of those which have been examined above include the improvements of transport infrastructure. But the most significant has been the position of the government as both the central strategiser of local development and as the sorting-house for articulating the broader discourse of 'cultural reconstruction' (*wenhua chonggou*), which in Nakeli's case has focused on exploiting its connections to the Ancient Tea Horse Road.

If we take a strict definition of authenticity as our point of departure, we may conclude that Nakeli is more of a theme park than a heritage park (perhaps the two are one and the same with just slight variations). However, if we focus on the formation of the new Ancient Tea Horse Road, we can see that Nakeli's experience with contemporary heritage discourse and tea tourism is sufficient in its own right for us to take the views of government, villagers and tourists seriously. That is, we can understand authenticity as existing on a range from hard to soft, or what is noted above as a form of relational authenticity. As Du and Yang (2018, 91) note, the tourists who visit Nakeli are in essence experiencing a modern form of leisure tourism with a focus on pleasant surroundings, good food and local hospitality dressed up in the veneer of the tea road. This is in essence what the new Ancient Tea Horse Road is all about, and in so doing Nakeli continues its connection to the past as a 'road service provider'.

Conclusion

This chapter has examined the development of tea route tourism in Southwest China taking the village of Nakeli, a historic staging post on the Ancient Tea Horse Road, as a case study. It shows how the past is itself a specific kind of resource that can be used to reimagine a location's connections with the broader world and market itself as a site that tells part of the bigger picture of tea mobility. This is supplemented by the mobility of the tourists themselves who visit Nakeli not so much to learn about the past but to experience leisure and entertainment in the context of a built environment that uses the past as a framework for toponymic branding. As in many contexts across the globe, the tensions between preserving and reviving an authentic past play out with the imperatives of servicing the developmental plans of government and the desires of tourists. This is a continuation of Nakeli as a service provider. But whereas that service provision once focused primarily on the immediate needs of caravans and trucks for rest and sustenance, in the contemporary context tourists also demand to experience the past as part of their visit. This shift to a focus on the past as a commodity in its own right is a new twist in Nakeli's socioeconomic history. Whilst its experience is similar to other sites on tea heritage routes across China, how it compares to similar sites in other countries remains to be explored. The author hopes that this chapter will be a catalyst for some to begin this research journey.

Discussion questions

1   How has route heritage, in this case relating to tea distribution, been used in the contemporary context to develop tea tourism? Find examples in China and elsewhere.
2   What does the author mean by positing a distinction between the old Ancient Tea Horse Road and the new Ancient Tea Horse Road?
3   In what ways is the case study of Nakeli indicative of the experience of a touristic modernity?

References

Bei Chamagudao Yanjiuhui (Northern Ancient Tea Horse Road Research Society), ed., 2011. *Zhongguo Bei Chamagudao yanjiu (Research on the Ancient Tea Horse Road in Northern China)*. Beijing: Shijie Zhishi Chubanshe.

Du, G. and Yang, Y., 2018. Xiandai xiangcun lüyou xia de lishi wenhua chonggou yu lüyou tiyan: Yi Yunnansheng Ningerxian Nakelicun wei gean (Tourist experience and historical cultural reconstruction under the conditions of modern rural tourism: the case of Nakeli, Ninger County, Yunnan Province). *Qujing shifan xueyuan xuebao (The Journal of Qujing Normal University)*, 37(1), 88–92.

Ling, W., 2020. *Chamagudao yu qianniuhua (The Ancient Tea Horse Road as Morning Glory)*. Kunming: Yunnan daxue chubanshe.

Ling, W., Luo, Z. and Mu, J., 2018. Chamagudao yanjiu zongshu (A summary of Ancient Tea Horse Road research). *Yunnan shehui kexue (Yunnan Social Science)*, 3, 97–106.

Liu, X., 2009. Shijie yichan shiyexia de Chamagudao lüyou kaifa: Jianlun Chamagudao de shijie yichan jiazhi (Tourism development of the Ancient Tea Horse Road from the perspective of World Heritage). *Lüyou Luntan*, 2(2), 199–204.

Mu, J., Chen, B., Li, X., Xu, Y., Wang, X. and Li, L., 1992. *Dianzangchuan dasanjiao wenhua tanmi (Cultural Exploration in the Triangular Region of Yunnan, Tibet and Sichuan)*. Kunming: Yunnan jiaoyu chubanshe.

Qi, K., 2020. Yunnansheng Nakelicun baohu yu fazhan guihua (Yunnan Province Nakeli village preservation and development plan). *Jianzhu yu wenhua (Architecture and Culture)*, 6, 22–27.

Shen, J. and Antolín, J., 2019. Chinese tourism in cultural routes: the Camino de Santiago and the Ancient Tea Horse Road. *PASOS: Revista de Turismo y Patrimonio Cultural*, 17(4), 811–826.

Sigley, G., 2010. Cultural heritage tourism and the Ancient Tea Horse Road of Southwest China. *International Journal of China Studies*, 1(2), 531–544.

Sigley, G., 2013. The Ancient Tea Horse Road and the politics of cultural heritage in Southwest China: regional identity in the context of a rising China. *In:* Blumenfield, T. and Silverman, H., eds., *Cultural heritage politics in China*. New York: Springer, 235–246.

Sigley, G., 2020. *China's route heritage: mobility narratives, modernity and the Ancient Tea Horse Road*. London: Routledge.

Smith, L., 2006. *The uses of heritage*. London: Routledge.

Su, X., Sigley, G. and Song, C., 2020. Relational authenticity and reconstructed heritage space: a balance of heritage preservation, tourism, and urban renewal in Luoyang Silk Road Dingding Gate. *Sustainability*, 12, 5830. https://doi.org/10.3390/su12145830.

Xiao, K., 2019. From highlands to lowlands: the Pu'er tea trading network and ethnic group interactions in the frontier of Yunnan 1662–1796. *In:* Lipokmar, L. and Baruah, M., eds., *Objects and Frontiers in modern Asia: Between the Mekong and the Indus*. London: Routledge, 93–108.

Zhang, J., 2014. *Puer tea: Ancient Caravans and urban chic*. Seattle, WA: University of Washington Press.

Zhang, J. and Gao, M., 2015. Cong shengsi xiangyi dao jianbei liqu: Yunnan Kunman gonglu yanxian Nakelicun de lu renleixue yanjiu (From a matter of life and death to gradually abandoned: Road ethnography, Nakeli village and the Kunming to Bangkok expressway of Yunnan). *Yunnan shehui kexue (Yunnan Social Science)*, 4, 98–104.

Zhao, X., 2019. *Xiangcun lüyou jingji fazhanzhong shehui guanxi bianhua yu shehui ziben xingcheng yanjiu: yi Nakelicun wei yanjiu duixiang (Research on the change of social relationship and the formation of social capital in the development of rural tourism economy: Taking Nakeli village as the case study)*. Masters Thesis, Yunnan University, School of Ethnology and Sociology.

3

TEA AND SPIRITUAL TRAVEL

Panchen Tuo Tea

Jianming Li, Min (Lucy) Zhang and Yanting Gu

Introduction

Tea is produced in many countries and consumed all over the world. It is intrinsically linked to many tourism forms such as culinary, cultural and agro-tourism (Jolliffe 2007). Although tea tourism is not as well established as wine tourism, many tourists in numerous countries have begun to see opportunities for tea-related tourism. Similarly, pilgrimage travel in different religions has been a common phenomenon since ancient times. However, spiritual and religious tourism is one of the most understudied areas (Timothy and Olsen 2006), although the growing interest in spiritual travel or spiritually oriented tourism has raised some interesting questions and discussions, such as travel patterns of religious tourists, economic effects of religious tourism and the management of holy sites, among others. The connection between tea and spiritual and/or religious tourism has not been much explored within tourism discourse. This chapter seeks to explain how tea and spiritual travel could be connected through embodied practice.

Literature review

Theory of embodiment

Since Descartes put forward the ontology of dualistic opposition between spirit and material and the epistemology of dualistic opposition between subject and object, the body and mind of human beings have been separated from each other. Cottingham *et al.* (1984) believed that only rational thinking of spiritual consciousness could lead man to the truth. Later on, philosophers such as Bacon (2000) questioned the way human beings understood the world through their senses; thus, the body became a burden and an obstacle to reach the truth. Influenced by this philosophical thinking, the research of social sciences tended to divide society and individuals into two levels, that is, body and mind (He 2013).

In recent years, in the process of reflecting on research objectification, the social sciences have drawn on the view of embodiment and emphasized the role of the body in the cognitive process. In particular, attention has been paid to the interactive relationship between human body and consciousness (Ye *et al.* 2020). The theory of embodiment suggests that the body is a basic tool that allows individuals to interact with others in a society, through which the specific understanding of the world is attained. In this vein, human cognition is not a priori logical ability but a continuous

DOI: 10.4324/9781003197041-5

and situational process (He 2013) formed in the interaction between body, mind and environment. Therefore, anthropological study should make efforts to investigate embodied experiences which affect individuals' self-construction by understanding how they interact with the external social culture.

Spiritual travel

Spiritual travel is often based on religious tourism (Novelli 2005). The relationship between and definitions of spiritual and religious travel too often overlap in tourism literature (Cheer *et al.* 2017). There have been several interesting studies conducted in various geographical locations focusing on spiritual travel. For example, Heidari *et al.* (2018) and his colleagues from Iran stated that spirituality aims to understand the deep meaning of religious beliefs and to address the interdependence of physical, emotional, mental and spiritual activities such as meditation. Wilson *et al.* (2013) observed that tourists feel *energized, inspired* and *uplifted* in spiritual travel. Although the previous research is undoubtedly important, it fails to elaborate individuals' subjective motivations for spiritual tourism and the process as to how it might take place (Cheer *et al.* 2017).

A seminal work by Turner and Turner (1978) that first identified three key elements of pilgrimage tourism – namely, motivation, journey and destination – has been widely used in understanding religious and spiritual travel (Shinde 2015). In detail, spiritual travel can be motivated by wellness and healing, personal quest and self-improvement, socialization, recreation and leisure (Cheer *et al.* 2017), as well as by religious pursuit at sites including Islamic mosques, Buddhist temples and Christian churches. According to Shinde (2015), a spiritual journey is driven by devotions, which provide a spiritual and moral renewal beyond the scope of a particular religion or faith. A journey can be also a personal religious practice through mindful activities which may contribute to a spiritual recovery (Heffernan *et al.* 2016). In this sense, spiritual destinations, such as mosques, churches and temples, can be defined as those places where people pursue religious purity (Shinde 2015). Conversely, Singh (2006) believed that a spiritual destination can be an abstract spiritual *home* where people are embraced with inner peace, the almighty and forgiveness. This chapter supports the view that no place is intrinsically sacred, but rather a process and the result of social construction through personal experiences and interpersonal interactions within a certain sociocultural environment.

Studies on tea-related travel and tourism has been connected mostly to history, traditions, culture, consumption (Jolliffe 2007; Jolliffe and Aslam 2009), and agriculture (Kajima *et al.* 2017), among others. In contrast, researchers from China investigated topics in a wider range, such as the connection between tea and religion, particularly in ethnic regions (Ding 2012). Interestingly, Yang (2020) observed that the combination of tea and Zen practice was a strong motivation for Chinese tourists to travel to sacred destinations. Chinese Zen Buddhism (*Chanzong*禅宗) was considered to facilitate the understanding of one's mind by immersing into the nature. In this sense, tea was an essential medium between body and spirit. Other examples from the literature included the role of tea in pro-poor projects (Li 2019) and tea cultural itineraries like the Ancient Tea Horse Road (Shi 2002). However, little is known about the connection between tea and spiritual travel with a theoretical analysis of how they are intrinsically linked. This chapter attempts to contribute to address this research gap by adopting the theory of embodiment as a guiding conceptual framework.

Method

The theory of embodiment emphasizes the role and function of the human body through experiences in various situations or contexts. Thus, the human cognitive process is characterized as embodied, experiential and situational (Shinde 2015; Ye *et al.* 2020). Furthermore, the view of interaction in the theory of embodiment pays attention to the significant role of interpersonal and social interactions in understanding others. Therefore, this chapter uses qualitative and interpretive data because it involves studying people in a certain social context (Denzin 1994).

A case study strategy is adopted, which is deemed appropriate because such studies are used to develop new insights and knowledge about contemporary phenomena within a real-life context (Yin 2003, 13). The selected case refers to a typical tea industry representative, Ms. Cai, who has been running a tea business within a well-organized network of the Xiaguan tea brand and its franchise in Kunming, the capital city of Yunnan Province. Previously, she worked as a tour guide in the (religious) tourism industry. Based on her personal experiences in both tea and travel industries, she might further consider opportunities and the potential to integrate these two businesses in the future. Semi-structured interview questions concerning three aspects – namely, motivation, journey and destination of her spiritual travels – aim to interpret how tea, as a significant medium of social interactions, played a crucial role in her transformation of way of life and individual value in the society.

Case study context

Panchen Tuo Tea, a type of tightly pressed tea, was first produced in Xiaguan, Dali Prefecture of Yunnan Province back in 1902. Evolving from the Tuan Tea of the Ming dynasty (14th–17th centuries), it has maintained the traditional art of tea making for hundreds of years. The producing craft of the Panchen Tuo Tea follows complicated steps and most of them are manual operations, such as raw material preparation, screening, picking, weighing, steaming, rolling, pressing, moulding, drying and packaging. Thus, various values of the Panchen Tuo Tea in economy, culture, traditional local knowledge and aesthetics have been highly appreciated and widely recognized. As such, it was designated as one of China's Intangible Cultural Heritage items in 2011, aiming for its better protection and promotion.

The origin of Panchen Tuo Tea is southwest China, in Yunnan Province, a multi-ethnic region composed of the Hanzu and 25 ethnic minorities. Yunnan is abundant in both natural and cultural resources. It is adjacent to Tibet in the northwest and Myanmar, Laos and Vietnam in the south. The climate belongs to the subtropical plateau monsoon type. Many ethnic groups mastered (and still master nowadays) the specific technology of tea planting, which made Yunnan the most important tea production area in China and notable in the world. In ancient times, the tea from Yunnan was transported by horse caravans to Tibet and surrounding countries through a commercial network named the Tea-Horse Ancient Road (Shi 2002).

The tea produced in Xiaguan was named after the Tibetan Buddhism Master Panchen due to his visit to the Xiaguan Tea Factory in 1986. Two years later, in memory of Master Panchen and the friendship that he built between Yunnan and Tibet, Panchen Tuo Tea became a symbol of compassion that the great Master left to the Tibetan and Yunnan people.

Tibetan Buddhist temples might be the most important places for Panchen Tuo Tea consumption for three reasons. First, tea sharing is inseparable from daily Buddhist chanting activities. Vitamins provided by tea are essential for monks' daily nutrition. In addition, tea helps to digest greasy food like beef and mutton at the Tibetan plateau, where fresh vegetables are rare. Thus, tea provides the material support to religious practices like daily chanting. Second, tea functions as a sacred object to Buddha and as a gift to temples. For instance, donors often pray for the blessing of the Buddha by offering tea to temples and masters. Third, tea-drinking is a symbolic rite to integrate new monks into the social organization of the temple.

Case study 3.1   Panchen Tuo Tea

Motivation

Influenced by her grandmother's practice of Buddhism since her childhood, Ms. Cai has always held respect for religions. In her early career, she worked as a tour guide and acquired abundant knowledge about Buddhism. Gradually, during her trips and experiences with the Buddhist tour groups from Southeast Asia, she began to feel that it was difficult to go further in her understanding of Buddhism, as she described:

> During ten years of taking tour groups to Tibet, I gradually felt the appeal of religion. However, faith is difficult to judge with rational knowledge. I feel that I'm standing at the edge without knowing how to truly get into this seemingly wonderful spiritual world.

When Ms. Cai was suffering from her confusion, she happened to find that the Panchen Tuo Tea produced in her hometown was frequently consumed in the daily diet of Tibetan monks. The unexpected discovery led her to learn about the Panchen Tuo tea, and since then, she began to run a tea business as an amateur while enjoying tea tasting with her friends. Transferring from tourism to the tea industry, as most friends came into contact with tea and Buddhism, she was encouraged to practice Buddhism, starting from reading classic scriptures.

However, practice is never easy. Scriptures were so difficult to understand that Ms. Cai almost wanted to give it up. Every time she felt that she could not bear the suffering any more, a familiar scene always came to her mind, reminding her of the persistence of young monks in Tibet. Young lamas needed to learn how to make Panchen Tuo milk tea (made from the Panchen Tuo Tea and fresh Tibetan yak milk) and serve it to the older ones when they were chanting sutra. For these young lamas, it seemed that they were doing nothing important; however, in the process of tea making and serving, they were totally immersed in the religious atmosphere, in which young monks devoted themselves with body, mind and soul. With this scene deeply imprinted in Ms. Cai's mind, she became aware that it was rather impossible to practice alone. Instead, she needed to find a way to immerse herself in the Buddhism context for a breakthrough. Coincidentally, a friend working in the Panchen Tuo Tea business was practicing at Wuming Buddhist College in Ganzi Tibetan Autonomous Prefecture, Sichuan Province. Ms. Cai decided to join in and took 20 pieces of Panchen Tuo Tea with her on her journey of practice.

Journey

Religious journey

Wuming Buddhist College was founded in 1880 and rapidly expanded after the 1990s. Practitioners came from all over China increased its enrolment from fewer than 100 to tens of thousands. On her arrival, Ms. Cai was arranged to practice in a small class with ten mates under the guidance of a Lama Master. The Panchen Tuo Tea became an important medium for Ms. Cai to communicate with other practitioners in her spare time. Many Han practitioners were curious about the history and stories behind the Panchen Tuo Tea, through which Ms. Cai built strong friendship ties. As the difficulty of practice increased, half of the practitioners retreated while ten pieces of the Panchen Tuo Tea were finished. Ms. Cai piously extended her practice to the end of the 20th piece of Panchen Tuo Tea. Her practice was much appreciated by her Master:

> Your practice is just like the Panchen Tuo Tea that you are about to finish. Buddhism advocates emptiness; like your tea is about to finish, your practice will soon enter the realm of emptiness as well.

Up to this point, the Panchen Tuo Tea is not only a symbolic appearance of material, but also a transformer and facilitator for Ms. Cai's self-cultivation and improvement.

Social journey

Since 2000, several journeys from Yunnan to Tibet have been organized by the Xiaguan Tea Factory to show respect and gratitude to Tibetan Buddhism Masters and monks for their support of Panchen Tuo Tea. Ms. Cai participated in these activities by offering the Panchen Tuo Tea to temples, as well as donating to poor families along the journey.

In the process of presenting the Panchen Tuo Tea as a gift, Ms. Cai came to understand the social life of things related to tea. As a commodity, the process of tea being transported from Yunnan to Tibet was a social re-embedding process, during which the social life of tea was alienated from a biological vitality to an undifferentiated commodity, and then it was endowed with religious significance and sanctified into a sacred object. From the hands of tea-picking women to tea-making workers, tea was produced in the specific shape of Buddha's hair bun before it flowed to Tibetan areas and became a necessary living object for monks and the wider public. In this sense, Tibetan monks and the public endowed the Panchen Tuo Tea with a strong religious meaning, through which the nature of the Panchen Tuo Tea as a commodity gradually declined as the symbolic and religious meaning gradually strengthened. The sanctification process of the Panchen Tuo Tea provided Ms. Cai great insights and inspirations; thus she started to concentrate on the Panchen Tuo Tea business while practicing Tibetan Buddhism and looking for her spiritual destination.

Destination

Spiritual and religious destinations do not merely refer to some physical spaces like mosques, churches and temples but to any places as long as they are endowed with spiritual and/or religious meaning. Since no place is intrinsically sacred, it means that any place could be transformed and constructed into a spiritual destination through certain experiences and social interactions. Ms. Cai returned to Kunming, Yunnan, after six months' collective practice in Wuming Buddhist College. It seemed that she finished the spiritual journey, but in fact, she just arrived at the real destination, as she described:

> I thought Wuming was my destination of the spiritual travel. However, after coming back to Kunming and starting to practice in my daily activities, including the Panchen Tuo Tea sales, I came to realize that my tea shop is truly the destination of my body, mind and soul, because the Panchen Tuo Tea is no longer merely a commodity, but

a unity of spirit and material, which continuously provides me with peace of mind and religious experience of secular everyday life. Of course, thanks to the journey to Wuming, I was "awakened" by getting to know wise Buddhism masters, nice friends and enjoyed the specific experiences in the unbelievable atmosphere.

Ms. Cai's tea shop is a franchise chain store of the Xiaguan Tea Factory. The most prominent feature of the tea shop is the spatial layout of its Tibetan Buddhist culture. A Tibetan Buddhist thangka is hanging on the wall in the centre of the tea shop. In the portrait, the statue of Tibetan Buddhist Green Tara is worshiped. On the right side of the thangka is the golden mandala, while on the left side of the thangka is another smaller thangka and the sutra tube. A brand trademark of "Panchen Tuo Tea" in black is hanging below the thangka on a red background. The award label of the time-honoured brand of Panchen Tuo Tea and the sign of Chinese Intangible Cultural Heritage appear on both sides of the trademark. In front of the statue, Ms. Cai has placed a table, where she offers tributes such as fruits and candies to the Buddha statue.

Analysis

Embodied practice: the transformation from economic capital to cultural capital

Since the 10th Panchen Lama visited the Xiaguan Tea Factory and expressed the demand of the Tuo Tea in Tibetan areas, the Tuo Tea was gradually given the name of "Panchen Tuo Tea". This symbolized that the tea, as a commodity, had become a precious object with certain significant religious meaning.

Taking Ms. Cai as an example, those who were involved in the Panchen Tuo Tea business gained easier access to the Tibetan Buddhist system through long-term tea marketing activities and close contacts with monks in religious surroundings. Gradually, the Panchen Tuo Tea became a significant symbol in the gift circulation which linked individual practice and the Tibetan Buddhist world. In this process, the religious significance of Tibetan Buddhism was embodied in the attitude of tea dealers towards commodities through the medium of the Panchen Tuo Tea. The religious symbolism of the Panchen Tuo Tea in Tibet, in turn, inspired the spiritual world of tea entrepreneurs. For instance, the profit-focused economic activities have gradually shifted to religious and social experiences to promote personal cultivation and social welfare. Thus, tea played an important role in the process of the transformation from economic capital to cultural capital, which might further become a resource for tea-related cultural tourism and spiritual travel.

Spatial practice: overlapping the commercial and religious spaces

In ordinary tea shops, a tea table is usually located in the central position, which is used to serve potential consumers to taste tea. However, with the deepening of Ms. Cai's religious practice, the symbols of Tibetan Buddhism, like the thangka, incense and altar, occupied the central position of her tea shop. She *converted* the tea table to an everyday practice space where religious and spiritual issues were discussed with friends or guests who were interested in religion and tea culture. In this way the religious space was overlapped with the commercial space, forming a compound space with dual functions and significance.

The Panchen Tuo Tea has experienced the transformation from an alienated commodity to an embodied religious symbol with new social vitality. In the process of reimagining the Panchen Tuo Tea, great changes have taken place in Ms. Cai's spiritual world. She deeply realized the meaning of emptiness in Buddhism and applied this mentality into her understanding of commodity exchange. Having such a mindset, she began a new way of life. Reflected on the arrangement in space, religious elements were integrated into the regular space of commerce in the process of practice embodiment.

Conclusion

Guided by the theory of embodiment, this chapter investigated how the tea merchants in Yunnan Province of China transformed the religious significance of tea as a commodity into their religious practice and social activities through travel. Firstly, the case study context introduced how the Panchen Tuo Tea, produced by the Xiaguan Tea Factory, became an indispensable living object in the religious practice of monks in Tibetan Buddhist temples. Then the analysis of a typical case revealed that the Panchen Tuo Tea was identified as a transformer through the description of personal experiences and social interactions, from the three aspects of motivation, journey and destination. In particular, the religious practice during Ms. Cai's personal spiritual journey was facilitated by tea sharing, whereas her social journey further extended the range of individual tea-related spiritual travel. Unexpectedly but naturally, it enacted social responsibilities of a specific merchant group from the tea industry. Meanwhile, tea businesses and tea-related travel as a carrier of religious and social practice incubated a novel opportunity for entrepreneurship in the region. The chapter finally discussed how the businessmen could subtly change their value of life from instrumental rationality (profit orientation) to value rationality in pursuit of the meaning of life value through the spiritual and religious travel experiences related to the tea business.

Discussion questions

1   In what ways is tea connected to spiritual and religious travel and tourism?
2   How have Ms. Cai's two careers of tour guide and tea company representative prepared her for future involvement in tea tourism?
3   What strategies could be adopted to enact the social responsibility of regular tourists through tea-related spiritual travel?

References

Bacon, F., 2000. The new organon. *In:* Jardine, L. and Silverthorne, M., eds. *Cambridge texts in the history of philosophy*. Cambridge: Cambridge University Press.

Cheer, J.M., Belhassen, Y. and Kujawa, J., 2017. The search for spirituality in tourism: toward a conceptual framework for spiritual tourism. *Tourism Management Perspectives*, 24, 252–256.

Cottingham, J., Stoothoff, R. and Murdoch, D., 1984. *The Philosophical Writings of Descartes*, Vol. II. Cambridge University Press: Cambridge.

Denzin, N.K., 1994. The art and politics of interpretation. *In:* Denzin, N.K. and Lincoln, Y.S., eds. *Handbook of qualitative research*. Thousand Oaks, CA: SAGE, 500–515.

Ding, J., 2012. Research on tea custom culture of De'ang Nationality from the perspective of religion. *Journal of Yunnan Minzu University (Philosophy and Social Sciences Edition)*, 3, 96–98.

He, J., 2013. *Body image and body schema: a study of embodied cognition*. Shanghai: East China Normal University Press.

Heffernan, S., Neil, S., Thomas, Y. and Weatherhead, S., 2016. Religion in the recovery journey of individuals with experience of psychosis. *Psychosis*, 8(4), 346–356.

Heidari, A., Yazdani, H.R., Saghafi, F. and Jalilvand, M.R., 2018. The perspective of religious and spiritual tourism research: a systematic mapping study. *Journal of Islamic Marketing*, 9(4), 47–798.

Jolliffe, L., ed., 2007. *Tea and tourism: tourists, traditions and transformations*. Clevedon: Channel View Publications.

Jolliffe, L. and Aslam, M.S., 2009. Tea heritage tourism: evidence from Sri Lanka. *Journal of Heritage Tourism*, 4(4), 331–344.

Kajima, S., Tanaka, Y. and Uchiyama, Y., 2017. Japanese sake and tea as place-based products: a comparison of regional certifications of globally important agricultural heritage systems, geopark, biosphere reserves, and geographical indication at product level certification. *Journal of Ethnic Foods*, 4(2), 80–87.

Li, H., 2019. Tea and tourism integration and targeted pro-poor strategy: a case study of Hainan Province. *Social Scientist*, 6, 68–75.

Novelli, M., 2005. Niche tourism: an introduction. *In*: Novelli, M., ed. *Niche tourism: contemporary issues, trends and cases*. London: Routledge, 1–11.

Shi, S., 2002. The Tea-Horse Ancient Road and its historical and cultural value. *Tibetan Studies*, 4, 49–57.

Shinde, K.A., 2015. Religious tourism and religious tolerance: insights from pilgrimage sites in India. *Tourism Review*, 70(3), 179–196.

Singh, R.P.B., 2006. Pilgrimage in Hinduism: historical context and modern perspectives. *In:* Timothy, D.J. and Olsen, D.H., eds. *Tourism and religious journeys*. London: Routledge, 220–236.

Timothy, D.J. and Olsen, D.H., 2006. *Tourism and religious journeys*. Routledge.

Turner, V. and Turner, E., 1978. *Image and pilgrimage in Christian culture*. New York: Columbia University Press.

Wilson, G.B., McIntosh, A.J. and Zahra, A.L., 2013. Tourism and spirituality: a phenomenological analysis. *Annals of Tourism Research*, 42, 150–168.

Yang, X., 2020. Zen-tea: Chinese Buddhist practice. *China's Religion*, 6, 74–75.

Ye, H., Guo, L. and Ma, Y., 2020. Enaction and dynamics: the theory of interaction in embodied cognition. *Psychological Exploration*, 6, 483–488.

Yin, R.K., 2003. *Case study research: design and methods*. 3rd ed. Thousand Oaks, CA: SAGE.

4

WILD TEA AND INDIGENOUS TOURISM

A case from Vietnam

Cuong Duc Hoa Nguyen and Mai Chi Nguyen

Introduction

Drinking tea became a part of Asian culture nearly 2000 years ago. This cultural heritage manifests in various aspects of life, including tea cultivation, processing, cultural rituals, festivals, cosmetic products and foods, or merely as a daily beverage. Tea heritage tourism takes advantage of distinctive dimensions of the tea culture in designing and delivering tea-themed tourism activities and services for the benefit of the local communities.

In Asia, wild tea trees are usually grown in remote highland forest areas where ethnic minority people reside. While wild tea is less popular, with a much smaller cultivation area and lower productivity, it is also conceivable that wild tea provides natural aroma and taste and is perceived as a chemical-free beverage and food ingredient. Also, it helps preserve important biotic genes and is used to produce precious nursery plants for larger scale cultivation. Shan Tuyet tea (*Camellia sinensis* var. *Shan*, or Snow Shan tea) was imported into Vietnam long ago and has been grown in many mountainous areas. This wild tea variety has been transplanted widely in Moc Chau District, contributing to making the destination a premier tea cultivation centre in Vietnam. Although tourism has been developing sharply over the last two decades in Moc Chau District, the local ethnic minority communities, which are the owners of great Snow Shan tea fields, can only engage in very simple tourism activities such as small tea shops or cheap local food services, hence gaining slight benefit from the industry.

After fully discussing relevant concepts and methodologies, this chapter will illustrate best practices and challenges in developing types of wild tea–related tourism by demonstrating lessons drawn from a case study. The chapter has the two following main purposes: first, to transfer knowledge of the history and culture of heritage tea production and tea tourism development and tourist involvement with and attachment to tea tourism heritages in the ethnic minority–inhabited areas of the northern mountainous region of Vietnam; and second, to promote indigenous tourism for improved livelihoods in local tea plantation areas.

The four specific research questions include the following. First, what are the unique historic and cultural dimensions of Snow Shan tea in Moc Chau District? Second, how has tea heritage tourism evolved in Moc Chau District? Third, what are the benefits and challenges for the tea tourism industry in local destinations? Fourth, how can tea heritage indigenous tourism be developed and benefit indigenous people living in local tea production areas?

DOI: 10.4324/9781003197041-6

Literature review

Concept of wild tea versus cultivated tea

Given that most tea comes from plantations, definitions and classifications of cultivated tea varieties or plantation methods are diverse. Darjeeling tea from West Bengal in India is grown in the Kurseong area at altitudes of 600–2000 m under a cool and moist climate with a minimum annual rainfall of 50–60 inches. The main types of Darjeeling processed teas include black; green (natural aroma, using water below the boiling point); Oolong (semi-fermented, fragrant flavour and fruit, sweet aroma); white (finest varieties, unmatched subtlety, complexity, natural sweetness); and blended, flavoured and scented teas (Mahua and Parthajoy 2015). The term "wild tea" applies to a wide variety of tea-growing situations and includes tea grown on plantations with minimal pruning, fertilization and pesticides; tea grown in forests, also with minimal pruning and no fertilization and pesticides; and tea grown in abandoned plantations covered by the natural forest ecosystem. Wild tea also refers to ancient tea trees growing in the forest (Chowaniak *et al*. 2021).

Concept of tea heritage tourism

As Aslam and Jolliffe (2015) stated, the long history of tea embraces a number of heritage aspects. This includes tea as a traditional beverage, culture and livelihood, plantation and processing. Heritage tourism explores distinctive elements of tea culture for the purposes of tourism: providing authentic tourist experiences and products such as visits to tea houses and tea-tasting customs; participating in tea planting, picking and processing; or staying at historic tea-themed lodgings. As a niche, tea tourism includes tea festivals and events, tea tourism experiences, tea tourism marketing and management, and responsible and community-based tea tourism.

Tea tourism can contribute to significant benefits of the host community and tea-related industries, in line with the UN Sustainable Development Goals in general and specifically to the Year 2020 theme (tourism and rural development) of the World Tourism Organization. Guo (2016) studied the effects of tourism on the tea industry in Wuyi Mountain (China) in terms of planting, processing, sales and new operating modes, finding it is important to promote the tea culture for the sustainable development of the tea industry. Guo (2016) concluded that to satisfy different demands of tourists for tourist products, Wuyishan City should integrate elements of tea into the activities of eating, walking, accommodation, amusement and shopping, extending the application of tea products.

Guo and Rato (2019) analysed the development of tea planting and production in the Thai Nguyen Province of Vietnam from a historical perspective. Nowadays tours to experience tea culture in this destination have become quite popular with tourists seeking knowledge and scenic landscapes, and for some, the indigenous cultural identities of the local ethnic minority people. Packages usually include sightseeing (combined with nearby tourist attractions), experiencing tea production and harvesting, and tea tasting.

Indigenous people and tourism

Based on the above understanding of wild tea, it is obviously connected with indigenous people inhabiting the naturally forested areas and their ethnic culture. According to the UN, an individual is recognized as "indigenous" when they are self-identified as such and are accepted by the indigenous community as a member. These people have distinct social, economic and political systems with a distinct culture. Indigenous peoples are the holders of unique languages, knowledge systems and beliefs, and they possess invaluable knowledge of sustainable natural resources management practices. Indigenous tourism refers to tourism development within those indigenous communities that wish to open up to tourism. Also, the UN has provided specific recommendations on developing sustainable indigenous tourism in which emphasis must be given to human rights, preservation of traditional culture, community participation and empowerment, and bringing equitable benefits to indigenous people through design of indigenous tourism products and services, investments into support facilities, or transparent and accountable mechanisms for revenue distribution (UNWTO 2019). Guidelines from UNWTO as a response to the continuing crisis caused by COVID-19 recommended to place indigenous communities at the centre of recovery plans, create an enabling policy environment for indigenous-led tourism and strengthen skills to leverage indigenous peoples' resilience (UNWTO 2021).

Social networks and online travel agency platforms in the tourism industry

With the advent of information technology, social media networks and online travel agency (OTA) platforms are increasingly becoming effective tools in both marketing tourism destinations and selling local tourism offerings. Social network sites (SNS) enable users to make visible their social networks, engage in networking or look to meet new people and communicate with others. When joining an SNS, users are prompted to identify others in the system with whom they have a relationship, labelled as friends, contacts or fans. Most sites allow users to leave messages on friends' profiles as comments or to send a private message. Some SNS have photo-sharing and video-sharing capabilities and/or have built-in blogging and instant messaging technology (Boyd and Ellison 2008).

Besides SNS, OTAs are also an effective marketing channel. Tourists tend to prefer new experiences and book their accommodation through non-traditional channels, for example online booking with Airbnb for hostels, villas and homestays. Airbnb is an online platform where people can upload rental information about their properties for tourists, although these accommodations do not feature the standard amenities of a hotel (Guttentag 2016). In addition to accommodation, there are Airbnb Experiences, which are activities with the objectives of (1) being led by local people who love their cities, (2) in small groups and (3) of high standards with unique access. This is a great channel to promote tea farm visits associated with sustainable tourism and authenticity.

Research gaps and framework

Although a great effort has been made by academia to identify tea as a unique tourist attraction and assess the potential for local people to get involved in the tourism sector, there is limited research about designing specific tea tourism experiences and services, promoting and selling tea tourism offerings. This includes solutions to maximize participation and benefits to indigenous people living in tea production areas. These are important rationales to determine the purposes and the questions of this research, as mentioned above, and to devise a framework as follows (Figure 4.1).

Methodology

This chapter mainly employs qualitative research methods. First, it reviews relevant academic literature about wild tea, tea tourism, indigenous tourism, and social network and online distribution in the tourism industry in order to clarify theoretical concepts and issues related to the research topic. Second, the authors collected secondary data and information related to background knowledge of the history and the cultural dimensions of Snow Shan tea, tea production and consumption, and tea tourism development in the studied areas. Sources of secondary data and documents come from local government agencies, company reports, media agencies and previous research documents. Third, regarding the primary data, the authors could only conduct two remote in-depth interviews

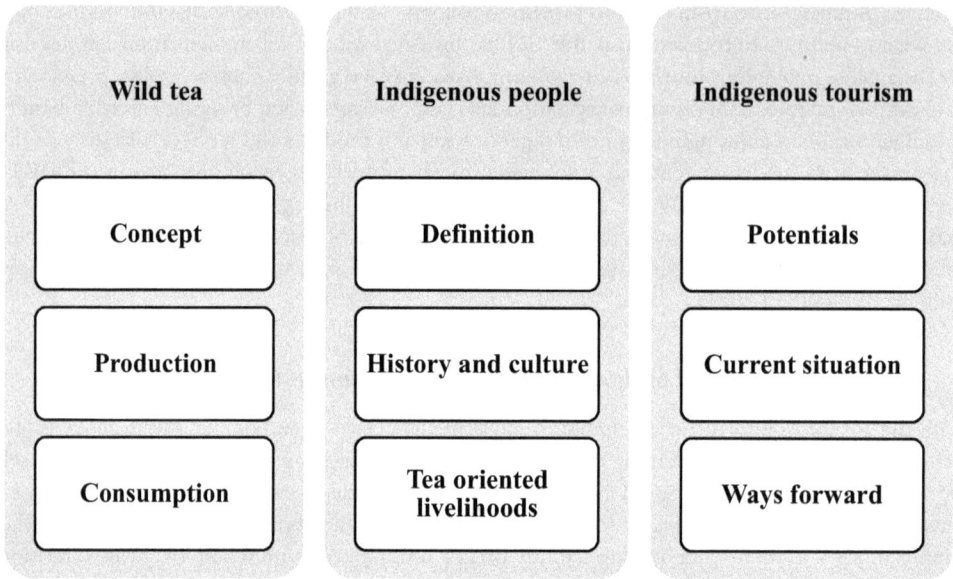

Figure 4.1 A framework for wild tea and indigenous tourism development
Source: Authors

with a tea producer and a tourism service agency in the studied area given the COVID-19 travel restrictions imposed during the study time. This is considered to be the most challenging obstacle and limitation of this research. Therefore, the authors also relied on their field experiences and findings from a previous research project about agro-tourism development in Moc Chau District undertaken in 2019.

Case study context

Moc Chau is a mountainous highland and border frontier district located in the southeast of Son La Province, 180 km northwest of Hanoi, with a natural area of 1082 km2 and an average altitude at 1050 m. This region has four distinct seasons, with the outstanding feature of a temperate and cool highland climate year round. The soil in Moc Chau is reddish-brown feralite with much humus and is very suitable for cultivating industrial crops like tea and coffee. This land has become the residential area for local people and makes agriculture a key economic sector of the province (Moc Chau District People's Committee 2017).

Snow Shan tea grown in Moc Chau District is a type of wild tea imported to Vietnam by the Shan ethnic people from China (the Shan or Shan Tai) over 300 years ago. The tree has large grey-white buds, and the tea leaves are covered underneath with a layer of fine, white fluff; therefore local people call it Snow tea. Snow Shan tea has a mild, nugget aroma and a cool and sweet green taste. Tea is processed through the traditional methods of the local Hmong and Dao ethnic groups. According to ancient Vietnamese bibliographies, tea trees have existed since time immemorial under two types: household garden tea trees in the Red River delta and mountainous tea trees in the northern region. In 1882, French explorers surveyed tea production and trading routes between the Mekong and the Da (Black) Rivers, especially in the northern mountainous regions of Vietnam, and further still to the Xishuangbanna region (Yunnan Province, China), where ancient tea trees existed. In 1970, in his biochemical research, a scientist of the former USSR Academy of Sciences

42

concluded that Snow Shan tea was discovered in the high mountains of Vietnam sometime between 1918 and 1930 and is very close to the Assam black tea of India (Science and Development Newspaper Online 2017).

While tea trees, including Snow Shan tea, had existed in Moc Chau for centuries, the local tea industry only started developing since the establishment of the Moc Chau Agricultural Farm in 1958. This state-owned and state-run enterprise is the largest tea producer and trader in the district, providing substantial livelihoods for a significant number of the local people. At present, there are 16 tea enterprises with a total tea cultivation area of 1800 ha in the district. The total production volume is about 20,000 tons of fresh tea buds per year. The main consumer markets for Snow Shan tea are Japan, Taiwan and Pakistan. The two villages of Cho Long and On, located only 6 km away from Moc Chau Township, currently conserve the largest population of about 500 ancient Snow Shan tea trees from which nursery plants can be produced, which are then widely transplanted in local tea farms or gardens. Besides traditional black tea products, local tea producers use all parts of the tea tree to create more product lines, including green tea, tea bags, matcha powder and instant milk tea.

Moc Chau changed its geopolitical position and destination image in 2014, when the Vietnamese prime minister approved the Master Plan for Tourism Development in Moc Chau National Tourism Area up to 2020, with a vision to 2030. Two years later, the destination welcomed more than one million tourist arrivals for the first time, satisfying an important criterion of recognized national tourism areas. This is an important milestone for unlocking the destination's potential for tourism development in different sectors of the tourism industry, including indigenous tourism.

Case study 4.1   Tea heritage tourism in Moc Chau District

Tea industry's impacts on the livelihood of the local people

Tea cultivation has been providing jobs and stable income for the ethnic minority people in Moc Chau. Interviews made with local tea producers and traders during a field trip to Moc Chau District in 2019 revealed interesting findings about impacts of tea production in the studied area. First, local tea enterprises provide local farmers with tremendous capacity building support. A Hmong ethnic farmer in Cho Long Village said:

> Moc Chau Tea JSC Company or Moc Suong Tea Ltd. Co. dispatched their technicians to raise awareness on tea production as sustainable livelihood and gave technical advice. Moreover, they provided financial loans, nursery plants, pesticides and fertilizers, purchased tea output and settled full payment for us. As a result, many ethnic minority households in our village have had a stable job and earn a living through tea production.

Second, local people seem more confident in tea production jobs when tea plantation production proves to be an important economic activity in the region. A veteran from the Regiment 280 who migrated to live in the District since the establishment of the Moc Chau Agricultural Farm in 1958 admitted:

> Our initial thought was that tea cultivation could merely help reduce poverty. Gradually, we could see high productivity and stable selling prices from tea products, and

thus decided to make this a permanent job. We have now 2 ha of tea fields, generating 30 tonnes of tea output worth of 300 million VND [or 13,200 USD] a year.

Finally, some tea farm owners find alternative livelihoods by providing tea-related experiences and travel services:

My son had built a tourist lodge with 10 cottages and 20 rooms on a tea field. His lodge provides not only accommodation service but also tours to local tourist attractions like Dai Yem Waterfall or Heart-shaped tea fields, foods and drinks, and taxi services. He could create regular employment for 8 people on an average income of 350 USD a month.

(Song, A.L., Vinh, D.D., and Lien, N.T., personal communication, April 21, 2019)

Tourism development in Moc Chau

The local tourism industry is growing rapidly, with over 1.25 million tourist arrivals, including over 1.18 million domestic and 67,000 international arrivals in 2019, reaching an average annual growth rate of 21% in the period 2016–2019. Revenue reached over 47 million USD in 2019. Table 4.1 depicts the main outcomes of the tourism industry in Moc Chau.

Figure 4.2 A tea hill in Moc Chau

Source: Authors' photo, 2016

Table 4.1 Moc Chau District key tourism performance statistics

Indicators	2016	2017	2018	2019
Overall tourist arrivals (thousands of trips)	1050	1150	1200	1250
International tourist arrivals (thousands of trips)	50	55	60	67
Domestic tourist arrivals (thousands of trips)	1000	1095	1140	1183
Revenue (millions USD)	38.18	41.37	44.23	47.30
Accommodation units	139	150	180	228
Number of rooms	1256	1347	1650	2200
Tourism employees	1200	1500	2800	4750

Source: Adapted from Management Board of Moc Chau National Tourist Area, 2021

Moc Chau has a rich cultural diversity of 12 local ethnic minority groups, each with distinctive local crafts, housing architecture, ethnic cuisines, indigenous customs, lifestyles and festivals. Taking advantage of its unique cultural diversity, Moc Chau District has determined community-based cultural tourism as one of the five key tourism development strategies of the Moc Chau National Tourist Area. Community-based cultural tourism programs have been developed, including homestays in stilt-houses, traditional dance and folk singing, ethnic foods and brocade souvenirs.

Marketing tea heritage tourism offerings

There are few tea heritage tourism products currently offered to visitors, including visits to heart-shaped or fingerprint-shaped tea hills, farm stays on tea fields, tea showrooms and souvenir products. An agro-tourism program named Visit Moc Chau Tea Area has been developed to draw tourists to Moc Chau National Tourist Area. A tea farm visit can be considered an authentic tourism experience when tourists participate in the process of harvesting and making tea while learning about the stories and traditions of the ethnic minority people.

Tea tourism has been promoted in certain local major events and through social networks. The Tea Festival, held annually in Moc Chau District, is one of the most important events to introduce and promote Moc Chau tea products and to honour tea producers and traders. The festival helps to enhance cooperation between businesses, tea production and trading cooperatives in Moc Chau District. In this vein, a search for tourism promotion information, particularly tea-related tourism activities and services, on social networks such as Facebook, Instagram or YouTube found few results and identified no effective channels. Moc Chau Tourism, the official fan page of the Moc Chau Tourism Management Board, has a low traffic volume with only 27 likes and 27 followers, outdated content and almost no comments. In addition, there were some postings on other pages such as Review Moc Chau (355,000 members), Vivu Moc Chau (58,000 members) and video clips on YouTube with a limited number of views.

Analysis

Wild tea management issues

Apparently, wild tea has fewer economic advantages than its cultivated tea counterpart given the nature of rare distribution, difficult access and low productivity. However, Snow Shan tea is associated with the lives of indigenous ethnic minorities; sometimes local communities consider the trees as sacred forests that protect local people. Apart from going daily to Snow Shan tea groves to pick leaves for drinking or selling raw materials in local markets, the ethnic minority people have gradually come to better understand the value of ancient tea trees, symbolizing the intense vitality of nature. They have a drive to preserve the tea trees and have begun to pay attention to exploiting added value, such as seed nurseries or associated indigenous tourism activities. Nevertheless, there are still many challenges that need to be solved in terms of raising awareness for local people, providing technical assistance and/or credit capital, or building supporting infrastructure.

Promotion of local indigenous tourism

In addition to selling in stores and promoting through events and festivals, tea tourism products and services in Moc Chau District need to be marketed and sold through more proactive and effective channels, including OTA platforms and SNS. Tea farm visits are the main type of tea heritage tourism product in Moc Chau currently commercialized on some OTA platforms. Seemingly, travel businesses are not yet keen on promoting this type of product through OTA channels. This can partly be attributed to a lack of stakeholder awareness and consumer demand at this early stage of the product life cycle. Return on investment can be lower than expected when the number of bookings from the OTA channels cannot pay off the costs of setting up and maintaining those products on sale. In contrast, SNS are the biggest source of potential customers. Facebook, Instagram and YouTube are the most popular channels to reach out and attract a large number of prospective customers.

To take advantage of SNS, companies and tourism development boards of Moc Chau District need to have a strategic plan to develop tea heritage tourism on SNS. First, they need to build a social network community with main interests in tea heritage and indigenous tourism. It is important to build the trust of prospective customers so that they can feel secure and share information about tea heritage tourism in Moc Chau District. In addition, articles must be attractive, contain adequate information and be entertaining. It is also necessary to maintain long-term relationships with past customers by encouraging peer-to-peer review, which also helps to reach new customers. Next, social networks are not simply one-way advertisements but also a general forum with high engagement. The reliable and useful information about the advantages of tea and tea heritage tourism should be shared frequently in order to interact with more customers.

Finally, it is essential to create engagement by replying to comments and maintaining conversations so that customers will share memorable moments and events from their travel, as well as things they may want more of in their future trips.

Conclusion

Snow Shan tea has both natural and indigenous cultural value. It is grown in the northern mountainous areas of Vietnam, and its value lies in its rare genetic resources, healthy seedlings and resistance to many diseases. Therefore, Snow Shan tea trees have been propagated and grown widely in many tea plantation areas in Vietnam. In Moc Chau District, Snow Shan tea has been planted on a large scale since 1958 and has contributed to creating jobs and a stable income for many local farmers, mostly the ethnic minority people. Apart from benefiting from tea cultivation and tea-based products, the ethnic minority people in Moc Chau are initially building new livelihoods, including providing local tourism services such as homestays, farm stays, food and drink, tours and taxi services.

For the development of sustainable tea-based livelihoods for the local people and local indigenous tourism, it is necessary to have tremendous supporting policies from the state, as well as cooperation and contributions from both local businesses and local people. The state government should formulate favourable mechanisms and policies on capital, land, seed resources, agricultural production materials, research and transfer of post-harvest processing technologies whilst also developing roads and other technical infrastructure in tea plantation areas and local residential areas. The local tourism industry should make efforts to market the destination and gain access to the markets for local tourism service suppliers whilst designing and investing in the construction of tea-themed tourist attractions and accommodations in the local agroecological landscapes. Local people, particularly the ethnic minority people in tea farming areas, need to have a sense of protecting the natural environment, the rare Snow Shan tea genetic resources and their indigenous cultural values.

There are some limitations of this study due to lack of field surveys in Moc Chau that are recommended to be addressed in future studies in this field. First, potential sites for building models of tea heritage tourism products have not been identified. Second, customer market demand for tea products and tea heritage tourism have not been investigated. Finally, operational resources, investment and business management capacity of local tea producers and traders and local tourism service suppliers have not yet been surveyed.

Discussion questions

1   What are the differences between wild and cultivated tea, and how do these relate to the potential for tea tourism in wild tea-growing areas?
2   In the areas where wild tea grows, how could indigenous populations there use tea resources for tourism-related activity?
3   In what ways could tea tourism in indigenous communities contribute to improving livelihoods?

References

Aslam, M.S.M. and Jolliffe, L., 2015. Repurposing colonial tea heritage through historic lodging. *Journal of Heritage Tourism*, 4(4), 331–344.

Boyd, D. and Ellison, N., 2008. Social network sites: definition, history, and scholarship. *Journal of Computer-Mediated Communication*, 13, 210–230.

Chowaniak, M., Marcin, N., Zhiqiang, Z., Naim, R., Zofia, G.S., Anna, S.S., Jakus, S., Maciej, K., Salimzoda, A.F., Usmon, M.M., Agnieszka, J., Andrzej., L. and Frorian, G., 2021. Quality assessment of wild and cultivated green tea from different regions of China. *Molecules*, 3620, 1–26. https://doi.org/10.3390/molecules26123620

Guo, G., 2016. Effects of the development of tourism in Wuyi mountain scenic spot on local tea industry and their correlation. *Journal of Food Science and Technology*, 8(1), 31–39.

Guo, Y. and Rato, M., 2019. The development of tea planting and tea culture tourism in Thai Nguyen, Vietnam. *Journal of Mekong Societies*, 15(3), 121–136.

Guttentag, D., 2016. Why tourists choose Airbnb: a motivation-based segmentation study underpinned by innovation concepts [Ebook]. Available from https://uwspace.uwaterloo.ca/bitstream/handle/10012/10684/Guttentag_Daniel.pdf?fbclid=IwAR0RE6RacHUU1L33ukVClO-M7fZ4AeqyDppk-7VmUm4YqTn_siPTHXm4LmY [Accessed 30 June 2021].

Mahua, B. and Parthajoy, B., 2015. Prospect of sustainable tea tourism in West Bengal. *International Journal of Current Research*, 7(8), 19274–19277.

Moc Chau District People's Committee, 2017. Overview of Moc Chau District. Available from: https://moc-chau.sonla.gov.vn/1306/31789/63244/gioi-thieu-chung-huyen-moc-chau [Accessed 15 March 2021].

Science and Development Newspaper Online, 2017. Origin of Snow Shan Tea in Moc Chau District. Available from www.khoahocphattrien.vn [Accessed 10 March 2021].

UNWTO, 2019. Recommendations on sustainable development of Indigenous tourism [ebook]. Available from: www.e-unwto.org/doi/book/10.18111/9789284421299 [Accessed 30 June 2021].

UNWTO, 2021. *UNWTO inclusive recovery guide – sociocultural impacts of Covid-19, issue 4: indigenous communities*. Madrid: UNWTO. https://doi.org/10.18111/9789284422852.

5

THE HERITAGE RAILWAY AND TEA TOURISM

The case of Senzu, Japan

Amnaj Khaokhrueamuang, Akari Takeguchi, Kohei Nagaoka and Koichi Kimura

Introduction

The value of heritage transportation as a recreational resource and tourism product, such as rail trails, is a relatively neglected field of research (Reis and Jellum 2014) particularly in tea tourism. The railway is a heritage mode of transport used to access tea tourism destinations and is the central focus of the visitor experience. These attributes raise the possibility of linking the value of the heritage railway as an attraction on its own to develop the tea-growing region.

The case examined here is that of Senzu, a tea-growing community in Kawanehon town in Japan's Shizuoka Prefecture and the location of the terminal station of the heritage tourist steam locomotive. Although Senzu and surrounding communities are home to plenty of tourism resources, taking the steam locomotive operated by the Oigawa Railway is the priority purpose of the trip, and other attractions are secondary. The overwhelming attractiveness of the steam locomotive journey is presumed to explain travelling along the railway line (Tsuchitani *et al.* 2014). As a result, areas along the railway line are not benefiting from the tourist trains, and the tourism resources are not fully utilised as tourist destinations.

The Oigawa Railway opened in 1927 in Shizuoka. Steam locomotives operate between Shimada City's Shin-Kanaya Station and Kawanehon Town's Senzu Station, which was built between 1930 and 1942. The line, which uses only steam locomotives, runs almost every day as a tourist heritage train accessing the tea fields. Some areas of tea fields have earned the Globally Important Agricultural Heritage System (GIAHS) designation. Steam locomotive travel is clearly superior to other offerings, such as hot springs baths or the tea tourism experience, in terms of tourist appeal. The study in this chapter, therefore, aims to optimise the value of rail transportation as a community asset and tourist attraction for creating tea tourism offerings.

Literature review

Heritage railways and tea-producing community development

Historical transports have been preserved and reutilised as tourism resources. Heritage railways in particular are a mode of transportation that uses limited resources to meet their objectives (Tillman 2002). These assets benefit the community by linking communities both socially and economically (Bird and Conlin 2014). This notion implies that liveability for inhabitants and attractiveness to

DOI: 10.4324/9781003197041-7

visitors and investors are influenced by how cultural heritage is maintained and how transport challenges are solved (Tonnesen *et al.* 2014).

In tea-producing communities where heritage railways operate, connecting the transportation asset (such as historical railways) and community resources is crucial to the policy and planning of sustainable tourism. According to Jolliffe and Aslam (2009), tea destinations were identified as exhibiting characteristics of tea-related history, traditions, ceremonies, cultivation and production, manufacturing, services, festivals and events, and retailing. Moreover, some tea-growing regions included railways that served tea and provided other services before serving tourists. Examples include the Udarata Menike in the Hill Country of Sri Lanka, the Darjeeling Himalayan Railway in northern India, the Alishan Forest Railway in Taiwan, and the Oigawa Railway in Shizuoka, Japan.

Accessing tea regions by heritage transport has increased in popularity due to tourist images circulated on social media, such as Sri Lanka's tea heritage train (Macan-Markar 2017). Although the heritage railway becomes integrally intertwined with the individual, the community, and the state (Thiranagama 2012), tea regions where heritage trains pass through have not paid much attention to benefitting from this demand by developing tea communities. As seen in the case of the Alishan Forest Railway, the Taiwan Administrative Railway has a plan to draw more visitors by establishing partnerships. However, related and supporting industries in the local community, such as travel agencies and hotels, put in only minimal effort to support the Alishan Heritage Railway (Chong 2019). In Taiwan, tea products are often promoted alongside cultural activities (Liou *et al.* 2020), but using the image of heritage trains and tea fields to draw customer attention is not a focus, even on tourism websites.

While the railway functions as both a heritage mode of transport to access tea tourism destinations and a self-contained attraction, these two attributes have not been combined to increase the value of the heritage railway by developing the tea-growing region. As Jolliffe and Aslam (2009) mentioned, tourism could enable the tea industry to rejuvenate and enhance the livelihood of tea communities. The linkage of railway tourism and tea tourism could play a significant role in sustaining the tea industry and its communities.

As railways and tea culture are relevant in this discussion, marketing heritage requires a focus on commodification and branding. According to Alexander and Hamilton (2016), replacing place identification through community heritage marketing involves three themes: community involvement in heritage activities, community heritage marketing and place identification, and facilitating community participation through Adopt a Station, a community engagement scheme in Scotland.

In Japan, developing and marketing heritage communities underpins the 'Renovation Machizukuri' scheme, meaning town development. This scheme, which aims to promote new industries and revitalise local communities, involves renovating existing buildings and land – such as vacant houses, vacant stores, and traditional buildings – to regenerate idle real estate and create high-quality employment. In addition, by increasing the value of the area, land prices will improve, and by increasing the property tax the local government's finances will be closer to a sounder business condition. This concept has been developed from the Machizukuri, which emerged in the 1960s as a way for residents to solve local community problems (Satoh 2019). Thus, linking the values of heritage railways and tea culture through tourism to develop tea communities in Japan is based on this framework.

Case description: Senzu and the Oigawa Railway

Kawanehon town, a famous tea tourism destination in Shizuoka, comprises many tea-producing communities. Senzu is one of these, and it is where the terminal station of the heritage steam locomotive, which operated for over a century, is located. Running 39.5 km for one hour and 20 minutes, it is the last station for steam locomotives departing from Kanaya Station. Due to the splendid landscape of tea fields along the Oigawa River and the charming experience of taking the steam locomotive, the trip is very popular with both Japanese and foreigners. Booking seats in advance via the Oigawa Railway's website is required. During the trip from Kanaya Station to Senzu Station, the Japanese interpreter explains the historical background of the railway and draws the attention of tourists to such attractions as Mt. Fuji, the Oigawa River, and the tea fields. Passengers on the train can buy souvenirs representing the image of the Oigawa Railway and tea region, such as a steam locomotive–shaped green tea cake. Lunch boxes and local foods are available at the Kanaya Station's store, where models and information about the Oigawa Heritage Railway are displayed. At Senzu Station, visitors to the museum can learn the history of the Oigawa Railway. Pictures and heritage objects related to the steam locomotives used in the old days are exhibited. Although there is no entrance fee, visitors are requested to donate to the museum. The information provided in the museum is in Japanese. From this station, visitors can take ordinary trains to visit other famous attractions in Kawanehon, such as the hot springs, dam, and lakes.

Research methodology

This research involves approaches used in an ethnographic study of the Senzu railway station in Kawanehon town. This study was conducted between November 2020 and February 2021 to evaluate the tourism phenomenon influenced by steam locomotive trips from Kanaya Station to Senzu Station and the surrounding area. The primary focus was the field survey and model hiking courses around the Senzu Station area introduced by the Kawane Honcho Machizukuri Tourism Association. Research instruments included local interviews (tourism organisation officer, railway staff, and residents) and participant observation alongside visual and archival data. The collected data was classified into three sections: tourism resources of the area, the heritage railway and tea tourism connection, and the Cross-SWOT analysis for strategy consideration. The authors discussed proposed guidelines for linking the Oigawa Railway and the Senzu tea-growing community to create a dynamic and thriving tea tourism destination.

Results

Examining tourism resources around Senzu Station

Senzu is the central interchange station ending the steam locomotive trip on the Oigawa Line and starting the journey with the Ikawa Line to access other tourist attractions in Kawanehon. As of 2017, Senzu Station was used by an average of 573 passengers daily. Tourism resources around the station can be divided roughly into three zones, each of which depends on one of three recommended hiking courses: the Senzu Station area, the Kawane Ryokoku Station area, and the Yangbai area.

The Senzu Station area is surrounded by commercial buildings, such as shops and restaurants. Inside the station, souvenirs are available for purchase in the station shop. Beside the station is the Oigawa Railway Museum. The tourist information office organised by the Kawane Honcho Machizukuri Tourism Association is located in front of the station, while souvenir shops, tea shops, sweet shops, restaurants, and cafés line the riverside street. Unique tea-related products on offer include Kawanehon green tea soft ice cream and green tea leaves in a cigarette-shaped package, called Chabako, a souvenir box of tea, available from the vending machine. However, no tea-related cuisine is provided in the restaurants.

The Kawane Ryokoku Station area is a residential and tea farming zone. Tourists can walk along the railway surrounded by tea gardens and some tea stores. The highlight of the nature trail is crossing the 145-metre-long suspension bridge to the natural park, where there is a monument of a teacup at the gateway. The course takes about an hour from Senzu Station.

Figure 5.1   The tea garden view in the Yangbai area
Source: Author

The Yangbai area is on the opposite side of the Senzu Station area and is separated from the Kawane Ryokoku station area by the Oigawa River and its beautiful landscape. One can start a walking trip at Senzu Station and cross the long bridge to the site, where tea gardens dot the mountain slopes. Tourists can enjoy the splendid view of the village at the tea gardens (Figure 5.1), shop at the tea stores, visit the Shinto shrine, bathe at the hot springs, and stay overnight at a Japanese-style inn. Along the walking trail, tourists will find abandoned traditional houses and vacant stores.

Investigating the heritage railway connection with tea tourism around Senzu

Tourism in the three zones of Senzu has been promoted by hiking courses and shopping, neither of which focuses on tea tourism. Domestic tourists are the main target for the hiking courses because the information and route maps are provided only in Japanese. However, inbound visitors are accommodated, as the shopping map around Senzu Station is available in English.

According to an interview with the manager of Kawane Honcho Machizukuri Tourism Association, tea tourism has been promoted in different areas of Kawanehon, such as Chameikan, a place for Kawanehon tea tasting and learning experience located near Suruga-Tokuyama Station. Images promoting tourism in Kawanehon show pictures of the steam locomotive passing through the tea landscapes (Figure 5.2). Although the steam locomotive trip ends at Senzu Station and many tourists visit here, linking tea tourism with the heritage railway to promote tea-growing areas around the station has not been implemented (Figure 5.2). Tea gardens are small in number and, due to

Figure 5.2   The steam locomotive at Senzu Station
Source: Authors

depopulation and the migration of the younger generation, tea farmers are few. Furthermore, there was intense competition with other tea-growing areas that welcomed tea tourists. Therefore, tea tourism development in the Senzu Station area is complex.

A tea seller in the Yangbai area mentioned that few tourists visit this site because many stores are closed. Most visitors taking the steam locomotive buy souvenirs in the Senzu Station area. One older gentleman in the Yangbai area was surprised to see the authors as we investigated the community, and he welcomed us. He said that young people visited the village only infrequently. However, when observing the Senzu Station area, the authors found that various tea leaves and tea-related products were sold in the station. The staff invited us to taste their tea as part of a promotion sale. We knew that Kawanehon tea is a high-quality tea and that it had been awarded first prize in a tea competition by the Ministry of Agriculture, Forestry and Fisheries. Moreover, Kawanehon teas are certified as being produced by the GIAHS designated by the Food and Agriculture Organization of the United Nations. This information means little to tourists, who are more interested in taking photos with the steam locomotive than experiencing the tea. A group of young Japanese in the cosplay attire of a famous animation enjoyed taking pictures with the steam locomotive and in the general station atmosphere. This tourism activity has been influenced by the Japanese animated dark fantasy action film 'Kimetsu no Yaiba Mugen Ressha-hen' (also known as 'Demon Slayer: Mugen Train' or 'Demon Slayer: Infinity Train'), which was released during the COVID-19 pandemic, premiering in October 2020 in Japan and in late 2020 to mid-2021 internationally. However, no tea tourism programs or tea-related activities were introduced by the Oigawa Railway in the station.

Table 5.1 Cross-SWOT analysis

Internal Factors	Strength	External Factors	
		Opportunity	Threat
		1. Strength (S) + Opportunity (O) Kawanehon tea + first prize in tea competition and the GIAHS site = premium tea products served in steam locomotives and Senzu community Steam locomotive trips to gateway of Kawanehon's attractions + Japanese animation boom = contents tourism	2. Strength (S) + Threat (T) The destination for steam locomotive trips/gateway to famous tourist attractions + competition of surrounding tourist spots = branding the community
	Weakness	3. Weakness (W) + Opportunity (O) Abandoned houses + Renovation Machizukuri scheme = abandoned houses renovated for tourism	4. Weakness (W) + Threat (T) Depopulation + disease (COVID-19) = balancing under-tourism and over-tourism

Source: Authors

Considering tea tourism development for Senzu

According to the field survey and document review, a Cross-SWOT analysis of strength (S), weakness (W), opportunity (O), and threat (T) has been conducted to propose the development guidelines (Table 5.1).

The internal factors affecting the strength and weakness of the community are considered. The strength is the novelty of the steam locomotive trip and the Kawanehon tea, which boosts awareness of Senzu Station among tourists. The weakness is the lack of tea farmers (a result of depopulation and migration), which has increased the amount of vacant property. However, the external factors concern grasping the opportunity and avoiding the threat; Senzu can have an advantage. Senzu can take advantage of its government award-winning tea and the fact that the town is located within the certified GIAHS area. COVID-19 and competitiveness with other famous tourist attractions are threats; however, these can have a positive impact when considering the strengths and weaknesses.

As shown in Table 5.1, the Cross-SWOT analysis began with S+O to find strategies that use strengths to maximise opportunities. The Kawanehon tea is a strength to integrate with opportunities, the first-prize tea competition award and the GIAHS site designation. One proposed strategy is to upgrade Kawanehon tea as the premium tea product served aboard steam locomotives and in the Senzu community. Another strategy is to promote content tourism by combining the strength of steam locomotives with the opportunities provided by the Japanese animation boom. The strategy emerging from S+T is to brand the community to compete with other tourist spots by creating the image of a destination for steam locomotive trips that is also the gateway to famous tourist attractions. As for W+O, the policy of the Machizukuri scheme to renovate abandoned houses for tourism is a proposed strategy to manage the weakness. Concerning W+T, there is the need to create a strategy that minimises weaknesses and avoids threats. Here, depopulation and an ageing society is the weakness, while the decline in tourism due to COVID-19 is the threat. Therefore, the proposed strategy considers Gowreesunkar and Thanh's (2020) notion of how a community balances under-tourism, an under-visited area receiving fewer tourists, and over-tourism.

Discussion

According to the Cross-SWOT analysis, which proposes ideas to exploit the value of steam locomotives for the touristic development of Senzu's tea-growing community, the study proposes five strategies with the following guidelines.

1. *Linking Kawanehon tea with steam locomotives by creating more tea-related products for steam locomotive trips, particularly gastronomy*

This suggestion concerns the use of Kawanehon tea's strength in attaining the tea competition award and the GIAHS site designation (S+O). Tea products can be created more than just as beverages but to increase the importance of gastronomy as a food product. As seen in the field survey, lunch boxes and green tea cakes were available for the steam locomotive trips. This popular service could be developed further to define the steam locomotive trip's uniqueness, such as tea cuisine lunch boxes and a variety of tea sweets served on the train. The Oigawa Railway and Senzu should take advantage of the fact that Kawanehon tea has received an award and been certified as a tea produced in the heritage farming system. Kawanehon is one of five regions in Shizuoka designated as a GIAHS site practising the traditional tea-grass integrated system to conserve the heritage tea-growing method and biodiversity. This can be used in valuing and branding the tea products and contacts with outsiders through tourism because it reflects the area's unique characteristics, providing tourists with meanings and stories (Kajima *et al.* 2017). Here local interpretations of heritage play an important role in place marketing (Alexander and Hamilton 2016).

Interpretation is particularly important for travellers' enjoyment and understanding of local heritage and destinations and for promoting attitude change (Hardy 2003 as cited in Camargo *et al.* 2014). As noted in the case description, the Japanese interpreter explains the history of the railway and draws tourists' attention to attractions during the steam locomotive trip from Kanaya Station to Senzu Station. This is one stage of the tourism interpretation process, which can be provided before departure or during the journey using brochures, videos, or talks by train staff (Camargo *et al.* 2014). Pre-departure tourism interpretation can be presented by referring to the package design of tea products provided at the tourist generating region, such as the label on the tea bottle (Khaokhrueamuang *et al.* 2021). Linking steam locomotive trips and the Senzu tea-growing community can thus communicate tea tourism interpretation through tea-related products.

2. *Taking advantage of the contents tourism trend by attracting Japanese animation lovers to experience tea tourism through steam locomotive trips and Senzu's tea tour*

Another strategy regarding the S+O involves applying *contents tourism*, a Japanese-English term that the Japanese government first defined in 2005 for a tourism development policy. The term is derived from popular culture intended originally for Japanese youth, especially manga and anime, but including aspects of religion, mythology, folklore, popular literature, TV and internet drama, and creative beliefs, which are frequently broadcast on electronic media and the internet (Graburn and Yamamura 2020). Contents tourism is a strategy to attract young travellers or fans interested in steam locomotives as attractions used in animation. To develop this strategy, it is crucial to identify unique tourism activities to connect steam locomotive trips and the Senzu community's tea tourism program. For example, creating model tea tourism courses or routes links the steam locomotive museum and the Senzu community with a tea trail, integrating artistic activities such as tea art festivals into tea tourism programs.

3. *Branding the tea tourism destination to differentiate Senzu from other regions*

This strategy resulted from the S+T matrix, taking advantage of the strength to manage the threat. The proposed idea is to brand Senzu to compete with other tea communities and nearby tourist attractions. Compared with other communities in Kawanehon, Senzu is not a massive tea-producing area. However, regarding place branding, it does share the image of Kawanehon tea with steam locomotive trips. Thus, Senzu should create a distinctive appearance to differentiate the area from other, similar areas. regarding strategy 1, Senzu can create an image that focuses on creating gastronomy-related tea products for steam locomotive trips. For example, it uses the slogan and logo of 'Steam locomotive tea' or 'Heritage railway tea' due to the tea tourism destination and the fact that it is a gateway to other attractions in tea regions accessed by the heritage steam locomotive. This idea applies to the product-place co-branding concept, as Ranasinghe *et al.* (2017) states it is an attempt to market a physical product by associating it with a place that is assumed to have attributes beneficial to the image of the product.

4. *Rejuvenating the tea community by renovating abandoned traditional houses to serve tourists*
The strategy of renovating heritage buildings, such as traditional houses, was derived from the W+O analysis, strengthening the weakness by taking advantage of the opportunity. The railway station's community should benefit from the significant revenue and economic benefits generated by tourist trains (Camargo *et al.* 2014). This is especially true of Senzu, which is the destination of steam loco-motive trips and the gateway to other attractions. However, tourism income seems to be allocated to enterprises around the Senzu Station area. Few tourists visit the Yangbai area, where stores and houses are vacant. Implementing the Japanese concept of the Machizukuri scheme, the heritage buildings could be renovated to create tourism income, and employment in tea workshop locations, accommodations, shops, restaurants, cafés, and museums would help promote new businesses and revitalise local communities. In Japan, the renovation of traditional buildings for tourism in some tourism destinations has managed to underpin the concept of community-based tourism, which receives investments from real estate companies (Murayama 2021). In the case of Senzu, the Oigawa Railway may consider cooperating with a real estate firm and the community to take advantage of the heritage buildings.

5. *Promoting the tea community through rural tourism as the place for balancing over-tourism and under-tourism*
Regarding the problem found in the Yangbai area mentioned in strategy 4, the strategy to minimise the weakness of small tourist numbers and avoid the threat of COVID-19 (W+T) is a focus on tack-ling the balance of over-tourism and under-tourism. Because the steam locomotive trips to Senzu are popular journeys, attracting more than 500 visitors per day, over-tourism tended to be a tourism phenomenon prior to the coronavirus pandemic. However, post-pandemic, this risk should be man-aged by allocating visitors to other areas, particularly Yangbai, through tea tourism trails. Potential tea tourism–related resources in this area include tea field slopes, tea shops, and tea rooms in the cul-tural centre. These resources can be utilised with the renovated abandoned properties, establishing them as places for rural tourism promotion through education and entertainment, such as volunteer work with tea farmers, learning Japanese tea culture, and tea cuisine. This area, conceived as an under-tourism destination receiving fewer tourists (Gowreesunkar and Vo Thanh 2020), will help revitalise the weakened tea community.

Conclusion and implications

This chapter sought to offer guidelines based on five strategies for tea community development through the linkage of the heritage railway and tea tourism. Connecting steam locomotive trips and tourism in Senzu's tea community area involves three themes, as noted by Alexander and Hamilton (2016): community involvement in heritage activities, community heritage marketing and place identification, and facilitating community participation through a community engagement scheme. Community involvement in heritage activities relates to all strategies, particularly creating tea-related tourism products. Community heritage marketing and place identification concern strategies for branding the community as a tea tourism destination, including promoting contents tourism and rural tourism to recreate learning and entertainment activities that support tea tourism. The facili-tation of community participation through a community engagement scheme is in line with the corporate renovation of heritage buildings within the Machizukuri concept, involving the railway company, real estate firms, and the community.

However, these strategy guidelines are derived from the Cross-SWOT analysis, which lacks a tourism supply and demand study. Although the approach of developing proposed strategies based on the Cross-SWOT analysis is not new, marketing heritage transport and tea tourism by integrating community involvement, branding the destination, and implementing the engagement scheme is a challenging framework to adapt in developing global tea communities. Finally, future research should focus on the opinions of tourism stakeholders toward these adaptive strategies, such as the railway company, tea business enterprises, residents, tourism organisations, and tourists.

Discussion questions

1 Why does heritage transport, such as railways, play a significant role in developing the tea-producing community?
2 How can heritage railways connect with the tea tourism community?

References

Alexander, M. and Hamilton, K., 2016. Recapturing place identification through community heritage marketing. *European Journal of Marketing*, 50(7/8), 1118–1136.

Bird, G.R. and Conlin, M.V., 2014. No terminus in sight: new horizons for heritage railways. *In:* Conlin, M.V. and Bird, G.R., eds. *Railway heritage and tourism global perspectives*. Bristol: Channel View Publications, 279–287.

Camargo, B.A., Garza, C.G. and Morales, M., 2014. Railway tourism: an opportunity to diversify tourism in Mexico. *In:* Conlin, M.V. and Bird, G.R., eds. *Railway heritage and tourism global perspectives*. Bristol: Channel View Publications, 151–165.

Chong, K.L., 2019. Review of problems in the development of rail tourism in Taiwan, on the example of the Alishan forest railway. *GeoJournal of Tourism and Geosites*, 26(3), 943–955.

Gowreesunkar, V.G. and Vo Thanh, T., 2020. Between overtourism and under-tourism: impacts, implications, and probable solutions. *In*: Séraphin, H., Gladkikh, T. and Vo Thanh, T., eds. *Overtourism*. Cham: Palgrave Macmillan, 45–68.

Graburn, N. and Yamamura, T., 2020. Contents tourism: background, context, and future. *Journal of Tourism and Cultural Change*, 18(1), 1–11.

Jolliffe, L. and Aslam, M.S.M., 2009. Tea heritage tourism: evidence from Sri Lanka. *Journal of Heritage Tourism*, 4(4), 331–344.

Kajima, S., Tanaka, Y. and Uchiyama, Y., 2017. Japanese sake and tea as place-based products: a comparison of regional certifications of globally important agricultural heritage systems, geopark, biosphere reserves, and geographical indication at product level certification. *Journal of Ethnic Foods*, 4, 80–87.

Khaokhrueamuang, A., Chueamchaitrakun, P., Kachendecha, W., Tamari, Y. and Nakakoji, K., 2021. Functioning tourism interpretation on consumer products at the tourist generating region through tea tourism. *International Journal of Culture, Tourism and Hospitality Research*, 15(3), 340–354.

Liou, B., Jaw, Y., Chuang, G.C., Yau, N.N.J., Zhuang, Z. and Wang L., 2020. Important sensory, assesion, and postprandial perception attributes influencing young Taiwanese consumers' acceptance for Taiwanese specialty teas. *Foods*, 9(100), 1–15.

Macan-Markar, M., 2017. Sweet or bitter: the 'tea train' evokes colonial legacy [online]. Available from: https://asia.nikkei.com/Editor-s-Picks/Tea-Leaves/Sweet-or-bitter-The-tea-train-evokes-colonial-legacy [Accessed 4 August 2021].

Murayama, K., 2021. *Kanko saisei (Tourism revitalization)*. Tokyo: President.

Ranasinghe, W.T., Thaichon, P. and Ranasinghe, M., 2017. An analysis of product-place co-branding: the case of Ceylon Tea. *Asia Pacific Journal of Marketing and Logistics*, 29(1), 200–214.

Reis, A.C. and Jellum, C., 2014. New Zealand rail trails: heritage tourism attractions and rural communities. *In*: Conlin, M.V. and Bird, G.R., eds. *Railway heritage and tourism global perspectives*. Bristol: Channel View Publications, 90–104.

Satoh, S., 2019. Evolution and methodology of Japanese Machizukuri for the improvement of living environments. *Japan Architectural Review*, 1–17.

Thiranagama, S., 2012. A railway to the moon: the post-histories of a Sri Lankan railway line. *Modern Asian Studies*, 46(1), 221–248.

Tillman, A.J., 2002. Sustainability of heritage railways: an economic approach. *Japan Railway & Transport Review*, 32, 38–45.

Tonnesen, A., Larsen, K., Skrede, J. and Nenseth, V., 2014. Understanding the geographies of transport and cultural heritage: comparing two urban development programs in Oslo. *Sustainability*, 6, 3124–3144.

Tsuchitani, T., Takahara, J. and Hirabayashi, W., 2014. Issues of Oigawa railway as a tourism resource (Kannkou shigen toshite no Ooigawa tetsudou no kadai). *Komazawa Geography (Komazawa chiri)*, 50, 69–80.

6

CULTURAL HERITAGE AND TOURISM

Friesland tea

Lysbeth Vink, Annette Kappert and Hartwig Bohne

Introduction

The historic region of Frisia (*Frieslande*) is situated along the North Sea and is politically divided into four parts: West Friesland and Fryslân in the Netherlands and East Frisia (*Ostfriesland*) and Northern Frisia (*Nordfriesland*) in Germany. To cite Timothy (2021, 3), "what we inherit from the past and use and value in the present day" is pertinent to this chapter, specifically as it applies to the shared inter-dependency of Fryslân and Ostfriesland on tourism and farming. Yet very little has been explored regarding how their historical past can add value to current practices, and as such, this chapter seeks to redress this omission by highlighting how their shared cultural heritage of tea could offer both regions opportunities for culinary mapping, increase tourism expenditure and provide links between tea cultural heritage and tourism within the wider European context.

Initially, the chapter offers insights into Dutch and German historical tea routes, and more specifically those of Fryslân and Ostfriesland. Further, tenants of Frisia's tourism and characteristics are explored. This chapter concludes with some recommendations and practical applications on how these results may be implemented to create opportunities for culinary mapping on both sides of the political border.

Methodology

Overall, 38 semi-structured interviews were carried out with members of tea heritage sites, regional tourist information centres, tea specialists and hospitality employees. A further six in-depth interviews were conducted with representatives of the two biggest East Frisian tea museums (the Ostfriesisches Teemuseum and the Bünting Teemuseum), the head of East Frisia Tourism, the head of the cultural department of the East Frisia Cultural Heritage Board, the curator of the DE Heritage Center, and the Friesland Bureau of Tourism. Both sets of interviews were then coded and critically analysed to provide the initial recommended opportunities for culinary mapping.

Tea for Europe

Tea is an aromatic beverage of the *Camellia sinensis* plant, derived from either the *sinensis* or the *assamica* variety. It is believed that it was first consumed as a solitary practice sometime between 2700 bc and ad 220 in China (Heiss and Heiss 2007). However, the earliest mention of tea in Western

DOI: 10.4324/9781003197041-8

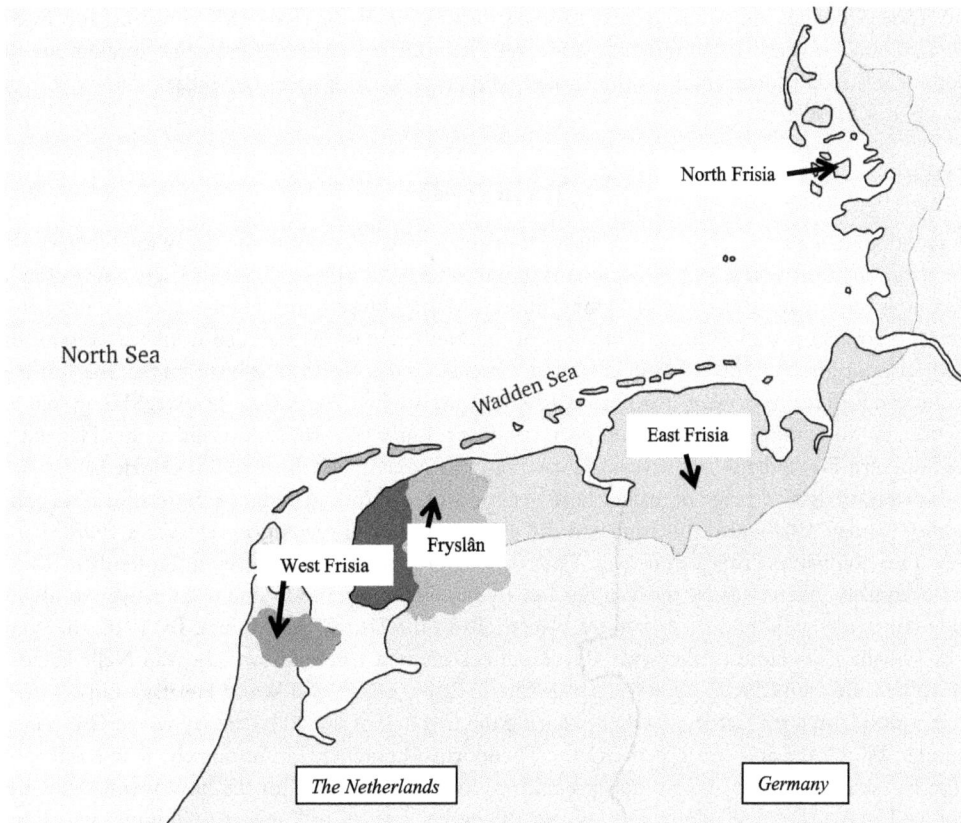

Figure 6.1   The Frisian states in the 1300s extended across parts of today's Netherlands, Germany and Denmark

Source: Made by David Cenzer for the History of the Netherlands podcast website (Source www.historyofthenetherlands.com)

literature dates to 1559 and refers to it as mainly being consumed for medicinal purposes (Martin and Cooper 2015). Tea was first introduced to the Netherlands in 1606 and was shipped from Dutch East Indies colonies to Amsterdam. It then made its way up to the north of the Netherlands before crossing into Germany. In 1768, the trading company *Verenigde Oost Indische Companie* (V.O.C.) began to export tea, such as Chinese pekoe, to other European countries and so to East Frisia (Liu 2007).

Ostfrieslanders became early adopters of tea, mainly due to canny marketing techniques which included royal patronage through the gifting of tea to the prince of East Frisia. That said, there is also evidence that the tea-drinking culture of East Frisia could have been established even earlier than this, because there are ship logs indicating that the first direct shipment of tea to the East Frisian harbour of Emden dates to July 1753, carrying green tea from China, porcelain, tea pots and other tea utensils. It has also been mentioned that East Frisian traders founded the Royal Prussian Asian Compagnie in Emden to Canton and China in 1751, using the Dutch V.O.C. as a role model (Klöver 2008; Haddinga 1977).

By 1660 the affluent Dutch had begun to accept tea as a household commodity, and we see the appearance of tea caddies fitted with special locks to avoid pilfering, handmade tea boxes made from silver and from paper, and tea became a symbol of wealth and status, with special rooms where tea was prepared, served and drunk. These rooms were soon to progress into actual tea houses (*theekoepels*) with wealthy families erecting them in their backyards. Many were influenced by Chinese and Turkish designs and situated next to small streams, canals or rivers (Meulenkamp and Boeve 1995). Some of these structures have been restored and are visible monuments throughout the Netherlands today, for example at the Kröller-Müller Museum Kasteel Rosendael, and extensively throughout Fryslân.

Tea in Fryslân

As mentioned, tea spread from its original port of entry of Amsterdam to the rest of the Netherlands, but rather than following logistical routes it travelled with the affluent Dutch society. For example, because it was the beverage of choice for Countess Maria Louise, van Hessen-Kassei, the mother of the then governor of the United Provinces of the Netherlands, tea travelled from the West of the Netherlands to the provinces in the East and eventually to the North. It is believed that the countess also introduced tea to Fryslân (Kooijmans 1997). However, by 1782 V.O.C. teas were being sold via auction and re-exported to other European countries (Liu 2007); Brabant, Flanders and Hainaut in the Southern Netherlands; the riverine areas along the Maas and Rhine; and to Ostfriesland.

Notably, the global price of tea began to decline around 1750, making it a more affordable commodity (Molen 1978), and this triggered the growth of a retail revolution (De Vries 1993). Large Dutch tea companies emerged, namely Van Nelle in 1782 at the Leuvehaven in Rotterdam, which was eventually taken over by the Van der Leeuw family, who in 1845 due to lucrative worldwide trade contracts established their own tea plantations in the Dutch West Indies. By 1916, the company bought a site close to the Schie River and constructed a new factory, the Van Nelle Factory, which became a UNESCO World Heritage Site in 2014 (UNESCO World Heritage List).

Another Dutch tea factory to emerge during the first half of the 20th century was in Groningen, founded by the tea and tobacco merchant Theodorus Niemeijer. Unfortunately, it was forced to close its tea line 60 years later when it could no longer compete with the now world-renowned Douwe Egberts (D.E.). A similar fate awaited the nearby factory of Kanis and Gunnink, which had established its coffee factory in Kampen between 1879 and 1885 and added tea to their line around 1920. By 1969, they became part of the coffee and tea consortium JDE Peet's.

Tea entrepreneurs

Indeed, it would be highly inappropriate to discuss the historical and cultural heritage of Fryslân tea without paying homage to Douwe Egberts. In 1753, Dutch Egberts Douwe and his wife Akke Thijsses opened their first store, De Witte Os, in Joure, Fryslân, selling tobacco, coffee, tea, and spices. When their son eventually took over, he renamed the company Douwe Egberts, the trade name currently used today (Groeneweg 2021). By 1937, D.E. had introduced their own tea brand, Pickwick, named after the Charles Dickens work *The Pickwick Papers*. By 2014, however, they suffered the same fate as the other tea conglomerates and their stock was taken over by the German investment company Joh. A. Benkiser (JAB) and in 2015, they too merged with Jacobs: JDE (Groeneweg 2021).

The cultural significance of D.E. in the Netherlands is extremely relevant to this chapter for several reasons. Firstly, their pioneering loyalty program, introduced in 1924, still binds customers to their products. Giving the illusion of good value, consumers can exchange loyalty points for tea and tableware, silver teaspoons, porcelain teacups, and so forth (Groeneweg 2021), "more than 4.5 million consumers save D.E. loyalty points . . . and have been saving loyalty points for almost one hundred years" (Douwe Egberts 2021). Sixty percent of Dutch households save these loyalty points, and two billion loyalty points in delivered products were spent in 2021 alone (Douwe Egberts 2021).

Furthermore, D.E. has managed to maintain a link between royalty and tea. Their *Burendag* has become an annual event, held on every fourth Saturday in September, under the patronage of the Dutch royal family; neighbours are encouraged to drink coffee and tea with each other. Today, it can be said that mainly due to the emergence of these Dutch tea conglomerates, the Netherlands experienced a widespread adoption of the domestic consumption of hot beverages, marking the beginning of the "hot drinks culture" (Shammas 1990). By 1955, tea drinking had become popular amongst the Dutch farming communities, afternoons were favourite moments to drink tea and subsequently the word *Theetyd* (teatime) was introduced (Molen 1978).

Tea consumption habits

According to Heiss and Heiss (2007), how one prepared and drank tea has always been unique to each country, as have the tea wares and the preparation methods. Historically, the Dutch were known for adding milk to their tea, because they had heard that this was how the Manchu emperor (Heiss and Heiss 2007) drank his tea. However, this changed during the 18th century, and tea was prepared by infusing very strong tea in a teapot and served in a porcelain teacup. At times they often added hot water from the *boullioire* to give the tea a less strong taste, and to give it a more vibrant colour, they added saffron and vast amounts of sugar (Voskuil 1988). Today tea is a global commodity for mass consumption in the Netherlands, but the merchant sailing vessels which transported it (the clippers) also carried social, cultural, economic and political transformations to Europe. The V.O.C. itself epitomizes a rich cultural heritage, with its fine examples of tea memorabilia, warehouses, auction halls and ship wharves. Their archives also offer a unique source of information about the 17th- and 18th-century history of many countries and cultures, and as such has become a part of the UNESCO project Memory of the World (UNESCO 1992), assisting in the preservation of the region's cultural heritage.

As such, all is not lost, and to see how Dutch tea history, traditions and culture are embedded into the society one only needs to look to the rural provinces, which have kept many of the old traditions, such as the serving of a single cookie with their tea, but also the reinvention of their rich tea heritages as tourist attractions. Instead of trying to shake off their historical past, the Dutch hospitality industry are using tea to reintroducing authenticity, cultural heritage, and "wow" experiences in order to sustain economic growth and thus creating opportunities for culinary mapping with bordering regions. One such province is Fryslân, where it has always been common in Fryslân to drink tea with a heavy breakfast, go for *tea visits* (Voskuil 1988) and take tea breaks (Knoop 1763).

Tea traditions

In sum, the province of Fryslân has always been a tourist destination, and today it is famous for its water sports and cultural experiences. Fryslân has kept its own language, and its unique landscape of dunes, dikes, forests, beaches, lakes and swamps has attracted tourists since the early 19th century. In 1910, the Compagnons Hotel was established, with its own terrace, playground for kids and an area of relaxation for adults. By 1912, Fryslân began to develop campsites with facilities such as washrooms and toilets. Unfortunately, the area saw a decline in tourism during the Second World War but recovered slowly with the introduction of agritourism, as local farmers, desperate to subsidise their dwindling income, cleaned out their cow sheds once the cows had migrated to the pastures in the spring (Dijkstra 2020).

By the 1960s, with domestic tourists becoming more discerning, expectations grew and the region introduced horse racing, amusement parks and cycling activities. Their hotels became more luxurious with hot and cold water, newspaper delivery services, and public tea houses that began cropping up sporadically (Dijkstra 2020). In 2015 a Tea Garden Tour, connecting nine Frisian tea gardens in Northwest Fryslân, was set up by the Tourism Office of Fryslân in cooperation with De Witte Os. Now there is a museum store featuring the heritage of D.E., the history of Frisian craftsmanship, and the opportunity to experience the blending and the tasting of the tea under the guidance of a master tea blender. A second museum dedicated to tea is the *De Theefabriek* in Hou-werzijl, which opened its doors in 1990. Set in an old, reformed church, this unique tea museum enables tourists to visit the tea museum, shop and the tea room and to experience a Fryslân/Grun-niger afternoon tea.

Today tourism is an important sector in the region, and with cycling still being a very common way for tourists to spend their holiday, the Fryslân *Fietsersbond* and Tourist Office, Visit Friesland, have created special Tea Garden cycling routes in the Northwest of Fryslân, and the regional Mar-keting and Tourism Office of Northeast of Friesland (SRMT) has created a magazine and website to promote the Northeast of Fryslân. A part of this promotion is dedicated to culinary tourism. Visi-tors are shown local places related to eating and drinking that are unique to Fryslân, and several tea gardens and restaurants dedicated to tea are part of this too.

East Frisian tea history

Similarly, the Ostfriesland Peninsula, located directly on the German-Dutch border and character-ised by agricultural and touristic structures, has residents who are reliable, enthusiastic tea ambas-sadors. For hundreds of years, the preparation and consumption of black tea have been symbolising the pride and traditional roots of the population. It is their tea culture and traditions that offer locals a source of identity and visitors an attractive opportunity to connect with the inhabitants (Stenger 2019).The history of these rituals is based on the stable rhythm of drinking tea five times a day and celebrating this as a social binding element and positive routine.

Unlike with the Dutch case, Ostfriesland's tea culture has been transferred via families, and the preference for certain tea blends has been passed down from generation to generation (Hangen and Kaufmann 1997). However, as per the Dutch population, at the beginning of the 18th century consuming tea was also a privilege of the upper class in Ostfriesland, but perhaps it did not hold quite the same royal regard since the then king of Prussia, Friedrich II, was not amused about the growing attraction of tea and coffee. This was due first to economic reasons and second because such beverages were imported from countries with colonies, and Prussia did not have any such colonies. Finally, this was down to his personal preference for beer, the local product. Consequently, in 1778 the king tried unsuccessfully to ban tea, but by the end of the 18th century the people of Eastern Frisia had already committed themselves to black tea (Klöver 2008). Initially, the black tea from Assam and Java was presented as a gift from Dutch trade men to the prince of Ostfriesland and then was developed as an exclusive and expensive product only to be offered by doctors or herb traders. It also became recognized as a medicine for a long time (Haddinga 1977). The trademarked blend of tea, the Original East Frisian Tea Blend, consists of ten or more varieties of black tea, mostly Assam, Ceylon and Java. Three tea companies are still operating in East Frisia (Bünting Tee, Thiele Tee, OTG) and they are allowed to sell the tea under the trademarked brand only if the last steps of blending the tea are done in East Frisia.

East Frisian tea culture

As tea consumption is an essential element of East Frisia's culture, there are several steps to celebrate the tea ceremony and its preparation. First, the people focus on certain time slots each day, where they meet automatically for tea, that is first thing in the morning, again at 11 a.m., in the afternoon and in the evening. As an East Frisian, you do not need to invite your family members because they know when it is time to drink tea. Guests and foreigners will get this structure soon and also tourists are involved in this scheme – if they get an authentic East Frisian "teetied" (teatime).

As a first step, a big teapot is prewarmed with hot water. Second, hot water is boiled, and the host calculates 8 to 10 grams of East Frisian Tea Blend per litre of water. The loose tea leaves are put into the teapot and covered by freshly boiled water. This mixture must steep for about four minutes, and then the remaining water is poured in the pot. Afterwards, this beverage is sieved into another pot. For an original teatime, the *resmer* porcelain is often used. It shows either the typical blue strawflower or the red East Frisian rose, the traditional tea-related design for porcelain in East Frisia.

Before pouring the tea into the wide porcelain cups (with handles), large pieces of white rock candy are placed in the middle of each cup. In the East Frisian dialect, this candy is called *Kluntjes*. As the final step, liquid cream is added to the tea by using a special cream spoon. This spoon is named *rohmlepel*. The goal is for the cream to show structures like clouds on the surface of the hot tea, which are named *Wulkje* (Reneburg 2018).

Tea drinking and spending time together while enjoying the Original East Frisian Tea blend has become a Frisian custom, and Ostfriesland is now known as the tea-drinking region of Germany. Further, Ostfrieslanders have managed to maintain many of their tea-drinking habits, illustrating their pride and traditions (Hangen and Kaufmann 1997). An example of the latter is the unique culture of preparation and consumption of tea, forming the basis of the now famous Ostfriesland tea ceremony. The process known as *Teetied*, is part of a custom that ensures all guests are welcomed with tea, and it does not matter if they stay one day or three weeks in East Frisia. *Teetied* is the symbol of hospitality, of feeling welcome and of showing interest in guests (Hangen and Kaufmann 1997).

Tea as a cross-border brand

Today, Fryslân and *Ostfriesland* still have similar attitudes towards tea, but there are also differences in regard to their traditions and habits. In East Frisia, tea is more than a simple beverage, but the original East Frisian tea blend is a symbol of pride; "it is their tea" (Haddinga 1977). Furthermore, in 2016, UNESCO listed their tea ceremony as an "intangible cultural heritage", making this the only tea-related UNESCO recognition in this category worldwide and firmly establishing the historical uniqueness of the East Frisian tea culture in the German part of Frisia. More specifically, based on centuries of tea consumption, numerous roles of tea as a drink, and as a currency and a symbol of social anchoring, requiring gatherings five times a day of family members or close people, to drink tea following a certain procedure, and using this time to recover, to exchange professional or private information and to get a feeling for each other's needs and requirements. It is a construct to come together, using the warm beverage as a ritualised rejuvenation (Bohne 2021).

In addition, East Frisians have been awarded the world championship as tea drinkers in August 2021. Guinness World Records analysed tea consumption worldwide and evaluated that the East Frisians are drinking more than 300 litres of black tea per person per year – more than every other social group or regional population worldwide (Thien 2021).

Consequently, Ostfriesland has three tea museums: Ostfriesisches Teemuseum (publicly supported), Teemuseum Norden (privately financed), Bünting Teemuseum (company's museum). They all have different collections of tea utensils, pots, ceramics, and branded tin boxes/caddies as well as information about tea production, imports, and the uniqueness of the East Frisian Tea Ceremony (Klöver 2008). These three tea museums have become symbols for tourism development, as well as lanterns of tea culture and its preservation. A rising numbers of guests and a higher demand for educational programs, with tea trainings, underlining the interest of local, regional and national as well as international tourists for the East Frisian Tea Culture, has safely embedded tea tourism in East Frisia (Bohne 2021).

Culinary mapping and tea tourism

Culinary mapping is a tool often used by the local inhabitants of a particular destination to market and showcase products, places and people. As per this chapter, it involves a process of collecting, recording, analysing, synthesising and visualising information to describe the culinary resources, networks and usage patterns of a specific group in a specific area. Interactive food maps are often a result of this mapping, as restaurants can be combined with other attractions and activities in order to stimulate local tourism. Tea tourism should be seen as a subcategory of cultural tourism and also has links to gourmet and heritage tourism. Within this concept of holiday creation, tea plays a role for choosing the destination, for cultural programs and for interaction with locals by following a beverage and its tradition based on specific regional customs (Jolliffe 2007). As such it offers strong potential for connecting regional brands for culinary services and products and could function as a harmonic link between both sides of the political border of two relatively underdeveloped regions. Within this falls the recognition as *intangible cultural heritage*. Awarding this status to any regionally identifying traditions raises the reputation and profile and provides alluring opportunities to establish a community of loyal guests and passionate cultural heritage actors.

More specifically, it has become evident that both Frieslands have very similar occupancy rates, and their RevPAR is also comparable. This suggests that a combined stay in both regions has potential, as guests' expectations would also be comparable. It can be seen that, by both parts being connected to the river Dollard, this natural attraction offers an alternative form of transportation or the potential for a combined sailing holiday. As such, in the concluding section recommendations are offered to a number of pertinent stakeholders.

Conclusion

In light of the case of cross-border Frisian tea traditions, it is obvious that political borders and language barriers cannot eliminate the same or similar preferences or customs, which have been developed for hundreds of years. Different cultural stakeholders, such as museums or cultural authorities, as well as tea companies, can and should use this heritage as a value for destination marketing based on unique, authentic and positively recognized consumption traditions. Also, the close relation between the population and its region is a big advantage for developing sustainable structures for touristic infrastructure or gastronomic offerings.

Hence, considering tea as a cultural cross-border brand for culinary services based on Friesland's rich cultural heritage offers multiple opportunities for tourism establishments to create authentic and wow experiences for the modern tea consumers. From a historical perspective, we see that tea began as a solitary event and progressed to small group gatherings at in an aristocrat's garden. That said, according to the interviews conducted there is still a distinct lack of knowledge surrounding tea. Very few respondents were aware of the historical relevance of the Dutch tea industry. They also lacked knowledge regarding the product itself, although there was some evidence regarding alternative uses for tea. This was in response to the question: *How can Frisian tea tradition become a connecting cross-border brand for culinary services and products, and therefore assist in the preservation of the region's cultural heritage?* Our penultimate recommendation is to the tea industry, which must reconsider the design of their customer experience by focusing on different ways to educate their customers and staff, as well as young people, regarding the legacy of the Frisian tea industry, for example to reinstate the 18th-century Dutch *Theekransjes*. Considering the area's many canals, lakes and rivers, the industry should consider the location of its tea houses and perhaps regain some of the authenticity from the *theekoepels* of former times.

The swift and widespread adoption of the domestic consumption of tea, along with a variety of accompanying utensils and consuming practices, has come to be embellished in Frisian traditions and culture today. As such, the making and serving of tea holds many traditions and is the ideal commodity from which to explore the concept of cultural cross-border branding for culinary services and offers direction to anyone who wants to invest in rejuvenating the Frisian tea industry. Tea could be a good anchor for both parts of Frisia and a stable topic in order to establish Frisia as the European Tea Region itself, based on long-term traditions showing contemporary adaptations of tea houses, consumption rituals and positive social effects.

Both regions could and should use more intensely the positive reputation in order to develop joint activities and tea-related authorities, to attract international guests – for which political orders are less important – showing the pride of cultural heritage and the quality of hundreds of years of harmonic social gathering, cultural anchoring and touristic handling of tea.

For further research and industry-related activities, it should be considered how the traditional tea companies in Friesland can contribute to the building of the tourism infrastructure of the region. For researchers, follow-up questions stemming from this chapter ask, how can the tea regions create opportunities from the receipt of UNESCO's recognition and the world championship award, and what impact could this have on regional gastronomy and the branding of the region? Finally, to reiterate to the Dutch and the German authorities, tea cultural heritage has the potential to become a sustainable tourism development, to bind and support the Frisian population regardless of language borders and language barriers. How can tea be used more successfully for the development of cross-border tea routes and tea attractions? Also, how political and or financial instruments can be used to support this touristic process, and which training elements are necessary to raise the service quality on touristic establishments and the knowledge about tea, are relevant for a future focus? In addition, how can mass tourism and a reputation as a cheap destination be avoided in order to save the social binding character and the qualitative approach of this cultural heritage?

Discussion questions

1 How can traditional tea companies in Fryslân and East Frisia contribute to the tea tourism network?
2 How can tea be used more successfully for the development of cross-border tea routes and tea attractions?
3 How can political and financial instruments be used to support this touristic process, and which training elements are necessary to raise the service quality of touristic establishments and the knowledge about tea relevant for a future focus?

References

Bohne, H., 2021. Uniqueness of tea traditions and impacts on tourism: the East Frisian tea culture. *International Journal of Culture, Tourism, and Hospitality Research*, 15(3), 371–383.

De Vries, J., 1993. Between purchasing power and the world of goods: understanding the household economy in early modern Europe. *In:* Brewer, J. and Porter, R., eds. *Consumption and the world of goods*. London: Routledge, 85–132.

Dijkstra, T., 2020. Hoe gingen we vroeger op vakantie?, *Leeuwarder Courant*, 15 July [Online]. Available from: https://lc.nl/Hoe-gingen-we-vroeger-op-vakantie-25849808.html [Accessed 20 June 2021].

Douwe Egberts, 2021. *Geschiedenis* [Online]. Available from: www.de.nl/geschiedenis/ [Accessed June 2021].

Groeneweg, L., 2021. *Thuis sinds 1753*. Unpublished Manuscript, Reinwardt Academie & JDE Heritage Center, 27, 30, 51.

Haddinga, J., 1977. *Das Buch vom ostfriesischen Tee*. Leer: Verlag Schuster, 29.

Hangen, H. and Kaufmann, T., 1997. 'Un drink'n koppje tee' – *Zur Sozialgeschichte des Teetrinkens in Ostfriesland*. Aurich: Museumsfachstelle der Ostfriesischen Landschaft, 85–90 and 142–145.

Heiss, M. and Heiss, R.J., 2007. *The story of tea: a cultural history and drinking guide*. New York: Random House Digital, Inc.

Janssen, E., 2007. *Tea Almanach*. 3rd ed. Hannover: Felicitas Hübner Verlag, 62–63, 74–77.

Jolliffe, L., ed., 2007. *Tea and tourism: Tourists, traditions and transformations*. Clevedon: Channel View Publications.

Klöver, H., 2008. *Tee in Ostfriesland*. Barßel-Elisabethfehn: Sambucus Verlag, 54–57.

Knoop, J., 1763. *Tegenwoordige staat of historische beschryvinge van Friesland: waarin deszelfs legging, gesteldheid, natuur, . . . vervolgens de wereldlyke en geestelyke regeerings-form . . . duidelyk aangeweezen worden: alles uit oude en laatere stukken, ...* Leeuwarden: By A. van Linge Comp. Boekverkoper, 471.

Kooijmans, L., 1997. Friese adel en het huis van Nassau. *Virtus | Journal of Nobility Studies*, 4(2), 57–59.

Liu, Y., 2007. *The Dutch East India Company's Tea Trade with China: 1757–1781* (Vol. 6). Leiden: Brill, 141.

Martin, L.C. and Cooper, R., 2015. From herbs to Medicines: a world history of tea – from legend to healthy obsession. *Alternative and Complementary Therapies*, 17(3), 162–168.

Meulenkamp, W. and Boeve, E., 1995. *Theekoepels en tuinhuizen in de Vechtstreek; overvloed & welbehagen*. Weesp: Uitgeverij Heureka.

Molen, J., 1978. *Thema thee: de geschiedenis van de thee en het theegebruik in Nederland: [tentoonstelling] 18 maart- 4 juni 1978*. Rotterdam: Museum Boymans-Van Beuningen.

Renebarg, T., 2018. *Tee-Tied*. Kiel: Grabener Publishers, 25–40.

Republic of Amsterdam Radio, 2019. Episode 15: fuelling the flames of Frisian Freedom. [online]. Available from: www.republicofamsterdamradio.com/episodes/historyofthenetherlands/episode-15-fuelling-the-flames-of-frisian-freedom [Accessed 21 November 2021].

Shammas, C., 1990. *The preindustrial consumer in England and America*. Oxford: Oxford University Press, 29–32.

Stenger, M., 2019. *Celebrating the 100th anniversary of the Heimatverein Norderland*. Norden: Ostfriesisches Teemuseum, 100–110.

Thien, M., 2021. Die Ostfriesen sind offiziell weltmeister im teetrinken [Online]. Available from: www.nwzonline.de/ostfriesland/ostfriesland-kultur-in-ostfriesland-erneut-weltmeister-im-teetrinken_a_51,3,744366625.html# [Accessed 10 October 2021].

Timothy, D.J., 2021. *Cultural heritage and tourism*. Bristol: Channel View Publications.

UNESCO, 1992. The archives of the Dutch East Indian Company (VOC) [Online]. Available from: www.unesco.org/new/en/communication-and-information/memory-of-the-world/projects/full-list-of-projects/the-archives-of-the-dutch-east-indian-company-voc-project/ [Accessed 19 June 2021].

UNESCO, 2014. Van Nellefabriek [Online]. Available from: http://whc.unesco.org/en/list/1441/ [Accessed 19 June 2021].

Voskuil, J., 1988. De verspreiding van koffie en thee in Nederland. *Volkskundig Bulletin*, 14(1), 68–93.

7

RECOGNITION OF THE CULTURAL HERITAGE OF TEA

An international perspective

Hilary du Cros

Introduction

Drink your tea slowly and reverently, as if it is the axis on which the world earth revolves – slowly, evenly, without rushing towards the future.
— *Thich Nhat Hanh, Buddhist monk (in Martin 2007, 193)*

Both global and local factors impact on how different tea traditions are viewed. This chapter will explore how some overarching themes such as religion, international trade, and past colonial practices, as well as local concerns, have aided in developing different tea traditions and tourism products. A case in point is how the Buddhist religion (as revealed in the quote above) is closely linked to tea in many parts of East Asia. International trade practices have been instrumental in the spread of the demand and supply of tea. Connected to this spread in the past are aspects of colonialism, which have had world-changing impacts. This is especially the case for what are now large tea exporting countries, such as China, India, Kenya, and Sri Lanka.

The main regional cultural heritage traditions associated with tea can be grouped approximately into five major geographic categories regarding how tea has spread across the world or been popularised:

- East – China, Korea, Japan, and other East Asian countries
- Middle East/Central Asia/Old Silk Road Countries – Russia, Turkey, Iran, Iraq, the Emirates, and the "Stans"
- West – UK, France, and other nearby tea-drinking countries
- Hybrid/postcolonial – India, Sri Lanka, Hong Kong, Singapore, Indonesia, Vietnam, North America, Australia, and Africa (e.g., Morocco, Botswana, Uganda, Senegal)
- Key herbal and/or indigenous tea producers – South Africa (Rooibos), South America (Yerba Maté).

These traditions and examples are not intended to be exhaustive. Their study raises a host of questions, some of which can be dealt with in this space; others will be flagged for later research. The questions explored here are:

- What kind of tourism do each attract? Are ex-colonial or religious affiliations ever a factor?
- Is there much demarcation between the traditions in tea tourism?

DOI: 10.4324/9781003197041-9

- What form would that take in terms of destination marketing organisation and specialist promotions?
- Why is popular culture relevant?
- What is the evolving process of demarcation of tea cultural traditions in relation to tea tourism? Is it tied to colonial/postcolonial factors or some other set of considerations?

Literature review

Most research into tea-related cultural heritage tends to dwell on the details of its intangible and tangible characteristics. Such work covers the intangible heritage of tea processing; traditional, medicinal, and ceremonial practices; and legends of origin. Additionally, associated tangible cultural heritage of tea utensils, gardens, houses, museums, shops, collections, and plantations is of interest (e.g., Tanaka 1973; Ling 2000; Ali *et al.* 2013; Abdusalama *et al.* 2020). Sometimes these works are accompanied by recommendations for its safeguarding and continuation with observations about measures to raise public awareness (Ali *et al.* 2013; Abdusalama *et al.* 2020; Seo *et al.* 2020).

Tea and Tourism (Jolliffe 2007a) was the first publication of its kind to concentrate on the nature of the relationship between the two, with contributions from a multidisciplinary team of researchers. Several significant journal articles on tea tourism have followed this pioneering edited volume. Some scholars have concentrated on items of tea cultural heritage and their appeal to tourists and the public; for example, the research into trade routes centred around tea (Tu and Kuang 1995; Avery 2003; du Cros and Lee 2007; Sigley 2010, 2021).

Sigley (2021) has observed how tea can be presented to domestic tourists in China as a narrative device. As such, it is particularly effective in linking the diverse peoples along the Ancient Tea Horse Road (ATHR) "into a single story, and by extension, into a single family of the Chinese nation" for the purpose of maintaining social unity (Sigley 2021, 5). Notably, concerns about impacts of development, particularly unsympathetic tourism development, are raised by many of these authors. Sigley (2010) is especially critical of the drawbacks of potential World Heritage inscription for elements of ATHR between provinces in Northwest China and Tibet. UNESCO's valorisation of the intangible heritage of tea culture is dealt with elsewhere in this book.

Some groundbreaking research has been carried out into colonialism and trade, where the sociopolitical context of the spread of tea drinking has been highlighted. It has generated a few published histories (e.g., Rappaport 2017; Merritt 2017) which could raise public awareness of the long-reaching impacts of colonialism. These histories have traced the full global circle of exchanges, which were responsible for putting tea produced in China and India into Western parlours. By a close examination of Britain's East India Company's debts in London, it was possible to track the train of exchanges all the way to India and China, which Merritt calls Edmund Burke's "mighty circle of commerce" (Merritt 2017, 79). Detailed studies are starting to appear (which flesh out this circle) and its importance for developing new postcolonial tea plantation hospitality products and tourism experiences. Accordingly, Chepwony in Karlsson (2021, 1) maintains regarding the history of tea production in Kenya:

> The world needs to know that that tea which is being exported from Kericho is blood tea. We need to bring a closure by demanding reparations and an apology from the UK government. Those multinationals are operating from stolen land.

Such articles observe that problems of fairness and local autonomy are still ongoing in many places receiving tourism, as "tea is intimately connected with power" (Karlsson 2021, 1). How this conflict affects their brand and its appeal is less certain without further research, although there are indications that it is more of an issue for domestic tourists than others. Nevertheless, some of this history is reaching a broader audience (Guerty and Switaj 2004; Martin 2007; Easton 2019). Accordingly, some aspects of how tea is dealt with in various forms of widely available popular culture are examined later in this chapter.

However, very little of this literature directly addresses the connected issues of tea tourism destinations and the demarcation of tea traditions by producers and marketers, particularly in relation to religious practices. Also, complementary products or the relevance of popular culture associations to tea tourism marketing are not commonly addressed. Hence, while supply and management of products/experiences for tea tourism are becoming popular research topics, the general awareness of tea cultural heritage and associated marketing is a less explored area. This chapter will briefly scope out a research agenda for both academics and marketers concerned about this gap in the literature.

Methods

The research methodology used for the chapter consisted mainly of a case study approach. Data was obtained in a desktop study combined with an opportunistic focus group. The latter was intended to roughly gauge perceptions about and depth of interest in, the heritage side of tea. The use of the focus group approach was meant as a test of a potential research tool in the time of COVID-19, which limited other ways of conducting face-to-face research.

Desktop study

The following chapter comprises an internet survey of academic and popular culture databases (e.g., the Internet Movie Database [IMDb]). Also, emails were sent to co-contributors and others on several issues. These individuals included researchers also involved in studying tea tourism or Buddhist religious tourism. However, it is not intended to be an exhaustive study of the topic, which deserves a much larger canvas than this chapter to fully encompass.

Focus group

A focus group (using a draft of the introduction and literature review of this chapter as a prompt) was held over Zoom to provide feedback on their perceptions of tea culture and tourism. The group of around ten people normally meets once a month in person (and over Zoom during COVID-19) to critique nonfiction pieces. They all live in Sydney, Australia, and are members of Writers NSW. The group attending the Zoom meeting comprised eight women and two men varying in age from late 20s to early 70s. Eight were native English speakers, mostly born in Australia, and two had English as a second language and were born overseas. This opportunistic sampling was undertaken because of limitations the restrictions associated with the COVID-19 pandemic have put on face-to-face focus groups. It was a roughly 25-minute session held on 15 August 2021; all participants had read the material in advance of the session and offered verbal comments during the session and some written ones afterwards by email.

Case study 7.1 Potential forms of demarcation in tea tourism

Most tea marketing and demarcation appears geographical and related to types of tea (e.g., green tea in Fujian, China, and matcha tea in Japan) or whether destinations are mainly known to produce,

---

consume, or re-export it. These distinctions are also supported by how international bodies view tea, such as the European Union's classification system and that of the International Tea Committee (Jolliffe 2007a). Meanwhile, the demarcation of tea tourism experiences is more complex. The key forms of distinction between tea traditions for tourism that are explored here are:

- tea experiences that are typically place-based;
- linked to a regional/national identity, religion, and/or principles of hospitality; or
- experiences being increasingly influenced by conflicts over cross-cultural and trade practices.

How tea tourism for these categories might translate into actual demand is explored briefly at the end of this section with a discussion of the results of the focus group.

Typical place-based tea experiences

A common form of tea tourism demarcation is place-based. For example, tea-growing countries offer mainly farm-stays, factory and harvesting experiences, while tea-consuming countries develop various tea-drinking practices and ceremonies for tourism. Though it may not be that clear-cut because there are countries and regions that offer both, it is one way to view demarcation. In some cases, there are attempts to distinguish these experiences from similar ones within a region that are potential competitors (Leung 2007). Alternatively, a quick internet survey of such offerings shows very little high-level commodification or marketing to link them together as part of a strong regional/national tradition, particularly in developing countries.

Tea experiences linked to regional/national identity, religion, or principles of hospitality

Meanwhile, more active demarcation can be seen in the way these practices and ceremonies reflect national identity (Jolliffe 2007a), religion and/or how this identity also features in the way tea traditions are shown to support a notion of hospitality for these places. More sophisticated forms of destination marketing have been found to link a nation's brand of hospitality as perceived in the way hosts offer their guests tea to their tourism branding. A recent example of this linkage can be seen in the Japan National Tourism Organisation's campaign to link the Japanese principles of hospitality behind their tea ceremony, *Omotenashi*, to its 2021 cultural tourism promotions (see JNTO 2021a). Meanwhile, tea rooms in the UK and Canada have been doing that for a while with their traditions of tea service (Hall and Boyne 2007 and Jolliffe 2007b).

It should not be forgotten that the religion of Buddhism has always been closely linked to the spread of tea. From China, the taste for tea was transmitted by the means of the Eastern Silk Roads to Japan and the Korean Peninsula (UNESCO 2021). In all these places, the beverage developed close connotations with religious and social rituals, because it was commonly consumed by Buddhist priests who disseminated this religious ideology. How this is handled in tea tourism experiences is still an area of emerging interest (Cora Wong, personal communication, 4 September 2021). A recent example of how it might be offered to tourists is the "Live like a Buddhist monk" experience in Japan (JNTO 2021b).

Tea experiences being increasingly influenced by conflicts over cross-cultural and trade practices

Yet another approach is to focus on the historical spread of tea where the aftereffects of colonialism can influence brand identity in a 21st-century context. A case in point is the Darjeeling plantation/non-plantation dichotomy in India. Again, the history of the colonial trade in tea starts to play a role as do later marketing initiatives, as well as sustainable development. Key in the Darjeeling example is how some non-plantation tea tourism experiences have been branded as artisanal, sustainable, and distinct from plantation ones. The former is more likely to be run by descendants of Nepalese imported plantation labour squeezed out of owning plantations post-independence (Sen 2019). They have also tried to include more work practices that are in line with the UN Sustainable Development Goals (SDGs; United Nations 2021).

These newer artisanal enterprises are arguing for inclusion as part of the group covered by international/national designation as Darjeeling tea producers (with associated tourism), hence the conflict (Sen 2019). It is unclear to what extent other ex-colonial plantation areas are currently adopting SDGs or are experiencing the same dichotomies or distinctions. Sri Lanka seems the closest to Darjeeling with some unofficial interest by a few of its more artisanal tea plantations in following SDGs, unlike more conventional plantations (M.S.M. Aslam, personal communication, 5 September 2021). Overall, this is an under-researched area.

There is also limited information on the extent to which different attractions within destinations struggle to compete with or complement each other in these ways for tourists' attention.

Results of the focus group and implications for further research

The focus group that studied information on the historical development of tea yielded two individuals out of the ten who were deeply interested in following up on the heritage of tea. Both supplied verbal comments during the session and then sent written comments and photographs afterwards to the author regarding their experiences of tea cultural heritage and interest in it. The others found the topic mildly interesting and noted that they would be happy to undertake tea day tours of plantations or ceremonies, if the chance ever arose. It is likely that much of the group would be satisfied with typical place-based tea tourism experiences, whilst the participant who provided the photos could be interested in those linked to hospitality or religion. The one participant fascinated by the sociopolitical aspects of colonial tea history would probably find the Darjeeling region of interest because of the contrasting tea production experiences it provides.

Case study 7.2   Popular culture and tea

Many people learn about the cultural heritage of tea by watching internet videos, films, and/or reading other media that feature examples of its main traditions. It is useful to explore the kind of portrayal of these instances of popular culture about tea's cultural heritage and their relative popularity. Also, this exercise helps to see where awareness could be enhanced without trivialising tea's cultural value.

After searching relevant keywords at the IMDb website, the results were analysed initially regarding whether tea is central to the feature or more peripheral (foreground or background). Tea can be in the foreground, as in the case of the television episode *Begin Japanology: Tea Ceremony* (NHK Educational 2007), or in the background, as in the film *Tea with Mussolini* (MGM 1999) and numerous British, Korean, and Chinese costume dramas. Then, a note was made of each feature's popularity, as measured by IMDb's algorithm, which provided a rough guide.

Overall, "tea" yielded 11 results and "tea ceremony" 35 results – the best results out of the terms searched. The list for "tea ceremony" was larger than the former, mostly because of Japanese movies that include it as a cultural signifier or focus and the North American movies that reference Asian culture. That is, the Japanese features tended to have it in the foreground more often and North American ones in the background. When a filter for popularity is added, the most popular of these entries was *The Karate Kid Part II* (Columbia Pictures and Delphi Pictures 1986), part of a Hollywood movie franchise. The protagonist, Daniel (the karate kid), accompanies Mr. Miyagi (his karate master) to the latter's childhood home of Okinawa in Japan, where a tea ceremony features in a scene. Another older example is *You Only Live Twice* (Eon Productions 1967), which shows part of a wedding ceremony with tea drinking by James Bond and his bride. These scenes occur in these movies as more of a cultural marker or signifier that the main characters are really in Japan and experiencing authentic local culture, rather than as a focus on tea. They are still popular today because both movies are part of film franchises that still attract audiences. The Karate Kid and James Bond franchises are available to view globally on internet streaming services or YouTube.

In the IMDb list, there are several documentaries where tea's cultural heritage or drinking tea in a ceremony are the key premises for placing tea squarely in the foreground. A case in point is the film documentary *All in This Tea* (Hoffman 2008), which follows the journey of a Californian tea importer around China. His objective is to find the best hand-picked and authentically processed tea. Significantly, it also references Robert Fortune clandestinely searching China in 1843 to steal plants to establish the colonial British tea industry in India and Sri Lanka (Ceylon). This is because the colonial historical event makes a great hook to garner attention with a bit of skulduggery, as well as an important part of cross-cultural relations surrounding the spread of tea around the world.

Understandably, many of the other entries on the list feature tea cultural heritage with considerably less attention to detail; however, they were more likely to create a generic or cliched image of the tea traditions depicted. The performance of a Japanese tea ceremony by a *Geisha* is a case in point. James May's *Hey Bim!*, an episode of *Our Man in Japan* (May 2020), shows a *Geisha* at a *Geisha House* in Kyoto carrying out the ceremony with more discussion of her status than the traditions associated with serving and drinking the tea. As such, it feels like the local tea cultural tradition and its authenticity is secondary to discussing whether *Geisha* were courtesans or not. It is the only time a tea ceremony is shown in the whole series, leaving the impression it is only performed by *Geisha*. More definitely needs to happen in this space about cross-cultural understanding and the role tea plays in Japan. After watching this episode, it is possible that even viewing random videos on YouTube would be more helpful than some travelogues, such as James May's.

Meanwhile, tea is in the background for many TV series and books, such as those for the *No 1 Ladies' Detective Agency* in Botswana (BBC *et al.* 2009). Here tea is needed to support exposition about a situation by the characters. The two main protagonists, Mma. Ramotse and Mma. Makutsi, often sit together around the kettle and tea utensils they commonly use while talking about their investigations.

Again, it is done less to feature tea and more to provide support to plot development, as well as the situational authenticity of being in a slightly nostalgic Botswana, as was experienced by English author Andrew McCall Smith.

Discussion and conclusion

When tea tourism was first considered in a global context (Jolliffe 2007a), it appeared that most of the research was centred on the major tea-producing countries (e.g., China, India, Sri Lanka, and Kenya) and significant tea-consuming countries (e.g., the UK, Canada, and China). The research in this current volume goes beyond this geographical focus, and so should any future research on forms of demarcation. Alternatively, nothing much had been attempted on the depiction of tea heritage in popular culture and social media before this book. This chapter has tried to deal with the former, and work on social media appears elsewhere in this volume. If read together, a picture emerges regarding how tea heritage might be perceived as being important for its own sake, a religious experience, as a cultural marker, or a way through posting experiences to signal the tastes and desires of a potential/actual tea tourist.

Next it is important to identify gaps in the literature. Creating a model of the three forms of demarcation of tea traditions in Case study 7.1 is one way to assist this process. It could even be viewed as an evolving process, like the Web:

* Typical place-based tea tourism (tea tourism 1.0, the first or most typical type of experience);
* Linked to national identity, religion, and/or principles of hospitality (tea tourism 2.0; more aware of a need to distinguish traditions culturally); and
* Influenced by conflicts over cross-cultural and trade practices (tea tourism 3.0; more aware of linkages to past and future social justice issues as a means of demarcation).

Plotting the rough geographic groups against the forms outlined above shows where more information is needed to ascertain whether there is an emerging process or not for demarcation across the world (see Table 7.1).

Whether all these three forms of demarcation really can be found globally is a big question. Other issues raised by this investigation requiring further investigation include how much demarcation is good for the sustainable development of tea tourism, and how can more of its heritage enter popular culture in a relatively authentic form?

Directions for future research

Looking at many of the questions raised in this chapter requires more research by academics and marketers. For instance, it would be timely (given the UN's promotion of the SDGs) to examine

Table 7.1 Geographical spread of forms of demarcation

Group	Form of Demarcation	Comments
East	Tea tourism 1.0 (all)	
	Tea tourism 2.0 (Japan, possibly South Korea)	Unknown if China has 2.0
Middle East/Central Asia/ Old Silk Road Countries	Tea tourism 1.0 (all)	Some parts of Middle East may have 2.0
West	Tea tourism 1.0 or 2.0	2.0 is more about hospitality and national identity than religion
Hybrid/postcolonial	All possible (only India and Sri Lanka [?] are known for 3.0)	Maybe Kenya has 3.0 too
Herbal	Tea tourism 1.0	Unknown if there is 2.0 or 3.0

more deeply tea tourists' aspirations for including authentic and sustainably derived tea heritage products in their travel. It could help marketers create products where demarcation allows for healthy competition, as well as ways to complement similar products/experiences where appropriate. One approach for understanding demand could be by arranging online focus groups (with incentives, where necessary) through community cultural groups, where it is possible to find a variety of typical cultural tourists, as these groups already contain people who regularly consume cultural products (Richards 2013; du Cros and McKercher 2020). If conducted in the main tea-consuming or even ex-colonial countries, these groups could provide excellent feedback for destination marketing organisations and other tea tourism promoters elsewhere. In addition, online surveys could be trialled where online focus groups are not possible.

Discussion questions

1 Choose one of the questions left unanswered in this chapter. How would you design a research project to investigate it?
2 Discuss how you would test the evolving process model proposed in the discussion section for tea tourism demarcation.
3 Review a selection of artisanal plantations' websites and comment on how much their stated practices follow the UN Sustainable Development Goals (SDGs). Discuss whether this could be a successful way for such attractions to distinguish themselves from more traditional offerings.
4 Select two examples of popular culture featuring the cultural heritage of tea (one foreground and one background), and compare tea's depiction in them with that in two similar examples in social media (e.g., TikTok, Facebook, Twitter). What conclusions can you draw about which are the best media for potential tea tourists (and what type of tea tourists would be most likely to access it)?

References

Abdusalama, A., Zhang, Y., Abudoushalamud, M., Maitusuna, P., Whitneye, C., Yang, Y.F. and Fu, Y., 2020. Documenting the heritage along the Silk Road: an ethnobotanical study of medicinal teas used in Southern Xinjiang, China. *Journal of Ethnopharmacology*, 260, 1–10.
Ali, A., Anwar, R., Hassan, H.O. and Kamuran, H.R., 2013. Significance of Japanese tea ceremony values with ceramic art interpretation. *Procedia – Social and Behavioral Sciences*, 106, 2390–2396.
Avery, M., 2003. *The tea road: China and Russia meet across the Steppe*. Beijing: China Intercontinental Press.

BBC, HBO and The Weinstein Company, 2009. *The No. 1 ladies' detective agency*. BBC, HBO and The Weinstein Company.

Columbia Pictures and Delphi Pictures, 1986. *The karate kid part II*. Columbia Pictures and Delphi Pictures.

du Cros, H. and Lee, Y.S.F., 2007. *Cultural heritage management in China. Preserving the Pearl River delta cities*. Abingdon: Routledge.

du Cros, H. and McKercher, B., 2020. *Cultural tourism*. Abingdon: Routledge.

Easton, E., 2019. History of high tea – history English afternoon tea – tea etiquette. Available from: https://whatscookingamerica.net/history/highteahistory.htm [Accessed 10 August 2021].

Eon Productions, 1967. *You only live twice*. Eon Productions.

Guerty, P.M. and Switaj, K., 2004. Tea, porcelain, and sugar in the British Atlantic world. *OAH Magazine of History*, 18(3), 56–59.

Hall, D. and Boyne, S., 2007. Teapot trails in the UK: just a handle or something worth spouting about? *In:* Jolliffe, L., ed. *Tea and tourism. Tourists, traditions and transformations*. Clevedon: Channel View, 206–223.

Hoffman, D.L., 2008. *All in this tea*. Flower Films.

JNTO, 2021a. Omotenashi [online]. Available from: www.japan.travel/en/au/experience/culture/omotenashi/ [Accessed 25 August 2021].

JNTO, 2021b. Live like a Buddhist monk experience Japan [online]. Available from: www.japan.travel/experiences-in-japan/en/1757/ [Accessed 25 August 2021].

Jolliffe, L., 2007a. ed., Introduction. Connecting tea and tourism. *In:* Jolliffe, L., ed. *Tea and tourism. Tourists, traditions and transformations*. Clevedon: Channel View, 3–22.

Jolliffe, L., 2007b. ed., Tea tourists and tea destinations in Canada: a new blend? *In:* Jolliffe, L., ed. *Tea and tourism. Tourists, traditions and transformations*. Clevedon: Channel View, 224–246.

Karlsson, B.G., 2021. The imperial weight of tea: on the politics of plants, plantations and science. *Geoforum*. Available from: https://doi.org/10.1016/j.geoforum.2021.07.017

Leung, P., 2007. Tea traditions in Taiwan and Yunnan. *In:* Jolliffe, L., ed. *Tea and tourism. Tourists, traditions and transformations*. Clevedon: Channel View, 53–70.

Ling, W., 2000. *Chinese tea culture*. Beijing: Foreign Language Press.

Martin, L., 2007. *Tea. The drink that changed the world*. Vermont: Tuttle Publishing.

May, J., 2020. *Our man in Japan* (Season 1 Episode 4. Hey Bim! c. James May. 2020. 51 mins). Plum Pictures/New Entity.

Merritt, J., 2017. The trouble with tea. The politics of consumption in the eighteenth-century global economy. *In: Studies in early American economy and society from the library company of Philadelphia*. Baltimore, MD: Johns Hopkins University Press.

MGM, 1999. *Tea with Mussolini*. MGM.

NHK Educational, 2007. *Begin Japanology: tea ceremony*. NHK Educational.

Rappaport, E., 2017. Tea revives the world. *In:* Rapport, E., ed. *A thirst for empire: how tea shaped the modern world*. Princeton, NJ: Princeton University Press, 264–304.

Richards, G., 2013. *Cultural tourism. Routledge handbook of leisure studies*. Abingdon: Routledge.

Sen, D., 2019. What makes Darjeeling tea authentic? Colonial heritage and contemporary sustainability practice in Darjeeling, India. *South Asia Chronicle*, 9, 177–203.

Seo, S.J., Jin, Y.R. and You, W.H., 2020. A study of sustainable conservation for tea farming in Boseong Region. *Journal of the Korean Institute of Traditional Landscape Architecture*, 38(3), 64–74.

Sigley, G., 2010. Cultural heritage tourism and the Ancient Tea Horse Road of Southwest China. *International Journal of China Studies*, 1(2), 531–544.

Sigley, G., 2021. Reimagining the 'Central Plains' (Zhongyuan) and 'Borderlands' (Bianjiang): the cultural heritage scholarship of the Ancient Tea Horse Road (Chamagudao) of Southwest China. *International Journal of Heritage Studies*, 27(9), 904–919.

Tanaka, S., 1973. *The tea ceremony*. New York: Harmony Press.

Tu, N.H. and Kuang, W.D., eds., 1995. *In search of the Ancient tea caravan route*. Hong Kong: Hong Kong China Tourism Press.

UNESCO, 2021. Cultural selection: The diffusion of tea and tea culture along the silk roads [online]. Available from: https://en.unesco.org/silkroad/content/cultural-selection-diffusion-tea-and-tea-culture-along-silk-roads [Accessed 25 August 2021].

United Nations, 2021. Make the SDGs a reality [online]. Available from: https://sdgs.un.org [Accessed 25 August 2021].

8

TEAICS AS A FRAMEWORK FOR KNOWLEDGE USE IN TEA TOURISM

Brian Park

Introduction

With a history going back over 4,700 years, the use of tea as a medicinal plant and beverage has influenced science, art and culture for centuries. Knowledge of tea and its influences on some 224 distinct disciplines has led to the development of *teaics*, a systematic, logical and objective framework encompassing all knowledge of tea – from shrub, to plantation, to caddy, to cup – together with all aspects of ancient and modern tea culture (Park 2009).

Many different industries have evolved around teaics, one of which is *tea tourism*. Tea tourism is defined as tourism that is motivated by the history, tradition and consumption of tea (Jolliffe 2007). Compared with other types of tourism, knowledge of tea and tea culture gleaned through research enables rich opportunities for cultural experiences, sociological interpretation of culture and education through tea tourism. This chapter discusses how, using the framework provided by teaics, a structured approach to organising knowledge of tea tourism could be applied to help develop and improve the tea tourism industry.

Literature review

A brief history of tea and tea culture

Legend has it that tea, known taxonomically as *Camellia sinensis* (L.) Kuntze since 1753, was discovered by the Chinese 'Divine Farmer' Shen Nong in around 2700 bc. According to various local folklore, the birthplace of tea is Szechuan, Yunnan, Burma and Siam. In Korea, records suggest that the Indian princess Hwang-Ok Heo brought tea seeds with her to the historic city of Gaya in ad 48, when she visited for marriage; the king of Gaya then ordered these seeds to be planted in Gimhae (Jung 1990). *Chajing*, the first book published about tea, was written by Lu Yu in 780, while in 828 Daeryeom is reported to have brought tea to Korea from the Chinese T'ang dynasty (Jung 1990).

Tea was established in Japan during the Heian period (794–1185), when tea seeds from the T'ang dynasty were imported and planted. Sen no Rikyu (1522–1591) established *wabi-cha*, a detailed set of rules for ceremonial tea-drinking traditions that take place in small tea rooms across Japan. Other examples of Korean tea culture include *Dabang*, an official institution on tea; *Dachon*, a kind of 'tea village' where tea was offered at temples; and *Dashi*, an everyday tea-time regularly observed by public officials (Park 2020). In colonial Great Britain, 'afternoon tea' became part of society that is

DOI: 10.4324/9781003197041-10

maintained today, combining Western culture with ceramic traditions from the East, as well as sugar from Brazil.

Methods

Development of a comprehensive knowledge framework would be useful for the rapid development of tea tourism and to increase the sustainability of this industry (Lee 2009). Teaics is helpful in this regard; it is a structured framework encompassing the many different themes within tea research, including knowledge about the tea plant, tea as a beverage material and the beverage itself, and knowledge of tea culture (Choi 2016).

Teaics was developed through a classical seven-step process: (1) experience, (2) value recognition, (3) fact, (4) information, (5) knowledge, (6) theory and (7) discipline. First, people have unique experiences, which develop into perceptions of value recognition and in turn become facts. Facts are only promoted to *information* when they are gathered based on the same goals and the same values, and when a certain threshold is reached such that that information is valuable to society. To become *knowledge*, information must meet further criteria: information has a certain direction and subject, and when that information can be used for educational purposes, it can develop and improve people's abilities. When knowledge is sufficiently logical, it becomes a theory, and when theory gathers sufficient objectivity, it becomes a discipline. All disciplines belonging to teaics were therefore developed through the process of experience, value recognition, fact, information, knowledge, theory and discipline (Lee 1989).

Study context

Teaics as a knowledge framework for tea tourism

Take botany, as it relates to tea, as an example of how the teaics structure was developed. First, tourists have the experience of looking at a tea shrub. As they repeat this experience, they realise that tea shrubs have flowers and fruits at the same time. As they recognise the botanical value of this phenomenon, it becomes a fact that tea shrubs simultaneously develop both fruits and flowers. Gathering sufficient other facts about tea shrubs – including that tea seeds can germinate by picking them from the shrub and planting them before the pods are dry, that these seeds can take 2–5 years to germinate, and that these seeds develop in a process of cross-fertilisation – makes them useful for society and promotes facts to 'information'.

When information is linked by specific rules or criteria, then it becomes *knowledge*. For example, information about the processes of vegetative propagation and seed propagation becomes knowledge under the banner of 'tea tree propagation'. Going further, combined with other related knowledge, such as that tea is a plant of the genus *Camellia*, and associated ideas in botany and agronomy, knowledge becomes *theory*. Finally, a *discipline* emerges when academic consensus is reached after critical review of theories arising from scholarly research at conferences and meetings, or through publication of academic theses and journal articles (Lee 1989).

Disciplines forming the structure of teaics

The 224 disciplines that have so far been developed under the framework of *teaics* can directly support tea culture, convention and production industries. The disciplines are mainly made up of structured knowledge, divided into five key interrelated categories. Knowledge that is useful for tea tourism can be found in each of these. Table 8.1 provides an overview of these disciplines, and their relationships with tea tourism are explained below (Post 2009).

Table 8.1 Structure of teaics

Category	Discipline	Subdisciplines	Further subdisciplines

Social science
The scientific and systematic study of various phenomena of human society related to tea

Sociology of tea
The study of human society as it relates to tea, and social behaviours such as growing, making and drinking tea

Social theory of tea culture:
- For the elderly, adults, young people, juveniles, children, infants, etc.

Education for tea-related behaviour
The study of educating people about how to behave through the processes of brewing and drinking tea
Educational theories of:
- General tea-related behaviour
o For the elderly, adults, young people, juveniles, children, infants, families, etc.
- Special tea-related behaviour
o For the elderly, adults, young people, juveniles, children, infants, families, etc.

Pedagogy as applied to tea culture

Pedagogy for children and adolescents as applied to tea culture, including:
- General pedagogy
- Special pedagogy, etc.

Andragogy as applied to tea culture
Education for adults participating in tea culture
- General andragogy as applied to tea culture
- Special andragogy as applied to tea culture, etc.

Tea economics
The study of the behaviour, balance and social optimality of economic agents in the fields of tea production/sales/tea-related environments/tea-related land
- Tea demand and supply
- Tea production economic theory
- Tea price and policy response, etc.

Statistics on tea
Statistics that use the range of tea industry management as a population

Tea business administration
The study of the structure and behaviour of tea-related organisations
- Tea brand management theory
- Tea shop management theory, etc.

Humanities
The study of fundamental problems, ideas and cultures related to tea and humans

Cultural theory related to tea
The scientific understanding, description, analysis and prediction of tea-related cultures

Literary theory for tea
- Verse-based literature about tea
- Creative theory for verse-based literature about tea
- Critical theory for verse-based literature about tea
- Prosaic literature about tea
- Creative theory for prosaic literature about tea
- Critical theory for prosaic literature about tea

Theory for books related to tea culture
- Translation theory for book about tea
- Criticism theory for book about tea

Comparative theory related to tea culture

Comparative theory related to:
- Tea-related behaviours
- Tea-drinking customs
- Music for tea
- Tea painting
- Tea ceramics

Theory on the content of tea culture

Conventional business theory related tea culture

History related to tea

The study of events, social changes, ideas and cultures related to tea in the past
- World history related to tea
- Regional history related to tea
- History of the Opium Wars
- History of Boston Tea Party, etc.

Philosophy related to tea and tea-related behaviours

The study of comprehensive concepts related to tea and exploring the origin of tea

Seon-teaics

The practice of seon using teaics and a knowledge system on the essence and methods of the practical implementation of teaics
- Seon-teaics related to asceticism
- Seon-teaics related to criticism

Natural sciences

The study of natural phenomena related to tea

Botany related to tea

The study of tea as a plant
- General botany for tea
- Theory of tea plant physiology
- Theory of taxonomy for *Camellia sinensis* (L.) Kuntze
- Theory of tea as medicinal plant

Crop science related to tea

- Tea cultivation
- Tea propagation
- Soil
- Fertiliser
- Pathology
- Entomology
- Sustainable agriculture
- Growth regulation
- Agricultural machinery
- Micro-meteorology

Theory of genetics related to tea (*C. sinensis* (L.) Kuntze)

- Taxonomy
- Heredity
- Variation
- Reproduction
- Chromosome
- Linkage
- Mutation
- Cytogenetics

(Continued)

Table 8.1 (Continued)

Theory of breeding related to tea

- Breeding technology for tea
- Genome
- Variety
- Progeny test of tea plant
- Combining ability of tea plant
- Sterility

Landscaping theory related to tea

Landscape history
Landscape architecture plan
Scenic planting
Landscape construction engineering
Landscape management

Sitology related to tea
The study of fields related to tea as a food

Theory on tea

- Theory on six kinds of teas (green, white, oolong, yellow, black, dark)
- Regional tea
- Korean tea
- Chinese tea
- Japanese tea
- Indian tea
- Sri Lankan tea
- Indonesian tea
- Turkish tea
- Iranian tea
- Kenyan tea
- Vietnamese tea
- Bangladeshi tea
- Malaysian tea
- Malawian tea
- Myanmar tea
- Nepalese tea
- British tea

Theory on tea blending

- Theory on:
  o Blending principle for tea
  o Tea and tea blending
  o Tea and plant blending
  o Tea and flower blending

Theory on tea brewing

- Theory on tea brewing skill
- Theory on brewing:
  o Green tea
  o Oolong tea
  o White tea
  o Yellow tea
  o Black tea
  o Dark tea

o Regional tea
- Theory of the relationship between tea and water
- Educational theory on tea brewing

Tea-processing theory

Theory related to functional tea

Theory on food hygiene in tea

Theory on tea tasting

Theory on regional tea

Theory on tea from:
- Asia
- Pakistan
- Russia
- Rwanda
- Taiwan
- United States/Canada
- Argentina
- Australia
- Azerbaijan
- Bhutan
- Brazil
- Cambodia
- Chile/Ecuador
- Colombia
- Ethiopia/Tanzania
- Georgia
- Germany
- Guatemala
- Madagascar
- Mali
- Netherlands
- North Korea
- Portugal
- Spain
- Switzerland
- Thailand

Nutrition as it relates to tea
The study of the nutritional phenomena of tea as a nutritional substance for humans
- Biological effects of tea
- Physiological effects of tea
- Tea and food pairing

Chemistry as it relates to tea
The study of the chemical composition of tea ingredients and the chemical reactions that occur when tea is absorbed in the human body, using chemical methods
- Biochemical theory of tea
- Chemical theory of tea
- Chemical theory of water as solvent for tea
- History of tea compounds
- Polyphenols/catechu/catechins/caffeine
- Theanine/theophylline
- Stimulants/theory of tea as a remedy

(*Continued*)

Table 8.1 (Continued)

- Fermentation/oxidation
- Amino acids/vitamins/sugar
- Aromatic compounds/chemical compounds
- Organic acids/saponin/minerals
- Chemical theory of the health benefits of tea
- Natural products in tea/flavanols and flavones

Art

A field related to human creative activities that form artistic works related to tea

Theory of tea music

Theory of music as an art that expresses thoughts and emotions felt in tea activities

Theory on flowers for tea

Theory on thoughts and emotions connected with flowers as a component of tea activities

Decorational theory of tea rooms

Theory on the places occupied by people and tea, or the places where humans drink tea and the area in which the act of drinking tea takes place

Theory on tea ceremonies

Tea activities include growing, harvesting, making, brewing and serving tea; also includes behaviours in which the body and mind can be trained during tea-related activities

Theory on tea utensils

Theory on utensils required in the process of brewing and serving tea

Structural theory for tea rooms

Theory on the structure of the places in which tea is brewed and served

Cross-Disciplinary Studies

Studies combining two or more separate categories of theories to create an integrated theory related to tea

Theory on tea tourism

Tourism that is motivated by an interest in the history, traditions and consumption of tea

Theory on sixth industry–related tea activities

Study of tea industry-related activities whereby new value and jobs are created in rural areas by converging the primary industry of agriculture, with the secondary processing industry and the tertiary service industry

Theory on artificial intelligence related to tea

Study of the human abilities to learn, reason, perceive and understand natural language through computer programs related to tea

Theory on Big Data related to tea

Study of the tea-related accumulation of structured, semi-structured and unstructured datasets, and the technology that can extract and analyse economic value from these

Social science

Tea culture, and the business administration of the tea industry, are important educational disciplines within the social sciences category of teaics. Indeed, tea culture often overlaps with business and education at a fundamental level. For example, in Korea and Japan, when drinking tea, a polite attitude, and cultured speech and behaviour are required. Regular tea drinkers are taught to maintain focus and to be organised and methodical – behaviours that induce motivation and are valuable in the business and educational environments (Jolliffe 2022). Courtesy, which is embedded in Korean and Japanese tea culture, is also encapsulated in the field of educational engineering and is a concrete example of the application of tea culture to education. Knowledge of tea industry management is important for managing and operating a corporate organisation cultivate, process and sell tea. Tea tourists in Korea and Japan, therefore, have the opportunity not only to drink tea but also to understand how tea culture influences education and business.

Humanities

Within the teaics category of humanities, the category of world tea history encompasses global historical events and traditional folklore as they relate to tea. Examples include legends, such as the discovery of tea by China's 'Divine Farmer' Shen Nong in 2700 bc; the introduction of tea seeds to Gaya in Korea in ad 48 (Lee 1918); the role of the infamous Boston Tea Party in the American War of Independence in 1773; and the relationship of tea with the Opium Wars between Great Britain, China and Brazil during the 19th century. Indeed, a typical British *afternoon tea*, which combines sugar originally imported from Brazil and black tea from Asia, was established thanks to development of the international shipping industry. Invention of the Russian 'samovar' water-boiling system was another key moment in world tea history, as are the teachings of Lee Sampyung, which influenced the beginnings of Japanese ceramics culture.

'Seon' teaics is a logical and objective knowledge system of the practice and effects of meditation ('seon' in Korean, or 'zen' in Japanese) using tea. 'Seon-cha', formulated by Korean Buddhist monks, is effective in promoting and maintaining mental health. Its concepts can be utilised in tea tourism, where seon teaics provide basic theories and knowledge of meditation practices in nature, such as in tea fields, or on the theme of tea relics.

Natural science

Teaics disciplines falling within the category of natural science include tea shrub landscaping, tea food science and tea nutrition. In tea garden tourism, tea tour planners can evaluate an area for its plants, civil engineering materials and sculptures to determine its suitability as a tea tourism destination, and promote the work of landscape planning, design, construction and management. A high-level understanding of artistry, the ecological environment and tree botany is required to develop and operate tea tourism products with environmental conservation value. The tour guide also contributes to interpretation of ecological, aesthetic and artistic values by understanding the landscape of the tea tourism area.

In terms of the food science of tea, a tour guide can explain the principles and processes of tea production, and the classification of tea according to the different processing methods, as part of a tea factory tour. From a nutritive and health point of view, they can also impart knowledge of the anticancer and antioxidant compounds found in tea, such as catechins, thus enhancing the tour from the perspective of healing tea tourism.

Art

Within the teaics category of art, tea ceremony theory provides logical and objective knowledge, as well as interpretation, of the tea ceremony. In tea tourism, this theory can be exploited to provide tourists with an in-depth understanding of repeated tea-drinking practices and to develop tea ceremony experiences as tourism products. Likewise, theories surrounding knowledge of the tea room – the indoor or outdoor space in which tea is produced, made and drunk – provide logical and systematic knowledge to enhance tea tourists' understanding of gardens, factories, tea houses and tea rooms.

Cross-disciplinary

The fifth and final main category of teaics encompasses cross-disciplinary themes fusing philosophical theories in humanities, social science, natural science and art. As applied to tea tourism, this category currently includes:

- tea tourism semantics, which explains the meanings of words and phrases used in tea tourism, how these terms are used in real life and how humans respond to them;

- tea tourism value theory (axiology), which describes the philosophy of how the various cultures created by tea tourism explore the consciousness of value in history;
- tea tourism ontology, an academic theory that explores the existence and fundamental and universal regulation of tea tourism as an independent form of tourism; and
- tea tourism aesthetics, which elucidates the interrelated essence and structure of beauty contained in nature, life, art and tea tourism, including the emotional perception of tea tourism and the emotional recognition of beauty that can be obtained from tea tourism.

Analysis

Interactions between teaics and tea tourism

As illustrated, the current body of knowledge on tea has considerable potential impacts on development of the tea tourism industry. From the tea tourist's perspective, in-depth learning about tea from various fields allows them to understand, connect with and enjoy the sites they visit, and enriches their experience of existing tourism products. From an industry perspective, however, tea knowledge can be used to enrich the quality and quantity of tourism products. Some areas of tea knowledge have already been developed and are widely used in tea tourism, but there are many other areas that have not yet been fully explored. For instance, new knowledge might help to develop specialist, personalised hospitality for tea tourism, and new, tourist-centred tea tourism products might be brought to market in areas such as ecotourism, experiential tourism, healing tourism and educational tourism. Some examples are provided in the following paragraphs.

Tea ecotourism

Subdivisions of teaics including tea botany, genetics and breeding, crop theory, and landscaping knowledge can be used in tea ecotourism. Tea plant palaeontology would be an interesting addition to this knowledge, but this remains a relatively unexplored subject, outside of museums, for future research.

Experiential tea tourism

Tourists in Korea or Japan often experience the drinking of traditional tea, but it becomes a much more valuable experiential tourism product when the tourist is dressed in traditional clothing and immersed in a traditional Korean or Japanese tea ceremony.

Health and wellness tea tourism

Aspects of teaics involved in health and wellness tourism include knowledge of the nutritive qualities of tea and tea flowers, music, tourist behaviour, philosophy, literature and 'seon' teaics (Korean meditation as it relates to tea). New areas of teaics research could stem from topics such as the methodology of tea-related health and wellness, the development of health and wellness tea tourism and the use of tea tourism for psychological healing, and the application of technology for social and physical health and wellness. A recent example of health and wellness tea tourism is 'care farming' or 'agro-healing', which is practised in the Netherlands, Germany, Japan and Korea. This field utilises rural areas and agriculture for human healing beyond the purpose of producing agricultural products. Cooperation with care agriculture provides an opportunity to expand the health and wellness tea tourism industry and is another new field that remains to be explored in teaics (Gyung and Kim 2017).

Educational tea tourism

Aspects of teaics that might be exploited for use in educational tea tourism, for example for visits to a tea factory, include the bromatology and chemistry of tea, as well as tea nutrition. New, unexplored aspects of teaics that would provide helpful material for educational tea tourism might include knowledge of the different processing characteristics of six types of tea grown in different tea-producing regions, the technology to produce functional tea, and the principles underlying this technology.

Example of the World Tea Tour

A tea tourism product being proposed in Korea is the World Tea Tour, a 21-day tour beginning at the ancient tea garden of Gimhae and ending with the tea plantations in the Languedoc region of France. Tourists would move from place to place by aeroplane, ferry and cruise ship; at each of the destinations, a local tea tourism guide would provide information about the tea fields, tea factories and tea museums that are visited by car. In the evenings, seminars would be held to answer tourists' questions along with opportunities for discussion and to enhance participant's tea knowledge.

Aspects of teaics used to develop this tea tourism product include theories of and comparisons between tea cultures; within and between countries, regions and eras; and comparisons of the use of different tea utensils. Future strands of teaics that might be useful include regional differences in tea characteristics, processing methods, extraction technologies and theories about the different connections between tea and food in different regions.

Structure of tea tourism theory

Using teaics as a framework, disparate strands of new tea tourism knowledge can also be organised into a defined academic structure. Table 8.2 lists the main themes of tea tourism knowledge thus far identified, and Table 8.3 demonstrates how these themes might be arranged into a structured framework.

Table 8.2 Themes of tea tourism theory and their definitions

Theme	Definition
Humanist	The study of language, literature and history as they relate to tea tourism
Historic	The study of past tea tourism activities, events occurring because of those activities, records of those events and analysis of those records
Analytic	Deconstruction of complex concepts or objects related to tea tourism into simple constituent elements, which comprehensively and definitively interpret tea tourism through analysis
Relational	The study of how tea tourism is related to a certain aspect or area, or interpretation of the important behaviours and interactions occurring in tea tourism
Cultural	Discussions on the culture deriving from tea tourism itself, and related activities
Folklore	Studies of religion, customs, legends, techniques and traditional cultures in tea tourism and how these are connected with private life
Comparative	The study of similarities, commonalities and differences between tea tourism and other forms of tourism and other activities
Social science	Elucidating the objective rules governing social phenomena caused by tea tourism
Education	Studies associated with the education necessary to operate and develop tea tourism businesses, as well as the essence, purpose, content, method, system, administration, etc. of education for tea tourism

(Continued)

Table 8.2 (Continued)

Social	Studies of tea tourism behaviours and theories related to the nature, cause and effect of social relationships, as well as interactions between individuals and groups
Industrial management	Use of engineering knowledge and scientific management techniques to manage tea tourism–related product production, tea tourism operation technology and information related to tea tourism, especially from a systemic point of view
Legal	Study of the concepts and types of laws pertaining to tea tourism, as well as those accompanying the acts caused by tea tourism; for example, there is the Tourism Promotion Act, a law that seeks to develop the tourism industry in Korea and the effects, applications and interpretations of these laws
Policy	Government-level plans and systems to achieve policy goals, including goal setting, policy development and policy-related processes to enable activities such as macroscopic development of the ideal tea tourism industry
Natural science	The study of natural phenomena related to tea tourism and fields related to those phenomena
Botany	The study of plants related to tea tourism, including plant morphology and structure, development, growth, reproduction, genetics, metabolism, disease, evolution and lineage, environment and ecology
Crop science	Study of cultivated plants related to tea tourism, including development, morphology and physiology, varieties, classification, cultivation, use and crop improvement
Sitology	Study of the physical, chemical, sensory and commercial properties of tea pertaining to the nutritional, symbolic, sanitary and economic value of tea ingredients, as well as tea processing and food ingredients related to tea tourism
Art	A field of human creative activity that forms aesthetic works (e.g., paintings, sculptures, architecture, crafts, calligraphy) to express beauty in spaces related to and within the realm of tea tourism
Music	Study of music related to tea tourism, for example beat, melody and voice, and the expression of thoughts or emotion through the voice or musical instruments
Performance	The study of music, dance and theatrical performance as related to tea tourism and its associated behaviours
Design	The study of formative realisation for purposes related to tea tourism and its associated activities

Table 8.3 Structure of tea tourism theories

1		The philosophy of tea tourism
	1.1	Concepts in tea tourism
		1.1.1 Etymology
		1.1.2 Conceptual structure
		1.1.3 Conceptual analysis
	1.2	The value of tea tourism
		1.2.1 Value structure
		1.2.2 Value systems
		1.2.3 Value analysis
	1.3	The existence of tea tourism
		1.3.1 Structure
		1.3.2 Systems
		1.3.3 Analysis
	1.4	The aesthetics of tea tourism
	1.5	The logic of tea tourism
	1.6	Ethics in tea tourism

Conclusion

This chapter has defined the disciplines of tea knowledge that make up the logical, objective and structural hierarchy of 'teaics' – covering all aspects of tea: from shrub, to plantation, to caddy, to cup, and including the many and varied cultural and social aspects surrounding tea. Teaics was inspired by the pharmacist Changrim Park, Soo-rin's words, 'Medicine is the crystal of science and conscience,' (Songpo 2001) and attempted to create the crystal of tea knowledge. As a broad body of knowledge, teaics comprises 224 identified disciplines categorised into five main themes: natural science, social science, humanities, art and cross-disciplinary theories. Knowledge of, and obtained from, tea tourism can be subdivided into 60 disciplines across four main fields, and a structure for how these disciplines may fit into the framework of teaics is proposed.

Taking teaics and tea tourism together, we discover the value that the study of these disciplines has for the qualitative development of tea tourism products, and the quantitative development of expanding or reinforcing existing content in tea tourism. The relationship between tea knowledge and tea tourism is two-way: tea knowledge is necessary for the continuous development of tea tourism, but at the same time, tea tourism can contribute towards new fields of study for teaics.

As tea tourism develops, new knowledge will feed into training and education programmes relevant not only for tea but for all forms of tourism. Such education will play an important role in development of the International Competency Standards for tea tourism–related departments, tea tourism–related curricula and international certifications. Teaics education is currently being conducted at the International Teaics Education Centre located in Boseong-ri, Jeju Island, Korea.

Discussion questions

1 What are the characteristics of the relationship between tea knowledge and tea tourism?
2 What is the function of teaics as a source of knowledge for tea tourism?
3 What is the role of tea tourism theory in the sustainable development of the tea tourism industry?
4 How can teaics and theories of tea tourism be used in human resource training for the sustainable development of the tea tourism industry?

References

Choi, O., 2016. Academic identity of Saemaylogy as a Korean discipline. *Journal of Local Government & Administration Studies*, 302, 82–86.
Gyung, A. and Kim, M., 2017. *Gwan-Gwang-Hak-Ge-Ron*. Seoul: Bak-Mun-Sa, 14–18.
Jolliffe, L., ed., 2007. *Tea and tourism: tourists, traditions and transformations*. Clevedon: Channel View Publications.
Jolliffe, L., 2022. Tea tourism. *In:* Buhalis, D., ed. *Encyclopaedia of tourism management and marketing*. Cheltenham: Edward Elgar, 338–340.
Jung, Y., 1990. *Korean tea culture*. Seoul: Nu-Ruk-Ba-We.
Lee, H., 1989. *Hak-Mun-Ron-Seo-Sul Korea*. Seoul: Kyungin-Munhwa-sa, 15–26.
Lee, H., 2009. Academic value and system of kinesiology. *Journal of the Korean Academy of Kinesiology*, 11(3), 1–11.
Lee, N., 1918. *Josun-Bul-Gyo-Tong-Sa*. Seoul: Min-Sok-Won, 461–462.
Park, B., 2009. *Dictionary of teaics*. Seoul: Seok-Hak-Dang, 134.
Park, B., 2020. Peace tea in Han Peninsula. *Proceedings of the Inter-Korea Forum for Cultural and Artistic Exchange*, Gangnung City, 21–22 November, 58–68.
Post, R., 2009. Debating disciplinarity. *Critical Inquiry*, 35, 749–770.
Songpo, 2001. *Bark*. Seoul: Sok-Song-Po-Hoe.

9

DEVELOPING TEA TOURISM IN THE GLOBAL SOUTH

An African perspective

Madiseng Messiah Phori, Lebogang Matholwane Mathole, Unathi Sonwabile Henama and Lehlohonolo Gibson Mokoena

Introduction

Tourism is one of the world's largest industries, producing vast employment and economic growth, which has a plethora of niche markets. Africa is a developing continent which still attracts less than 10% of international tourism receipts, as noted by Sharpley and Telfer (2004).

The vast majority of African tourism destinations offer themselves as sites for safari, sea and sand, not using tea tourism. Tea is one of the top three major beverages enjoyed worldwide and is consumed very widely on the African continent. Tea tourism has emerged as a new niche market for many of Africa's tea estates, which is still in its infancy in many African destinations located in the Global South. Tea tourism has been able to diversify the revenue streams of the tea estates and offer additional employment opportunities. Tea plantations have beautiful natural environments and special heritage over centuries and decades that is packaged for the tea tourism experience. Two African countries, namely South Africa in southern Africa and Kenya in eastern Africa could be noted for their divergent tea tourism experiences.

The chapter seeks to provide a Global South gaze into Africa's participation in the tea tourism economy, and this would be of use to policymakers in developing tea tourism further on the African continent. The African continent is one of the largest producers in the world, and tea tourism is an opportunity to transform tea estates into tourist consumption sites as a means to diversify income earning opportunities. According to Woldemichael *et al.* (2017), agriculture is the backbone of Africa's economy, accounting for 70% of the population as the primary source of livelihood and 25% of the continent's GDP. African countries are notorious for being sites of agricultural production with little or no local benefits, which limits the opportunity to acquire maximum economic benefit from agricultural products. Agro-processing is important for increasing economic benefit and tea tourism is an avenue to allow for tours of agro-processing of tea and its packaging for distribution to consumers.

Research methodology

This chapter applied a case study approach as a research method. According to Eisenhardt (1989), case studies are generally applicable for new research in which there is limited existing theory. The case study selection is dependent on the research purpose and questions (Rowley 2002). This study investigates the potential of tea tourism in Africa to capitalise on this growing niche market. There

DOI: 10.4324/9781003197041-11

is a lack of academic gaze on tea tourism on the African continent, and the case of Kenya and South Africa provides a construct for tea tourism potential. Data derived from the case study is collected through secondary sources such as websites, journals, e-journals and books. Secondary data can be a valuable source to provide answers to research questions (Saunders *et al.* 2007).

Special interest tourism and tea tourism

Special interest tourism (SIT), sometimes called niche tourism, has gained traction in tourism literature since the 1980s (Ma *et al.* 2020), and Swarbrooke and Horner (1999) allude that people's desire to develop new interests in a specific destination may be crucial for conceptualising SIT. Niche tourism has emerged as the opposite end of mass tourism. According to Novelli (2005), niche markets are divided into micro and macro niches. Macro niche, such as cultural tourism, is related to a larger group, a special category of consumers. On the other hand, micro niche is concerned with a narrowly defined market segment such as gastronomy tourism, which is a subset of macro niche; tea tourism ties in well with micro niche because it is a special interest under food and beverage tourism. The distinction between general interest tourism and SIT is that the latter is motivated by the development of new interests, whereas the former is motivated by destination features. Swarbrooke and Horner (1999), on the other hand, differentiate SIT from general interest tourism by asserting that it involves little physical exertion.

Hassan (2012) relates SIT to tourists' interest in a particular field of tourism. SIT as a form of tourism focuses on the needs of markets searching for experiences and activities that are not associated with the general interest type of tourism (Soleimani *et al.* 2019). Lee and Bai (2016) emphasise that SIT is rather "experiential." In general, this niche market consists of recreational experiences motivated by the interests of both individuals and groups (Derrett 2001). There has also been an increase in tourism of special interests in South Africa, such as health tourism, religious tourism, adventure tourism, avio-tourism, safari tourism, and ecotourism. A special interest in tea tourism is fairly on the rise as tourists are beginning to pay more attention to it. In many countries, tea consumption and tea culture are highly prevalent, which does not come as a surprise.

The tourism industry is by nature highly competitive. Therefore, destination countries are constantly seeking new niches to explore, using destination marketing organisations (DMOs), and this has also led to the growth rate of tea tourism destinations and experiences. The term "tea tourism" refers to tourism which is motivated by an interest in the history, tradition and consumption of tea (Jolliffe 2007). Zhou *et al.* (2016) noted the rise of tea tourism as a new form of cultural tourism, an area that has received relatively little academic attention. The high consumption rates of tea have also led to the high rate of production as an agricultural product in many countries. With plentiful fertile grounds of tea plantations around the world, more opportunities arise for further studies and explorations towards tea tourism consumption. Moreover, from a business perceptive, the attractiveness of an additional revenue stream by leaping on the tea tourism bandwagon has made many tea plantations curate tea tourism experiences on their plantations. Although News Dome (2021) points to Pakistan, China and India as places where tea tourism has been perfected, Africa is also playing a significant role in continuing to expand the scope, awareness and consumption of tea tourism as an alternative experience.

Tea destinations on the African continent

Tourism can be regarded as an economic messiah on the African continent. It is disproportionately important for African destinations as a major foreign exchange earner and diversifying the economy base into services. Tourism within an African context is dominated by a handful of countries. According to Dieke (2020), Kenya, Mauritius and the Seychelles are popular tourism destinations in the Indian Ocean, Morocco and Tunisia in the north, South Africa and Zimbabwe in the south, and Ghana and Senegal in the west. The tea tourism countries that would be focused on are Kenya, located in eastern Africa, and South Africa, located in south Africa. In terms of tourism competitiveness, the World Economic Forum (2019) ranked South Africa at 61 and Kenya ranked at 82. Two of the predominant tea tourism countries in Africa are Kenya and South Africa, which also happen to be some of the leading tourism destinations on the African continent.

Case study 9.1   Kenya-Bomet Country

The largest African exporter of black tea is Kenya, which is the third-largest tea producer in the world, as noted by Sarojini (2020). This milestone was made possible by the Kenya Tea Growers Association, founded in 1933. Sarojini (2020) noted that tea production benefits the economy of Kenya, and Kenyan tea is exported across the world to Europe, Asia and the United States. News Dome (2021) noted that Kenya is globally known for its agricultural and horticultural offerings, especially tea, coffee and flowers. In Kenya, over 200,000 ha of farms are used by smallholders (also known as *shambas*) governed by the Kenya Tea Development Authority (KTDA) and other tea producers in both private and public sectors.

The KTDA was founded in 1965 to help lift up and aid smallholders. This support has been rather effective, as smallholders process more than 60% of tea produced in Kenya. Today, Kenya has over 50 companies which develop the leaf from other smallholder (shambas) farms around the country. The main source of assistance comes from marketing Kenya tea as one brand. This marketing has also capitalised on Kenya's ultimate advantage – the weather. The main ingredient contributing to the success of Kenya's tea production can be found in its climatic conditions. Its tea grows year round due to rainfall totals and to its fertile soil. At high elevations there are tropical temperatures suitable for growing tea.

Kenyan tea has the advantage of being particularly bright and colourful, with a reddish, coppery shade and a phenomenally pleasant, brisk taste. The tea is blended into many other British tea brands as well. Given its major contribution towards other nations, the majority of tea produced in Kenya is exported as bulk crush-tear-curl (CTC)-style blends as additional means to add to its special identity and promote its difference. Approximately 443,500 tonnes of tea are exported by Kenya each year, as alluded to previously by Sarojini (2020).

According to Tourism Update (2020), with Kenya's tea industry being one of its top foreign exchange earners, the government is focusing on further development of the country's tea tourism packages. Therefore, the Kenya Tourism Board, on behalf of the Ministry of Tourism, is identifying and promoting a number of authentic and distinctive Kenya tea travel experiences. The proximity of the tea plantations to the Mau Forest complex and the Maasai Mara Game Reserve made Bomet County the ideal location to launch tea tourism in Kenya. Sarojini (2020) noted that Kenya has opted for Bomet as a tea tourism destination due to the presence of diverse tea plantations, as well as numerous multinational tea companies and privately owned and operated factories. There are many attractions located close to Bomet giving guests an opportunity to mix bush and wildlife experiences with tea at Bomet.

KTDA managed factories exist in Bomet County. Tea tourism will be promoted in Bomet through the unveiling of the tea tourism plan. According to Citizen TV (2020), tour operators should package their travel plans as soon as possible to include tea. "Local and international tourists who are interested in sampling tea are being offered farm tours to educate them about the tea experience from the bush to the factory to the cup" News Dome (2021). In Kenya, a variety of tea plantations have provided a means of developing tea tourism by using the plantations.

Kenya is one of the leading tourism destinations on the African continent, and tea tourism would diversify the tourism product offering which is currently dominated by the bush experience. Tea-growing regions have also experienced substantial growth in tourism recently, and the Ministry of Tourism is committed to promoting agritourism and promoting places where tourists can experience great tea, adding diversity to its offerings. "Many small bushes cover the surrounding highlands, which makes them scenic and a great backdrop for photos," according to James (2021). In terms of future plans, Kenya is investing in creating a national brand identity through value incentives such as fair trade and organic certifications to attract consumers around the word and to demand higher prices for Kenyan tea in the marketplace. Tea tourism is a new niche market that is intended to diversify the agricultural dominated economy of Bomet and to embrace tea tourism in addition to the bush and safari experience that already exists in Bomet. Kenya will work with the private sector within the tourism value chain to market both tea experiences and the high quality of Kenyan tea.

Case study 9.2   South Africa Rooibos tours

The Department of Agriculture, Forestry and Fisheries (2016) noted that South Africa produces approximately 10,500 tons of rooibos tea per year. South Africa consumes 4500 to 5000 tons, and the rest is exported. The BBC Good Food (2021) noted that rooibos tea has many health benefits because of its antioxidant content, and it contains some unique polyphenols, including aspalathin. The Rooibos producing area is located in the southernmost province in South Africa, the West Coast of the Western Cape Province, which has Cape Town as its capital city. The Cape Floristic Region (CFR), which produces Rooibos, was named as a UNESCO World Heritage Site in 2004, a momentous event for the people of South Africa marking that the world valued this region as a site of outstanding universal significance to humanity.

The Cederberg region has limited arable land due to the rugged mountains, and this makes it a climbers' and hikers' paradise, so adventure tourism has developed here. The unique geographical landscapes and the existence of rooibos fields and processing plants created the requirements for the establishment of the Rooibos Route. As noted by Brand South Africa (2014), two creative sisters launched the Rooibos Route in 2012 in partnership with a website and a team of hardworking people. Sanet Stander and Marietjie Smit approached rooibos producers to join the venture to add value to tourists that visit the Cederberg region. The official launch of the Rooibos Route took place in Clanwilliam, 200 km north of Cape Town in the Western Cape, on 14 August 2017. There is a related tourism website that acts as a one-stop-shop for planning a breathtaking holiday based on experiential offerings in the local rooibos industry Brand South Africa (2014).

The Rooibos Route offers a variety of experiences and opportunities to learn about tea cultivation, processing and packaging. The development of the Rooibos Route was intended to meet the needs of the tea tourist and package the Cederberg region for tourism consumption through marketing outlays to attract more tourists. There is a new attraction to its tourism offerings in Clanwilliam, which is popular for its rooibos and buchu. A three-day tour through Heuningvlei and the Cederberg Mountain areas, with accommodation facilities, has been added. The development of tea tourism in the Cederberg attracted government support from South African Tourism and Brand South Africa, two government agencies that are responsible for marketing the country.

Steyn (2017) noted that the reason for the Rooibos Route is to provide an environmentally friendly tourist route between Wupperthal and Nieuwoudtville. The route encourages visitors to use the local accommodation and hospitality of rooibos farmers and their families who can share their knowledge and environment. During the rooibos tour, visitors get to witness tea processing and walk the fields where the tea is produced, especially during the harvesting period between January and March.

A Rooibos farm visit is an experience one should not miss. For the more adventurous visitor, 4 × 4 riding, hiking and cycling are available options. Rooibos dishes and drinks are mouthwatering after experiencing nature. "Visit a tea tasting or purchase cosmetics and Rooibos treats" on the Namaqua West Coast (2021). Steyn (2017) noted that 16 January 2017 was the first celebration of World Rooibos Day in Clanwilliam, the most important Rooibos producing area in the world, to celebrate the tea and its economic significance. The yearly celebrations were intended to gather pace until World Rooibos Day can be celebrated internationally and become a day of significance on the South African calendar. Rooibos, an agricultural product, has been able to anchor the economy of the Cederberg, and the development of the Rooibos Route has brought in tourists that can now sample the Rooibos experience in the CFR. The growing global demand for Rooibos tea should be matched with marketing outlays to make visiting the Cederberg a bucket list destination.

Results and discussion

South Africa, located in southern Africa; Kenya, located in eastern Africa; and Egypt, Morocco and Tunisia in northern Africa are the leading tourist destinations on the African continent. Kenya and South Africa, which are the chosen case studies for this chapter, are primarily selling themselves for the Big 5 bush experience in a plethora of parks and nature reserves. The existence of unique tea production plantations was the means to an end in developing a thriving tea tourism industry, led by the government to diversify tea plantation earnings and also develop the new tourism niche of tea tourism. African countries are leaders in terms of tea production and are latecomers into the tea tourism trade as a means to add a new niche to complement the bush experiences they market to tourists.

Kenya

In the case of Kenya, which is the third-largest producer of tea in the world, the development of the tea industry was led by the state, and there is significant state ownership in the tea value chain. The Kenya Tourism Board sought to promote tea tourism experiences due to the existence of so many tea plantations located close to major bush tourism experiences. This provided a route tourism opportunity to linking the two. Tea tourism is regarded as an important niche to diversify agriculture in Kenya whilst increasing earnings from tourism.

South Africa

Rooibos tea popularity can be associated with the sought-after superfoods and the numerous health benefits that they possess. The development of the Rooibos Route and therefore the emergence of tea tourism in the Cederberg destination was a community conceptualised by two sisters and realised with the active participation of rooibos farmers. The inscription of the CFR as a World Heritage Site in 2004 preceded the establishment of the Rooibos Route. The private sector-led route attracted the support of government agencies that promote tourism. The South African case differs from that of Kenya in that the majority of tea estates in Kenya are owned by the government, whereas in South Africa they are in private hands. Tea tourism development in Kenya was initiated by the state because of significant tea plantation ownership by the state, whereas in the case of South Africa it was initially a private initiative, led from marketing support by the state agencies.

Conclusion

The development of tea tourism is building synergy between tourism and agriculture, diversifying primarily agriculture and adding a new tourism niche. African destinations are latecomers in providing tea tourism experiences, and this is an opportunity to grow this niche considering that Africa is now the world's largest producer of tea. The large tea plantations are a means to an end in developing tea tourism. Tea tourism has grown in leaps and bounds in China, Pakistan and India, and it is an excellent opportunity to earn foreign exchange from tea sales and attract tourists through tea tourism. The demand for tea consumption will remain robust, and this augurs well for tea tourism growth and development. Kenya and South Africa are major tourist destinations on the African continent, and by developing tea tourism, they help African countries catch up in contesting for the tea tourism market.

The success of Kenya and South Africa in tea tourism documented in this chapter would encourage other tea-producing countries on the African continent to develop tea tourism, and this could improve tourist arrivals to Africa, matching growth in this form of niche tourism. Ensuring shared growth from tourism, especially in rural and outlying areas, can be achieved through tea tourism located in those geographic areas. Tourism will only be sustainable if it benefits both urban and rural areas and is integrated with other economic sectors, such as agriculture in the case of tea tourism. The success of tea tourism is therefore the success of diversified agriculture, adding an additional income stream for tea plantations, and this can benefit employees and owners. The development of African tea tourism destinations would motivate other destinations to also jump on the tea tourism bandwagon.

This would increase the presence of African countries in this important niche market. The vast majority of African countries sell themselves for the bush experiences, and tea tourism experience would diversify what African countries have to offer to international markets. This body of knowledge will be of great importance to policy markets detailing how the success of tea tourism is always a collaborative effort between the public and private sector. The starting point is a collaborative effort to develop a tea tourism strategy that would have the buy-in of those in the tourism value chain and the support of the public service.

Discussion questions

1   What is the difference between special interest tourism and mass tourism?
2   Discuss why tea tourism could be an important economic sector for African countries.

3 Describe how Kenya's success on tea production has laid the foundation for the development of tea tourism.
4 Briefly explain why tea tourism is regarded as a niche tourism product.

References

BBC Good Food, 2021. Top 5 health benefits of rooibos tea. Available from: www.bbcgoodfood.com/howto/guide/top-5-health-benefits-of-rooibos-tea/ [Accessed 1 April 2021].

Brand South Africa, 2014. Rooibos refreshes the agri-tourism industry. Available from: www.brandsouthafrica.com/tourism-south-africa/rooibos-refreshes-the-agri-tourism-industry/ [Accessed 1 April 2021].

Citizen TV, 2020. Kenya to position tea tours as part of tourism offerings. Available from: https://citizentv.co.ke/business/ [Accessed 1 April 2021].

Department of Agriculture, Forestry and Fisheries, 2016. A profile of the South African Rooibos market value chain. Available from: www.nda.agric.za/doaDev/sideMenu/Marketing/Annual%20Publications/Commodity%20Profiles/field%20crops/Rooibos%20Tea%20Market%20Value%20Chain%20Profile%202016.pdf/ [Accessed 1 April 2021].

Derrett, R., 2001. *Special interest tourism: Starting with the individual. Special interest tourism.* Hoboken, NJ: Wiley.

Dieke, P.U.C., 2020. Tourism in Africa: issues and prospects. *In:* Baum, T. and Ndiuini, A., eds. *Sustainable human resource management in tourism. Geographies of tourism and global change.* Cham: Springer. Available from: https://link.springer.com/chapter/10.1007/978-030-41735-2_2/ [Accessed 1 April 2021].

Eisenhardt, K.M., 1989. Building theories from case study research. *Academy of Management Review*, 14(4), 532–550.

Hassan, A., 2012. Package eco-tour' as special interest tourism product-Bangladesh perspective. *Developing Country Studies*, 2(1), 1–8.

James, H., 2021. A blend of tea and adventure at Karatina Green Estates. Available from: www.pd.co.ke/lifestyle/a-blend-of-tea-and-adventure-at-karatina-green-estates-64824/ [Accessed 1 April 2021].

Jolliffe, L., ed., 2007. *Tea and tourism: Tourists, traditions and transformations.* Clevedon: Channel View Publications.

Lee, S. and Bai, B., 2016. Influence of popular culture on special interest tourists' destination image. *Tourism Management*, 52, 161–169.

Ma, S., Kirilenko, A. and Stepchenkova, S., 2020. Special interest tourism is not so special after all: big data evidence from the 2017 great American Solar Eclipse. *Tourism Management*, 77, 1–13.

Namaqua West Coast, 2021. Rooibos. Available from: www.namaquawestcoast.com/rooibos/ [Accessed 1 April 2021].

News Dome, 2021. Kenya: from the bush to cup-tea tourism. Available from: https://newsdome.co.za/kenya-from-the-bush-to-the-cup-tea-tourism-launched/ [Accessed 1 April 2021].

Novelli, M., ed., 2005. *Niche tourism: contemporary issues, trends and cases.* Abingdon: Routledge.

Rowley, J., 2002. Using case studies in research. *Management Research News*, 25(1), 16–27.

Sarojini, P., 2020. Kenya launches tea tourism in Bomet. Available from: www.commonwealthunion.com/commonwaelth_news/ [Accessed 1 April 2021].

Saunders, M., Lewis, P. and Thornhill, A., 2007. *Research methods for business students.* 4th ed. Essex: Pearson Education Limited.

Sharpley, R. and Telfer, D.J., 2004. *Tourism and development: concepts and issues.* Clevedon: Channel View Publishers.

Soleimani, S., Bruwer, J., Gross, M. and Lee, R., 2019. Astro-tourism conceptualisation as special-interest tourism (SIT) field: A phenomenological approach. *Current Issues in Tourism*, 22(18), 2299–2314.

Steyn, L.L., 2017. Hooray! SA's celebrates first ever official 'Rooibos Day'. Available from: www.news24.com/news24/travel/sas-celebrates-first-ever-official-rooibosday-20170116 [Accessed 1 April 2021].

Swarbrooke, J. and Horner, S., 1999. *Consumer behaviour in tourism.* Oxford: Butterworth and Heinemann.

Tourism Update, 2020. Tea for tourism? Available from: www.tourismupdate.co.za [Accessed 1 April 2021].

Woldemichael, A., Salami, A., Mukasa, A., Simpasa, A. and Shimeles, A., 2017. Transforming Africa's agriculture through agro-industrialization. *Africa Economic Brief*, 8(7), 1–12. Available from: https://www.afdb.org/fileadmin/uploads/afdb/Documents/Publications/AEB_Volume_8_Issue_7_Transforming_Africa_s_Agriculture_through_Agro-Industrialization_B.pdf [Accessed 1 April 2021].

World Economic Forum, 2019. Travel and tourism competitiveness report 2019. Available from: http://www3.weforum.org/docs/WEF_TTCR_2019.pdf [Accessed 1 April 2021].

Zhou, M., Hsieh, Y.J. and Canziani, B., 2016. Tea tourism: examining university faculty members' expectations. *Travel and Tourism Research Association: Advancing Tourism Research Globally*, 35, 1–5. Available from: https://scholarworks.umass.edu/ttra/2012/Visual/35 [Accessed 1 April 2021].

PART II

Sustainability in tea tourism

10
INTEGRATED MANAGEMENT OF COMMUNITY-BASED TEA TOURISM

Value through symbiosis

Bussaba Sitikarn, Kannapat Kankaew and Athitaya Pathan

Introduction

The northern region of Thailand, a forested mountainous region, is the largest tea plantation area covering more than 15,700 ha (Khaokhrueamuang and Chueamchaitrakun 2019). For centuries, Assam wild tea (*Camellia sinensis* var. *assamica*) has been traditionally cultivated in the forests for local tea consumption, which is called *miang* or picked tea leaves (chewing tea). Unlike others, Thai people originally consumed tea as a chewing snack to keep them awake. This tradition was blended into local ways of life and maintained for generations (Keen 1978). Not only is there Assam wild tea, but there is also Chinese tea (*C. sinensis* var. *sinensis*) planted in the north and covering 13% of the area. It was introduced by the remnants of the Kuomintang army, which arrived from the southern part of China and settled down at Doi Mae Salong in Chiang Rai in 1949; it was then that commercial drinking tea production began to develop (Winyayong 2008). In addition to miang, Chiang Rai also offers oolong tea, green tea, jasmine tea, and varieties of Chinese drinking tea. Later, Doi Mae Salong became a popular attraction for tea tourism in Chiang Rai, where tourists can experience Chinese Tea culture and visit the tea plantation.

While Chinese tea is becoming well known, demand for miang consumption is declining, especially among younger generations (Sasaki 2008). Even though some amount of miang produced in Thailand is being exported to Myanmar, it is not economically beneficial for local people. Huay Nam Geun village is situated in the mountainous region of Chiang Rai Province, where the production of miang was a main income for local people. In 2014, community-based tourism (CBT) was introduced by the royal project and played a significant role in revitalizing the village through the concept of ecotourism. In these cases, CBT is seen as a means of local revenue generation, as well as conservation of its ecosystem. Likewise, tea tourism can be used as a sustainable development tool to preserve tea culture and maintain the traditional tea cultivated in the communities (Jolliffe 2007).

Therefore, a form of community-based tea tourism (CBTT) could be the key to the communities generating income, as well as sustaining traditional tea cultivation. CBTT can be sustainable when it is developed and managed consistently with symbiosis among stakeholders (DASTA 2017). Accordingly, this study examines the (1) potential of CBTT supply in Huay Nam Geun village and (2) its tourism stakeholders' demand on management of community-based tea tourism, as well as involvement.

DOI: 10.4324/9781003197041-13

Literature review

Tea tourism has been studied over the years. Jolliffe (2003) examined tea history, tea heritage and the tea industry in relation to tourism. In her study, tea has been a part of tourism during trips or at destinations as a beverage, a meal, or even a tourist attraction. Tea can be experienced in tea plantations, gardens and factories, as well as purpose built and organized tea exhibits, tours and festivals. Tea attractions include tea gardens where tea is planted, tea factories where it is produced, tea shops where it is for sale or served, and institutions that preserve and interpret tea culture. Tea tourism was later defined by Jolliffe (2007) as "tourism that is motivated by an interest in the history, traditions and consumption of the beverage, tea". In her book, Jolliffe (2007) disclosed the microscale development in integrating tea with tourism could positively affect the economy of the local community. In the same way, the role of tea industry attracting tourists to Sri Lanka has been examined by Gunasekara and Momsen (2007). Empirically, the researchers identified that tea tourism has potential to be both community-based tourism or high-end and deluxe. Conjointly, tea tourism could be aligned with rural farm tourism, heritage tourism, ecotourism and wellness tourism.

CBT has increased its significance to be a driving force mechanism of Thailand tourism industry development. Thailand Community-Based Tourism Institute (CBT-I) pointed out some keys to sustainable CBT that are the local communities must truly understand it, be involved in every decision made, and possess a strong sense of ownership (DASTA 2017). To ensure the greatest long-term success, CBT must employ an inclusive development process with strong community support and stakeholders' participation (Sebele 2010). According to DASTA (2017), key stakeholders in CBT comprise four sectors: academic sector, government sector, private sector and public sector. While the academic sector promotes skills, labour forces and knowledge to prepare the workforce for the industry, the government sector sets out policy in the tourism industry to assure that related sectors are all heading in the correct direction. Some government sectors would conduct quality control in some tourism activities which affect the public, such as the usage of natural resources and protecting exploitation of its environment. To enhance CBT and tourism products to meet consumer needs, it is required to have integration and collaboration from private sectors. Ultimately, the local committees should take part in directing the tourism industry to develop and promote proper management. Similarly, Oka *et al.* (2021) shed light on local people's motivation to enhance CBT sustainability. Because local people's expectation is to be primary key resources of the CBT development, education is essential to a local community on related knowledge and hospitality skills which could strengthen the destination development. Yet beyond new knowledge and skill sets, the local sage wisdom of tea consumption is greatly important. It is a way of life that people have practiced for decades. It is a magnificent value which creates a "wow" experience for tourists. Certainly, tea consumption processes and methods differ geographically and culturally. Then, the blending of local ways of life, and know-how into CBTT are the real value for its management and development. The result of doing so leads to a multi-win relationship in the view of symbiosis theory (Dan 2016). Dalling (2007) stated that integrated management directs and aligns organizations to work effectively. It is a nonlinear process with coherence between a variety of business functions and problem connections, which confront the tensions in different perspectives on value to implement strategies that create unity among the organization, society and the natural environment (Bizikova *et al.* 2011).

Furthermore, DASTA (2017) proposed the concept of symbiosis which could bring in sustainability for the community and provide customer satisfaction. Generally, value symbiosis refers to collaboration between a community as a service provider and their customers, resulting in customers' satisfaction and benefit to the community (Somsukwang 2014; Komphanchai 2015; Suksamran 2016). Aaker (2001) explained that value creation is a method to distinguish their products from others to provide customers with the highest satisfaction. From a tourism perspective, customers' value and satisfaction could attract more tourists to visit the community, repeat visitation and create

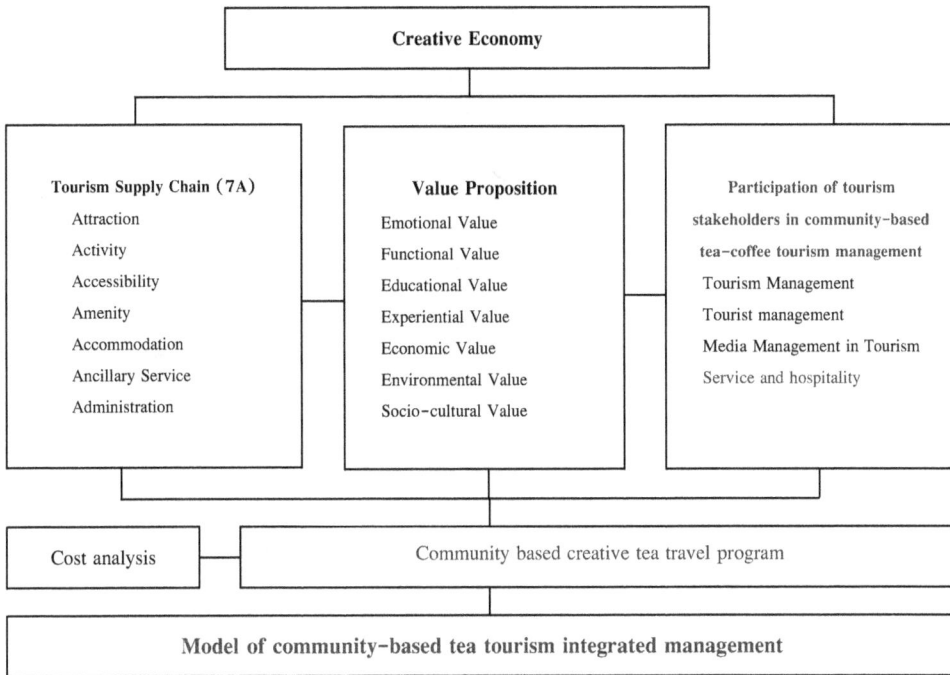

Figure 10.1   Conceptual framework

Source: Authors

word-of-mouth, as well as the willingness to buy more expensive travel products. In addition, the focal point of symbiosis requires local communities and their stakeholders to synchronize to achieve an ultimate mutual objective. This concept has been applied in the Thailand National Tourism Plan 2020–2022, which is launching an action plan for sustainable CBT under collaboration of partnerships. The plan integrates the symbiosis concept into community-based approach through value symbiosis among local community and key stakeholders, aiming at adding value to local resources and promoting supply-based marketing. The research framework is shown in Figure 10.1.

Methodology

A mixed methods design was employed that includes both qualitative and quantitative data collection and analysis. At the first stage, field surveys on the potential of CBTT supply at Huay Nam Guen village were taken by a group of ten researchers and ten external stakeholders. The assessment tool is developed based on CBT development guidelines (DASTA 2017) with additional tea tourism elements. This is to examine the current situation and its potential to be developed.

The second stage was to collect data on demand of stakeholders' participation in management towards CBTT. At this stage, qualitative data collection was administered by semi-structured interviews and focus group discussions. Purposive sampling was employed to key informants who were representatives of key tourism stakeholders at villages. These include seven public sectors, five private sectors, ten local authorities, 40 villagers, four NGOs and three educational institutes. Afterwards, a questionnaire survey was developed and collected to confirm the qualitative information. Accidental sampling was directed to 200 of both international and domestic tourists visiting the village during 2018–2019.

Finally, content analysis and descriptive analysis were employed to explain the data and integrated management phenomenon. After that, structural equation modelling was used for confirmatory factor analysis (CFA), complementing the qualitative results (Toland and De Ayala 2005). A goodness-of-fit analysis was tested including CFI 0.998, GFI 1.00, AGFI 0.999, RMSEA 0.010, and CMIN/d.f. 1.24.

Results and discussion

The results of the study are conveyed in two sections based on two objectives as follows.

The potential of community-based tea tourism supply at Huay Nam Guen village

Huay Nam Guen village is located at Tambon Mae Chedi, Wiang Pa Pao District, Chiangrai Province, Thailand. The village consists of 170 households with over 300 inhabitants. It is a highland village at an elevation of 1200 m and has produced miang or picked tea leaves for over a hundred years. It shows that tea has been a part of the history and economy of Huay Nam Guen village.

The village is under supervision by the agricultural extension area of the Huay Pong Royal Project. In 2014, the royal project with the support of the Highland Research and Development Institute has also started encouraging villagers to run agritourism businesses underpinning the principal management of CBT. However, as the village is situated in the Khun Chae National Park, tourism development has been limited.

In addition, the village positions tea as a main tourism product. They offer a CBTT package, which includes homestay, meals, organic tea tasting, and tea-related souvenirs. The tour program offers a variety of modern tea-making experience activities such as picking the wild tea leaves, roasting, and kneading tea by hand; walking in the organic tea garden; trekking in the wild tea forest; and having local cuisine made from tea leaves. At the time of data collection, there are eight homestays that are certified by the Home Stay Thai Standard to accommodate tourists. Additionally, five-coloured tea is used as a main theme to promote the CBTT package. *Five-coloured tea* is a signature product created to represent a sense of place and a story of wild assam tea. It is locally called *cha ha si* and made using five different methods for five different colours including white, green, yellow, black, and dark red.

Given the above, the village has potential on CBTT supply (Figure 10.3). The result, as shown in Figure 10.3 below, pointed out the ranking from the most to the least as accommodation (2.69), administration (2.69), attraction (2.56), activity (2.44), amenity (2.33), tourist service (2.31), ancillary service (1.80) and accessibility (1.75). Obviously, the ancillary service and accessibility showed low results because the village is situated on a high mountain which makes for difficult access, especially during the rainy season. In addition, it is controlled by the national park, therefore the development of ancillary service is limited.

The result showed that the active stakeholders for CBTT in the village are the *community*; the *public sector*, including Chiang Rai Agricultural Occupation Promotion and Development Centre (Highland Agriculture Extension), the royal project, the national park, the local authority, Ministry of Tourism and Sport, Department of Cultural promotion, and educational institutes; and the *private sector*, including Chiang Rai Tourist Association and tourism business–related entrepreneurs. As mentioned above, the village is mainly supported by Chiang Rai Agricultural Occupation Promotion and Development Centre (Highland Agriculture Extension) and the royal project to develop agricultural products and CBT. The areas are controlled by the national park, where land and facility development are restricted by law. Consequently, tourist activities are the only way to develop further, and the cultural tea experience is a key to trigger value symbiosis between host and guest.

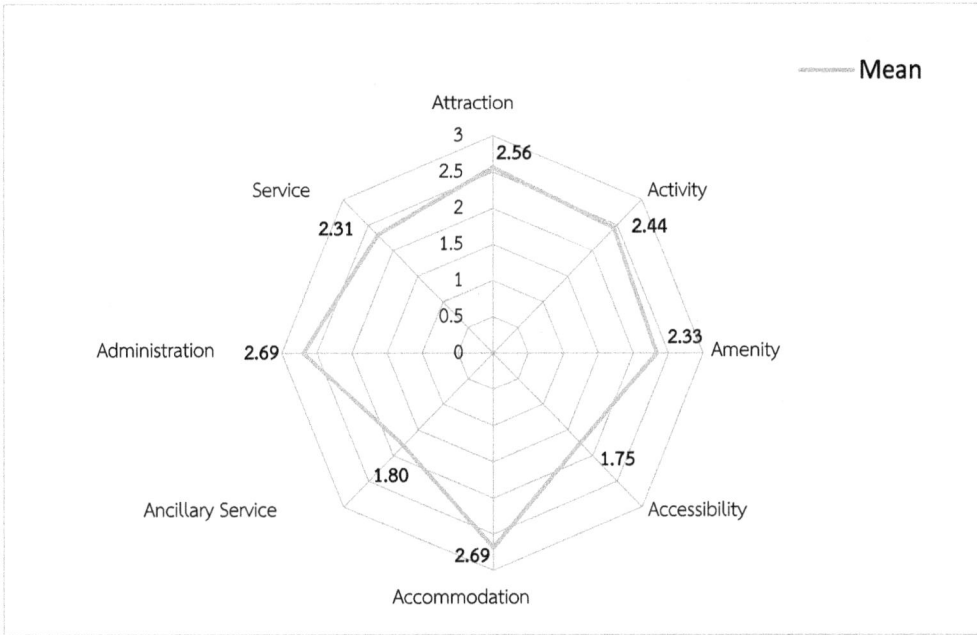

Figure 10.2 Potential of CBT–Tea Supply at Huay Nam Guen Village

Source: Authors

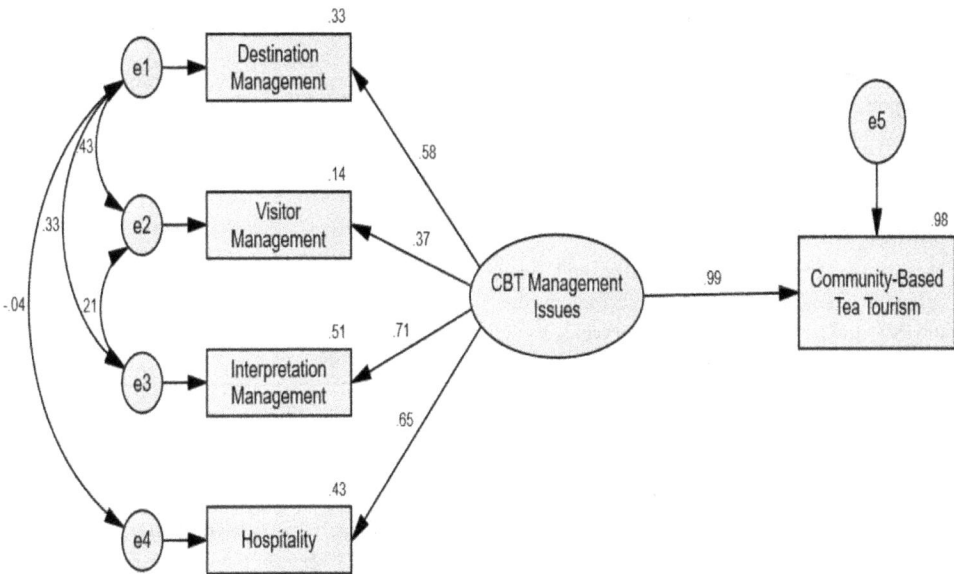

Figure 10.3 Confirmatory factor analysis

Source: Authors

Table 10.1 Tourism stakeholders' demand on management of community based tea tourism.

CBT Management Issues	Stakeholders			
	Community	Private Sector	Public Sector	Educational Institute
1. Destination Management				
Promote sustainable physical and environmental management	✓	✓	✓	✓
Promote sustainable sociocultural management	✓	✓	✓	✓
Promote participation of tourism stakeholders in CBT management	✓	✓	✓	✓
Local economic distribution for good quality of life	✓	✓	✓	✓
Activities and services that conserve cultural heritage	✓	✓	✓	✓
2. Visitor Management				
Accessibility: road condition, public transport, sign post and safety	✓	✓	✓	✓
Information Centre, map, interpretation and local guide	✓	✓	✓	✓
Banking services	✓	–	–	–
Tourist carrying capacity in according to destination carrying capacity	✓	✓	✓	✓
Enforcement of tourists' code of conduct	✓	✓	✓	✓
3. Interpretation Management				
Knowledge management process between host and guest	✓	✓	✓	✓
Awareness on conservation of natural and sociocultural resources amongst host and guest	✓	✓	✓	✓
Knowledgeable and skilled local guide	✓	✓	✓	✓
Activity enhancing perception and understanding on locals' way of life	✓	✓	✓	✓
Sign post, map, tourist information centre and distribution channels of CBT promotion	✓	✓	✓	✓
4. Hospitality				
Service and safety, criminality, drug, patent, overpricing	✓	✓	✓	✓
Activity operation with good hospitality and friendly	✓	✓	✓	✓
Sense of ownership amongst the locals, knowledge and skills and participate in tourism services, using local wisdom in all aspects to appreciate tourists	✓	✓	✓	✓

Conceptualizing ideas of the tourism stakeholders' demand on participation of CBTT can be categorized into four management issues including sustainable destination management, visitor management, interpretation management and hospitality. The management issues, as shown in Table 10.1, were then evaluated based on its impacts on CBTT together with the variables affecting the management issues by using CFA. The result, as demonstrated in Figure 10.4, indicated that interpretation management has the highest impact on CBT management issues (loading factor 0.71), followed by hospitality and destination management (loading factors 0.65 and 0.58, respectively). Also, the CBT management issues affect the CBTT with loading factor 0.99. Thus, CBT

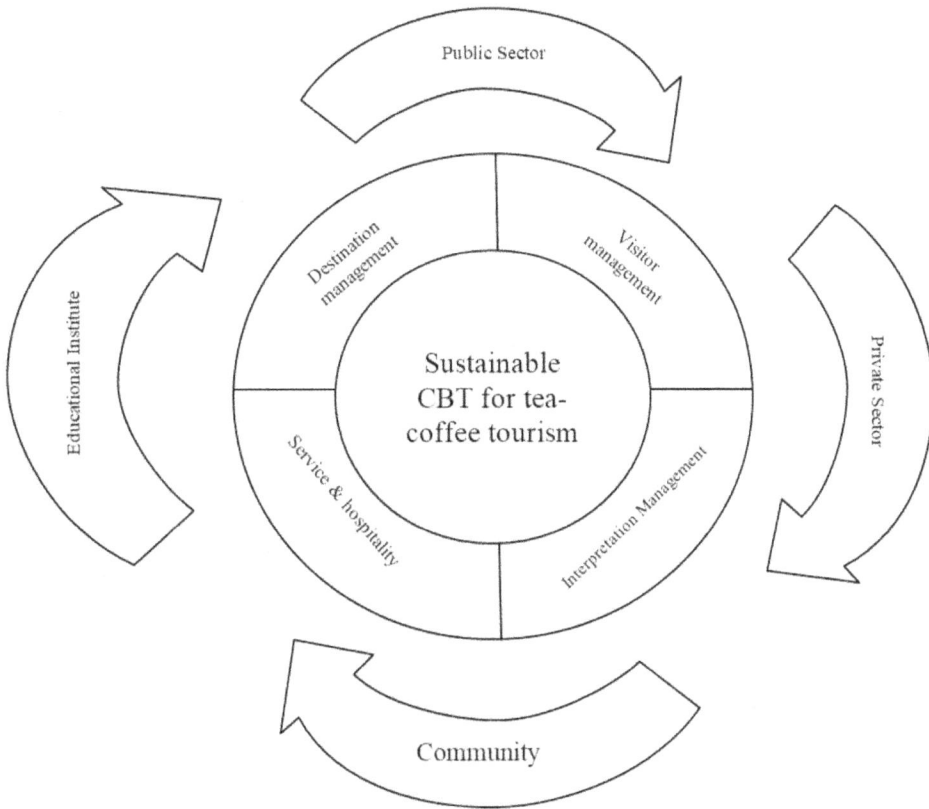

Figure 10.4 A model of community-based tea tourism integrated management

Source: Authors

management issues should be highlighted on interpretation management, hospitality management and destination management.

Ultimately, this study comes up with an overall image of an integrated model, as displayed in Figure 10.4. To manage sustainable CBTT, key stakeholders must co-create value through four integrated management issues: destination management, visitor management, interpretation management and hospitality. Their participation and involvement are important across management activities of the cycle with building, measuring and continuous learning.

Worth noting is that the ancillary service and accessibility have rather low CFA scores at 1.80 and 1.75, respectively. Certainly, these issues have to be improved, as Sitikarn and Kankaew (2021) defined CBT ancillary service as enhancing the products and services with informative promotion. Meanwhile, accessibility is dealing with infrastructure linking the tourists to the destination. Positively, accessibility and ancillary services enhance the tourists to acknowledge the destination, activities, products and services, and to be able to reach the tourist sites easier. On this account, the economic and life span of local people and communities would be improved. On the other hand, local people's life span in a traditional way and behaviour might be changed in response to their economic power, development and modernization. As previously mentioned, miang is not consumed by younger generations, and presently few people have knowledge about miang. The other issue is the environmental system has to be well aware. Hence, the integration of management of CBTT on providing knowledge and skill sets should cultivate the value of local originality and uniqueness.

Conclusion and implementation

Through the case study at Huay Nam Guen village, it can be concluded that the readiness of the local community in aspects of tourism supply concerning the CBTT is essential. The understanding of the roles and interests of involved tourism stakeholders, especially the local communities and local authority, is critical to the success of CBTT management. The management issues, therefore, have been divided into four categories, which are sustainable destination management, visitor management, interpretation management and hospitality management. All stakeholders must have a clear responsibility and integrated manners for involvement. In addition to management, value symbiosis among stakeholders is also important. To add more value to CBTT, the community should allow tourists to be a part of the value creation process. For example, the tourist should be able to actively design their own experiences such as personalized tea making and packaging.

Consequently, this study has identified integrated management for CBTT to create more value to stakeholders. It comprises existing resources, tourism stakeholder's collaboration, human resources, capital, time, ecological and sociocultural considerations, local goods and services, and the tourist activities. These, previously, existing resources are key benefits and challenges for tourism added value and symbiosis in the long term. It could be noted that it is a guideline for the development of sustainable CBTT management that combines non-destructive activities, ecological considerations and tourism stakeholders' collaboration. The result of this study indicates that the support or advocacy of the organization can drive some of the management for CBTT. Given the prior research, it was extended by the finding of current research illustrating integration of stakeholders' roles and participation to manage the quality of tourism, as well as increase more value for the community. It supports Bizikova *et al.'s* (2011) statement that it is essential for all sectors to work together in a coordinated way to overcome obstacles.

To implement CBTT development, policymakers should work closely with the local community, academics and other stakeholders in planning and leveraging the activities. The consideration of being non-destructive in terms of environment and traditional life span is crucial. Nevertheless, the incorporation of local ways of life, the benefit of miang and other types of tea consumption, must be embedded in the products and services of the community. The scholars should educate local people to conserve their uniqueness and cultural treasures. These would enhance CBTT sustainability.

Finally, this research has found a guideline for interested parties to apply for CBTT management and policy. The model could be used for increasing the value of tourism and systematic work. The positive aspect of this model is that it is easy to understand, manage, follow-up and consistently support. Despite the results of the CBTT management model, there are some limitations which need to be considered and improved in this study. Due to the restriction of time and place, the case study was limited to just a single field in Thailand, which is a small number compared to the overall number of tea fields in the global context. This limitation can be a possible direction for future research. The researcher might conduct a more varied case study to extend and elicit deeper information.

Discussion questions

1 How could you describe the effectiveness of a community-based tea tourism (CBTT) operation?
2 According to the potential of CBTT supply statistic results, accessibility has the lowest mean CFA value (1.76) among others. Due to its mountainous destination, specifically in the rainy season, it is difficult to visit. Do you agree or disagree with installing a concrete road to the destination so that tourists can visit year round? Or would you recommend leaving the road as it is in order to let the natural resources restore themselves for a specific time?
3 How could you explain the CBTT management model?

References

Aaker, D.A., 2001. *Strategic market management.* 6th ed. New York: John Wiley & Sons.

Bizikova, L., Swanson, D. and Roy, D., 2011. *Evaluation of integrated management initiatives.* Winnipeg: International Institute for Sustainable Development.

Dalling, D.I., 2007. Integrated management definition. *Chartered Quality Institute – integrated management special interest group.* Issue 2.1, 4–5.

Dan, Y., 2016. A study on the symbiosis between tourism scenic spot and local community development: a case study of Shawan Ancient Town. *International Business and Management,* 13(3), 22–26.

Designated Areas for Sustainable Tourism Administration (DASTA), 2017. *CBT: how to become sustainable? (A practical guide to community-based tourism).* Bangkok: Production House Accessories Limited Partnership.

Gunasekara, R.B. and Momsen, J.H., 2007. Amidst the misty mountains: the role of tea tourism in Sri Lanka's turbulent tourist industry. *In:* Jolliffe, L., ed. *Tea and tourism: tourists, traditions and transformations.* Clevedon: Channel View Publications, 84–97.

Jolliffe, L., 2003. The lure of tea: history, traditions and attractions. *In:* Hall, C., Sharples, L. and Mitchell, R., eds. *Food tourism around the world: development, management and markets.* Oxford: Butterworth-Heinemann, 121–136.

Jolliffe, L., ed., 2007. *Tea and tourism: tourists, traditions and transformations.* Clevedon: Channel View Publications.

Keen, F., 1978. The fermented tea (Miang) economy of Northern Thailand. *In:* Kunstadter, P., Chapman, E. and Sabhasri, S., eds. *Farmers in the forest: economic development and marginal agriculture in Northern Thailand.* Honolulu, HI: University of Hawaii, 379–402.

Khaokhrueamuang, A. and Chueamchaitrakun, P., 2019. Tea cultural commodification in sustainable tourism: perspectives from Thai and Japanese farmer exchange. *Rajabhat Chiang Mai Research Journal,* 104–117.

Komphanchai, S., 2015. *Scale development for measuring consumer perceived value towards products and services of social enterprises.* Thesis, Chiang Mai University, Chiang Mai.

Oka, I.M.D., Murni, N.G.N.S. and Mecha, I.P.S., 2021. The community-based tourism at the tourist village in the local people's perspective. *GeoJournal of Tourism and Geosites,* 38(4), 988–996.

Sasaki, A., 2008. Changes in the management system of the resources in the 'Miang Tea Gardens': a case study of PMO Village, Northern Thailand. *Tropics,* 17(3), 271–280.

Sebele, S.L., 2010. Community-based tourism ventures, benefits and challenges: Khama Rhino Sanctuary Trust, Central District, Botswana. *Tourism Management,* 31(1), 136–146.

Sitikarn, B. and Kankaew, K., 2021. Community based tourism logistic supply chain management in the Top North of Thailand: a key linkage to the Greater Mekong Subregion. *Uncertain Supply Chain Management,* 9, 983–988.

Somsukwang, A., 2014. *Factors affecting cross-cultural tourists' perceived value and loyalty toward Thai Spa.* Thesis, Kasetsart University, Bangkok.

Suksamran, T., 2016. *The influence of price perceived value and emotional perceived value on trust, customer satisfaction, word-of-mouth and revisit of SF Cinema City in Bangkok.* Thesis, Rajamangala University of Technology Phra Nakhon, Bangkok.

Toland, M.D. and De Ayala, R.J., 2005. A multilevel factor analysis of students' evaluations of teaching. *Educational and Psychological Measurement,* 65(2), 272–296.

Winyayong, P., 2008. Current status and future development of tea production and tea products in Thailand. *Proceeding of the International Conference on Tea Production and Tea Products,* 26–28 November, Mae Fah Luang University, Chiang Rai.

11

TEA COMMUNITY CULTURE AND TOURISM

The case of Turkey

Kadir Çetin and Emre Erbaş

Introduction

Statistics indicate that Turkey is the world's largest tea-drinking country, with 6.96 pounds of annual per capita consumption in 2016, followed by Ireland (4.83) and the UK (4.28; Statista 2020). Tea became an alternative to coffee after losing Yemen in the insourcing of coffee for the Ottoman Empire and starting the importing of expensive Brazilian coffee, and tea became an important national drink in 1923 (Wohl 2017). During this period, that one of the Japanese tea masters served Japanese tea to the Sultan accelerated tea-drinking habits (Honda 2012). Since then, tea has become a clock that determines the Turks' daily life dynamics. It has emerged as a strong boundary object. Gathering around tea, before getting into conversation at the table, people start to make commentaries about the tea being served, such as its origin, its way of cooking, its way of colouring, and the heat source used in its cooking.

The first thing tourists encounter about Turkey is the Turkish tea. Tea is a gateway for tourists wishing to travel into the authenticity of the visited destination. Just as a tourist is welcomed with a lei in Hawaii, they are welcomed with tea in Turkey. There is hardly a place in Turkey where a tourist is not offered a cup of tea. Locals start to volunteer as tour guides, using narratives about anything that tourists wonder about while sipping their tea. Tea is also a kind of strategic tool to keep tourists longer in their shopping experiences and ease the service recovery process in hospitality operations. In addition, because Turkish tea is grown and made by hand, there are many things for tourists to do from the time it is planted to when it is harvested and processed. Tourists are excited to go to the place where Turkish tea comes from after seeing all of these things (that is, the Eastern Black Sea Region).

Hence, a holistic look at tea as a practice developed in the socioecological world may provide a fruitful perspective for the planning of tea tourism and its sustainability. In this context, this chapter illustrates that the meanings of tea have given birth to the formation of the Turkish tea community, which will provide a point of departure for the future sustainability of tea tourism practices.

Theoretical background

Even though Turkish tea is an indispensable reflection of Turkish culture, there is no research drawing a holistic look into exhibiting its reflexivity in the context of tourism. For example, Wohl (2017) examined the "third space" that the Turkish tea garden (Çay Bahçesi) provides in terms of how

DOI: 10.4324/9781003197041-14

these spaces act as repositories of shared memory, mediating conflict that appears in other societal spheres. Honda (2012) and Tan (2018) explored the patterns of how the country switched from a coffee-drinking nation to a tea-obsessed one. Göken and Alppay (2019) examined a specific Turkish tea class called "ince belli" in terms of its design development potential. From a cultural perspective, Güneş (2011) has revealed tea preparation and service methods and tea-related tools.

Therefore, it is important to draw on the socioecological foundations of tea tourism to develop strategies for its design and sustainability in a tourism context. This will help to develop a background against which to evaluate the loss of tea culture, especially stemming from the increasing coffee consumption in tea communities. In this way, mapping the cultural structure of tea in a community is needed to come up with sustainable tea tourism strategies.

However, how socioecological measures can be a reference for sustainability in tea tourism is still one of the missing links. Therefore, the purpose of this chapter is to illustrate how the practices of tea can be the basis of tea tourism within a socioecological lens and its sustainability with evidence from Turkey. At this point, using a practice lens from the philosophical perspective will help us to develop an insight into the building blocks of tea's social reality. According to practice theory, everyday actions are consequential in producing the structural forms of social life (Feldman and Orlikowski 2011). Practices are social actions that produce and reproduce the structures that shape actions, which then enact social orders (Giddens 1984; Schatzki 2002). In brief, practices are "behavioral, cognitive, procedural, discursive, and physical resources through which multiple actors are able to interact in order to socially accomplish collective activity" (Jarzabkowski *et al.* 2007, 9). Revealing the tea practice in the case of the Turkish community as the outcome of the mutual interaction between specific instances of situated action and the social ecology in which the action takes place (Feldman and Orlikowski 2011) will give us the social ecology of tea tourism. A social ecological view (Clark 1997) provides an understanding in which to analyse social practice in a holistic and dialectical way. It reckons that individuals' behaviour is stimulated by a range of embedded contextual systems within their lived environments (Bronfenbrenner 1979).

Therefore, without understanding the social ecology of tea in its situated community, the transformation of tea-related practices into tourism will not be valid. In light of this information, this chapter examines the reflexive meaning of tea as a well-adopted practice that is interpreted and symbolized in Turkey. The data and social ecological theoretical knowledge indicate positioning tea as a spatial (or place) in which the folklore of a culture can be discovered and designed within the tourism system. Memory is spatial, so shaping space is a way to shape the way we think about things. A shared space such as a street can be a locus of collective memory in a double sense. It can express group identity from above, through architectural order, monuments and symbols, commemorative sites, street names, civic spaces, and historic conservation; and it can express the accumulation of memories from below, through the physical and associative traces left by interweaving patterns of everyday life (Hebbert 2005, 592). Hence, the themes revealed from the analysis indicate the socioecological pillars on which and how tea practices serve a valuable core tourism experience for tourists. This chapter develops a social-ecological picture of tea practices (i.e., praxis points for tourism) in Turkey.

Research method

This research adopts an interpretive research approach, because interpretive research helps researchers to view social reality as being embedded within social contexts, interpreting the reality through sense-making rather than a hypothesis-testing process. For this, the authors used a reflexive thematic analysis (RTA) approach to draw out the socioecological dimensions of tea practices within the Turkish community. RTA is an appropriate approach to examine the participants' true-life experiences, as well as the social structures that form experiences, meanings, and presumptions (Braun and

Clarke 2006). This approach helps to interpret the meaning of the studied contents to get a sense of the social world that builds the structuration process (Bauman 2010). Additionally, we have analysed around 150 minutes of content in total from nine different videos. For data analysis, we followed Braun and Clarke's (2006) six steps in the RTA process, which are familiarization, coding, theme development, revising themes, defining themes, and producing the report.

Data analysis and discussion

Turkish tea practices emerged under two main themes: ethnographic and folkloric, together with their sub-themes.

Ethnographic features

Ethnographic features have been examined under two main headings: (1) tools used, such as tea cups and glasses, tea saucers, tea spoons, teapots, samovars, and so forth, and (2) places such as tea gardens and coffee houses.

Tea cup/glass

The designer of the tea cup is unknown. The main design of the glass, which has a wide base, a thin middle part, and a wide mouth, hasn't changed for a century (Figure 11.1). It is named in several different ways due to its popularity among people. The most common ones are *slim waist*, which is

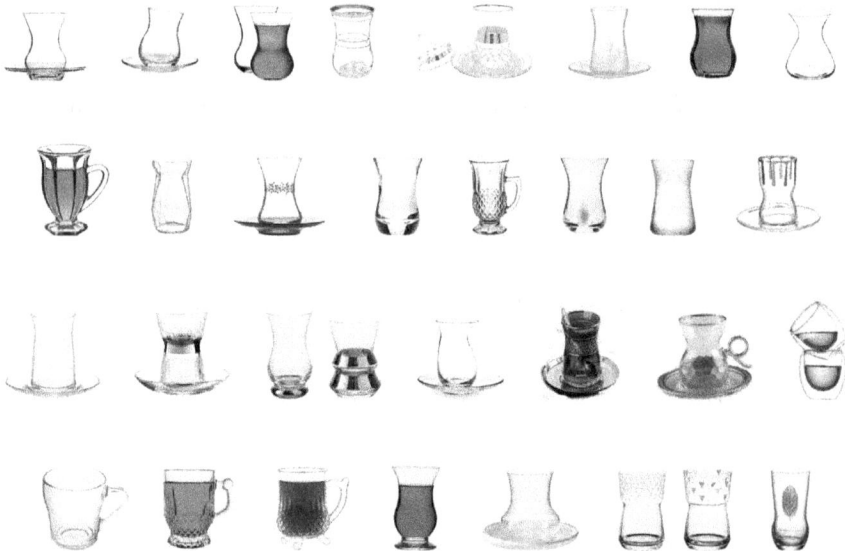

Figure 11.1 Turkish tea glass designs

Source: Authors

associated with delicacy and elegance; *tulip shaped*, which is a symbolic element of Turkish culture; and *Ajda*, which evokes the beauty of singer Ajda Pekkan.

It is assumed by most people that the thin-walled tea glass shape is more of a result of a scientific inquiry than a design. It is pleasing to the eye, because the transparent material used in the glass reflects the blood-red colour of the tea; it is pleasing to the nose, as it reveals the unique taste and smell of the tea thanks to the properties of the glass; and it appeals to the sense of touch, because it enables grasping with two fingers (forefinger and thumb) and feeling the warmth of the tea. Moreover, the sound of mixing sugar with a tea spoon appeals to the ear. Thanks to all these features, the tea glass increases the sensory benefits as well as the hedonic aroma. Besides, the fact that it warms both the hand and the palm in cold seasons and its being held with two fingers in hot seasons shows how suitable the design of the tea glass is for the Anatolian geography, where four seasons are experienced. Furthermore, it is used as a unit of measure in Turkish cuisine.

Tea saucers

A tea saucer is an inevitable accompaniment for a tea glass. With modernization, various design studies are being done to address some problems, like wet sugar and sticky tea saucers. However, the familiar red-and-white striped popular form, which is made of shatter-resistant melamine, has never changed. While the red stripes on the tea saucer reflect the "bright red" (dark brownish red) colour of the tea and symbolize a lovers' ruby lips, the white stripes show how well the tea has brewed. Additionally, it creates a space for sugar and tea spoons while serving and facilitates the serving of hot tea.

Tea spoon

The tea spoon complements the delicate and elegant look of the tea glass. It can be confused with the dessert spoon but is distinguished by its thin handle and small mouth part. These tiny stainless-steel spoons are used to stir the sugar in the tulip-shaped tea glasses.

Samovar, teapot, and cauldron

Samovar means self-boiling. It is produced in the shape of a cylinder or rectangular prism from copper, brass, or sheet metal. However, all samovars have the same technical features, even if their designs vary. The samovar consists of several parts. The samovar, at the base, has feet that serve as carriers and prevent the heat from damaging the surface. In the body section, there are a cauldron where the fire is lit, a cauldron cover, a cylindrical water tank, a metal pipe that heats the water by passing through the water tank, and a tap "kran". And in the extension of the body, there is a smokestack (*şeyka*) to ensure a draft and a *demkeş* to put the teapot in. The samovar is widely used to brew tea in Turkey, Russia, and the Central Asian Turkish Republics and Iran, where half of the population is Azeri Turks. The samovar has been associated with friendship and conversation among people. That is why there are popular sayings like "one samovar means a thousand conversations" and "if there is a samovar, there is a long conversation", as it keeps the tea warm for a long time. Sait Faik Abasıyanık, one of the leading writers of Turkish storytelling, defines the importance of the samovar for Turkish society in his book titled *Samovar* written in 1936 (Abasıyanık 1936):

> How beautiful that the samovar would boil in the room that smells of toasted bread. Ali likened the samovar to a factory in which there was neither misery nor strike nor boss.

A teapot is a kitchen utensil used in tea preparation, usually made of two pieces of various sizes and materials like copper, porcelain, and metal. Both parts have a handle and a spout with a narrow mouth that allows the liquid (tea and water) to be poured at the most appropriate angle. The small piece is called a "teapot" because tea is brewed in it. The teapot is a fixture of Turkish cuisine, because there are large or small teapots in every house to host crowded guests. However, the dominant belief among people is that the tea is more delicious on an open fire (stove), wood fire, coal fire, or in a copper teapot.

Tea cauldrons are similar to samovars. Nevertheless, they are bigger than samovars and are used in places where tea service is intense, such as tea gardens, coffee houses, tea houses, and businesses.

Tea gardens and coffee houses

Tea gardens are social surroundings where people gather to drink tea. Every city has its own tea gardens, which are mostly located under plane trees, around places with a direct water view, by a stream or among the greenery. That is because Turkish people enjoy sipping their blood-red tea while enjoying a beautiful view. In contrast to the calm and serene image created by Japanese tea gardens, tea gardens in Turkey are places where conversations are not only about neighbourhood gossip but also politics, and they are lively places where games such as backgammon are played while watching children have fun together. Therefore, it is possible to say that tea gardens are socializing centres for Turkish people.

Coffee houses are similar to tea gardens. However, they are places where retired men in particular come together and talk about politics and the economy, and play games such as backgammon, okey, and so forth while drinking tea, like water. The name suggests the image of coffee consumption rather than tea consumption. In fact, after World War II, tea started to be consumed more due to the widespread use of tea in the country, its ease of drinking, and the challenges of coffee imports.

Folkloric features

According to the findings of the analysis, the folkloric features generated by Turkish society on tea and tea culture can be divided into three categories: (1) rituals and traditions, (2) sociocultural values, and (3) tea dictionary.

Rituals and traditions

In the analysis conducted, it is concluded that the rituals and traditions generated by them are classified under three subthemes: (1) blending and preparation of tea; (2) tea service, consumption, and food pairing; and (3) social convention.

Blending and preparation of tea

As far as is known, one out of every three Turkish people makes their own blend by mixing various tea brands. In this blend, tea from Iran, Sri Lanka, or tea brands known as smuggled tea are used as the base product. It can be said that the characteristics of the blend prepared vary according to the geographical region. While the blends have a subtle colour, smell, and taste in Western Anatolia, the blends made in Central and Eastern Anatolia have a sharp and intense colour, smell, and taste. It is even known that sometimes only smuggled tea is consumed without blending.

The preparation of tea may differ from person to person, as in blending. However, the quality of the tea, the storage conditions, the properties of water and brewing temperature, time, and method are of crucial importance in this process. Quality tea shouldn't contain dust, stems, or fibres from the tea leaves. The water should be clean and rich in minerals. The brewing temperature should be between 95°C and 99°C. Brewing time and method change according to the type of tea and the level of bitter taste desired. A tea lover from Erzurum, where tea is consumed with a lump of sugar in one's mouth, describes the tea preparation as follows:

> Tea blend is poured into the upper teapot. The upper teapot is slightly shaken so that the fine tea powder can be filtered. Then, it is washed with cold water. In order to make the color more beautiful, one or two pieces of sugar are added according to preference. After this process, it is placed above the boiling water in the lower teapot. A raw smell comes from the tea roasted with the steam of boiling water in the lower teapot. After this scent, boiling water is poured over the tea and left to infuse for around 30 minutes. During brewing, the water in the lower teapot should be at a temperature very close to boiling, where the water "clicks" but doesn't boil. After the brewing process, the teapot is removed from the fire and left to rest for 10 minutes.

Tea service, consumption, and food pairing

Turkish tea is generally served in a "thin-waisted" glass with a cultic red-and-white special saucer. The glass is first rinsed with hot water. Tea and hot water are poured from the upper teapot with a strainer. It is not drunk with milk or lemon, but instead it is served with two small lumps of sugar. Although it doesn't have a certain size, a gap called *lip share* is always left for a sip. This gap is left so that the drinker can hold the hot glass with their two fingers safely and sip comfortably. Additionally, the size of this gap provides parody material for tea conversations. There are two important factors to pay attention to in tea service. The first one is the colour of the tea. Turkish tea should be bright red. If the colour of the tea is bad, expressions such as "go now" or "as black as pitch" are implied for the guest. Moreover, the colour of the tea can be *light, low-infused*, or *dark–very brewed*, depending on the preference of the person. The second factor is the temperature of the tea, which warms you in the winter and cools you in the summer months. The temperature of the tea is understood from the steam that appears after it is served. If the tea is not hot, the expression "imam's ablution water" is used. On the other hand, the tea prepared for children is called *pasha tea* and is served warm. The transition from pasha tea to the bright red one symbolizes the transition from childhood to adulthood among people. It is known that pasha tea dates back to the Ottoman bureaucracy. Then, due to the workload of the bureaucrats, their tea was served cold so that they could drink it quickly and comfortably.

Regional differences are observed in tea service. In Erzurum, tea is served without a spoon but with a lump of sugar called *kıtlama*. Tea is drunk like water with *kıtlama* sugar in this manner: a piece of sugar is bitten and placed under the tongue. Then, a sip of tea is taken. Then, the sugar is circulated in the mouth for one round and placed under the tongue again. This cycle is repeated with every sip. It is known that a person can drink a pot of tea with one or two lumps of sugar.

In Turkish culture, there is no specific time of day or season for tea drinking. It is drunk from the start of the Turkish breakfast until bedtime. Turks want to constantly mix the sugar added to the tea served in a tulip-shaped glass with an elegant and delicate teaspoon sound, grasp it with two fingers (forefinger and thumb) and slurp it regardless of time and place.

This warm drink, which is mingled with geographical and cultural elements of Turkish culture, is especially associated with breakfast and some foods. Within the scope of the Home of Turkey campaign, which aims at introducing Turkish culture to people coming from all over the world in a more transparent, sincere, and convincing way, tea is associated with chestnuts on embers. Tea is always consumed after every meal. That is why it is indispensable for restaurants to offer tea after the meal, and the consumers evaluate the service quality over tea.

Social convention

As tea has started to have a place in Turkish culture, some social convention rules have emerged among people regarding tea. It is customary to offer a cup of tea to a newcomer or guest as a polite way of welcoming them, and it is considered rude to refuse this offer. The oldest or most respected guest is the first to be served tea. If you are the host, the tea is refreshed until the guest says enough, and you should avoid saying, "We have no more tea". Generally, the teapots are not brought to the room if there is a guest. The host knows which glass belongs to which person and serves the tea accordingly. However, it is sometimes possible to bring the teapot to the room where the guests are in order not to interrupt the conversation. If you are a guest in Erzurum, it is said *arıldım* (I am full) for the last tea served by the host insistently after saying thank you. It is also called *hard tea*, and it is rude to refuse it. In other regions, a tea spoon is put on the glass, which means "that is enough, thank you".

Sociocultural values

Gerald Robbins, writing in the *Washington Times*, explains the importance of tea in Turkish life with this statement:

> Every business in Turkey is conducted based on traditional habits. The people who encounter bureaucracy know this situation very well. For instance, nothing starts before the tea is served in Turkish-style glasses in the capital of Ankara, for instance. Offering a cigarette with tea, which is followed by a long talk in a friendly or philosophical manner, has become the definition of business. Shopkeepers generally offer Turkish tea to the customers and then bargaining begins on the product to be sold. Tea plays a major role in Turkey's domestic economy. In work places, bazaars, and inns, there are tea shops and tea makers who are in charge of preparing tea and other drinks. The tea maker knows everybody, and everybody knows the tea maker. That is why it is thought that the absence of a tea maker in a workplace is noticed before the absence of a boss.
>
> *(cited in Hızlan 2005)*

Due to the fact that Turkish tea is an inevitable part of conversations in any social environment, it is said, "conversations without tea are like a night sky without the moon". A cup of tea served during busy working hours and a tea break provided a boost to employee morale. *Shall we drink a cup of tea?* means *a kind invitation. Shall we drink one more cup of tea?* means *I am still longing for you or I have a lot to tell. Tea is hot* means *Let's have a chat, long time no see. I will have a cup of your tea*, which means *I will visit you as soon as possible*". A cup of tea is an integrative and effective component of communication that arouses the feelings of humbleness, sincerity, helpfulness, sharing, pleasure, sensuality, grace, kindness, and enjoyment. Katharine Branning (2021) makes an analogy for Turkish tea: "Just like a Turkish person".

Tea dictionary

The rich mixture of tea culture has created a tea dictionary. Many words and phrases arise from the tea culture, bearing traces of the community: tea garden, tea field, tea sapling, tea sprout, tea seedling, bulk tea, scissors for cutting tea, tea basket (e.g., Figure 11.2), teapot (lower and upper teapot, spout), samovar (cauldron, cauldron cover, *truba, demkeş, kran, şeyka*), tea cauldron, tea glass (thin-waisted, Aida, Ajda, tulip-shaped tea, offering tea, refreshing tea, pouring tea, filling tea, bright red tea, sugar cubes, *ktlama* sucrose, pasha tea, slurping tea, tea lover, tea literature, and so on.

Figure 11.2 Tea harvest
Source: Kalender 2021

Conclusion

Implications

In Turkey, tea is regarded as an experience that enriches various tourism activities, and it is also a basic destination attraction. Tea serves as a strong mediator between the Turkish community and tourists. While sipping their tea, the locals share interesting information about their country. Based on this knowledge, this chapter is aimed to draw attention to tea as a socioecological practice in which the sustainability of tea tourism can be constructed. This perspective provides an opportunity to evaluate the social meaning of tea on the basis of cognitive, behavioural, motivational, and physical interactions of individuals and societies and its position in accessing the cultural forms of the relevant destination. Hence, in this present research, the sociological foundation necessary for the transformation of tea into an experience that enriches various touristic activities has been discovered within the case of Turkey, as well as the transformation of tea into a touristic practice.

Recommendations for future research

Herein, the question of how tea may become such a socio-material and or epistemic object embedded within communities arises. Furthermore, how such an embedded practice within communities is legitimized and achieves its final form in various contexts (e.g., behavioural, geographical, material) needs to be questioned. In this respect, it is recommended that future studies reveal the socioecological foundations of tea with a socioecological practice perspective to establish a foundation for the sustainability of tea tourism. The social background of tea practice will strengthen the strategizing activities of tea practice within the roots of society.

References

Abasıyanık, S.F., 1936. *Semaver küçük hikayeler*. İstanbul: Remzi Kitabevi.

Bauman, Z., 2010. *Hermeneutics and social science: approaches to understanding*. London: Routledge.

Branning, K., 2021. *Bir çay daha lütfen*. 3rd ed. Konya: Literatürk Academia.

Braun, V. and Clarke, V., 2006. Using thematic analysis in psychology. *Qualitative Research in Psychology*, 3(2), 77–101.

Bronfenbrenner, U., 1979. *The ecology of human development: experiments by nature and design*. Cambridge, MA: Harvard University Press.

Clark, J., 1997. A social ecology. *Capitalism Nature Socialism*, 8(3), 3–33.

Feldman, M.S. and Orlikowski, W.J., 2011. Theorizing practice and practicing theory. *Organization Science*, 22(5), 1240–1253.

Giddens, A., 1984. *The constitution of society: outline of the theory of structuration*. Cambridge: University of California Press.

Göken, M. and Alppay, E.C., 2019. A case study on Turkish tea glasses and Kansei Engineering. *In*: Fukuda, S., ed. *Advances in affective and pleasurable design*. Washington: Springer, 319–328.

Güneş, S., 2011. Türk çay kültürü ve ürünleri. *Milli Folklor*, 12(33), 234–251.

Hebbert, M., 2005. The street as locus of collective memory. *Environment and Planning D: Society and Space*, 23(4), 581–596.

Hızlan, S., 2005. Çaysız sohbet aysız gökyüzüne benzer [online]. Available from: www.hurriyet.com.tr/caysiz-sohbet-aysiz-gokyuzune-benzer-338508 [Accessed 8 July 2021].

Honda, Y., 2012. *History of Turkish tea drinking habits* (No. 2-33-6). Tokyo: Turkish Embassy.

Jarzabkowski, P., Balogun, J. and Seidl, D., 2007. Strategizing: the challenges of a practice perspective. *Human Relations*, 60(1), 5–27.

Kalender, A., 2021. Tea Harvest [digital image] reprinted with permission.

Schatzki, T.R., 2002. *The site of the social: a philosophical account of the constitution of social life and change*. University Park, PA: Pennsylvania State University Press.

Statista, 2020. China: coffee and tea consumption volume 2010–2020 [online]. Available from: www.statista.com/statistics/623672/china-coffee-and-tea-consumption-volume/ [Accessed 8 July 2004].

Tan, A.Ö., 2018. Turkish tea for liberty: changing the landscape of a region and drinkscape of a nation through political choice. *In*: McWilliams, M., ed. *Food landscape, proceedings of the Oxford symposium on food and cookery*. Great Britain: Prospect Books, 371–378.

Wohl, S., 2017. The Turkish tea garden: exploring a 'third Space' with cultural resonances. *Space and Culture*, 20(1), 56–67.

12

HOSPITALITY TEA CULTURE

Taking tea culture to the next level

J.A.R.C. Sandaruwani, G.V.H. Dinusha and R.S.S.W. Arachchi

Introduction

Tea production in Sri Lanka (formerly Ceylon) is vital to the national as well as the global economies. Sri Lanka is recorded as the world's fourth-largest tea-producing country, producing just under 300,000 tons per year or around 17% of the world's tea crop. This gives an annual income of $700 million. Apart from its trade value, Ceylon tea has been a pillar of Sri Lankan culture, heritage and identity. Tea is renowned for being one of the most enjoyed beverages among Sri Lankans. The average person enjoys at least three cups a day, and this habit of drinking tea has spread among the local commoners (Sumuduni and Piyumali 2015). A warm cup of tea is an integral part of Sri Lankan culture, reflecting the hospitality of the home towards guests. This 'home-based hospitality tradition transfers over to the offering of tea in commercial hospitality settings' (Jolliffe 2006, 165), which serves both as a symbol of the greatest Sri Lankan hospitality and a resource with noticeable potential in indulging the experience of tea-loving guests.

This blend of tea and culture captivates a growing number of tea lovers all around the world. Ceylon tea has become a phenomenon and part of destination tours for international tourists. Therefore, Sri Lanka is a main tea-producing country that also plays a role in tea tourism as one of its niche tourism segments. For Jolliffe (2007), tea tourism is defined as tourism that is motivated by an interest in the history, traditions and consumption of tea. Tea-producing countries are practicing tea tourism as a multifaceted concept with a centralized focus on sustainable development. Jolliffe (2007) identified the various aspects of viewing tea, from the agricultural, hospitality, art, religious and leisure perspectives. In many hospitality situations tea is a crucial aspect of food services.

When dissecting the scope of the literature on this industry, tea tourism research has focused more on destination marketing, potentials, issues and challenges rather than the role of tea culture in the hospitality industry. This chapter addresses how the hospitality industry crafts guest experiences by taking tea culture to the next level.

Literature review

Tea tourism has emerged as a burgeoning niche tourism segment. Thus, demand for the satisfaction of services towards tea tourism has increased, and catering to such requirements has become an important area of income generation. In particular, the tea industry has been vital to the economic

DOI: 10.4324/9781003197041-15

development of Sri Lanka, as it is the fourth-largest tea producer and the second-largest tea exporter in the world. Ceylon tea has grabbed global attention and has become renowned as one of the world's leading teas due to its unique taste and aroma.

In 1824, the British imported tea from China to Ceylon. Tea was introduced as a crop in the middle of the 19th century when coffee rust destroyed existing coffee plantations. James Taylor, a Scotsman, opened the first commercial tea plantation on the Loolkandura estate in 1867. Sri Lanka's tea industry legacy thus shows Sri Lanka is one of the world's oldest tea-producing nations. With its popularity increasing over the decades, the word 'Ceylon' has now become synonymous with quality tea (Herath and De Silva 2011).

Tea is both a local agricultural product for many provinces in Sri Lanka, as well as an export product. Sri Lanka specializes in producing black teas, which are classified into three categories: lower-growing, middle-growing and high-growing varieties (Jolliffe and Aslam 2009). Sri Lankans are addicted to drinking tea as a refreshment or herbal drink. They value the benefits of tea to their health. Thus, Sri Lankans have a strong tea-drinking culture, and the average person drinks three to four cups per day (Pearce 2016). Tea culture can be categorized into three dimensions: (1) material (producing and drinking tea); (2) behavioural (including tea houses, tea exhibitions, tea events); and (3) spiritual (aesthetic, religious and artistic) (Zhang 2004). This study shows how the hospitality industry considers taking tea culture beyond the traditional cup.

The concept of tea tourism differs from mass tourism in that it emphasizes hands-on experiences of tourism activities or events integrated with education and recreation rather than simply visiting tourism destinations (Jolliffe 2007). Ceylon tea and tourism are inextricably tied, as tea-growing areas are inarguably the most enticing tourist attractions in Sri Lanka (Ceylon Tea Land 2013). Koththagoda and Dissanayake (2017) explore the potential managerial applications of tea tourism in the Sri Lankan context. According to their findings, the concept of tea tourism could have the ability to build on different theoretical bases such as destination branding, farm tourism and attracting people from specific origins and be willing to pay for development to attract potential customers to the destination and expand customer choice revisited with the influence of customer satisfaction and loyalty. Gunasekara and Momsen (2007) proposed that tea tourism be developed into different niches including heritage tourism, ecotourism, health tourism, rural tourism, exclusive tourism and community tourism. Jolliffe and Aslam (2009) compared and analysed tea character accommodations, tea factories and local tea sale centres as supply components in Sri Lankan tea heritage tourism. In Sri Lankan travel itineraries, tea plantation visits, tea factory visits, tasting and buying different tea flavours, tea exhibitions, tea shops and tea garden train tours have attracted millions of tourists.

Today, tea has emerged as a distinct meal service in the hospitality industry, where they offer tea services reflecting local tea traditions. The World Travel and Tourism Council (WTTC 2020) identifies the hospitality industry as the main driver in global value creation. It brings trillions of dollars annually to the global economy and stimulates capital investment. Tea and hospitality have a hand-in-hand relationship. A cup of tea is a universal sign of hospitality in either a home or commercial hospitality setting (Walton 2001). In particular, offering a cup of tea is iconic for the tradition of welcoming guests in Sri Lankan culture. Moreover, it is a sign of Sri Lankan hospitality. Gradually, Ceylon tea is connected to commercial hospitality settings. The hospitality sector uses tea as an appealing product to market food. It offers an opportunity to develop appealing products and services that can also extend the use of dining facilities, attract new market segments and create new revenue streams (Jolliffe 2006).

In the lodging setting, tea is an experience, and it is not surprising that tea has evolved into a trendy product. Then, in the context of commercial hospitality, tea serves as both a symbol and a resource, with significant potential for commercial hospitality provision (Martin 2007). Whitmore (2000) observed that the hospitality industry has expanded its role in tea provision as a meal service, in particular by catering to business and corporate functions. Currently, afternoon tea or high tea service has evolved into a renowned tea culture in the hospitality industry (Bohne and Jolliffe 2021) from where it originated in 19th-century Victorian Britain (Jheng 1993). Afternoon tea is typically served on low tables in the middle of the afternoon. Tea etiquette, lace decorating, silver tea service and dainty foods like meat pies and cheesy casseroles are associated with it.

Tea tourism is defined as 'tourism that is motivated by an interest in the history, traditions and consumption of the beverage, tea' (Jolliffe 2003, 136). Therefore, tourists seek out authentic tea experiences during their visits. As in China, Sri Lanka also can use tea culture to create a comfortable and positive experience for guests. It can be a value addition to the guests. Jolliffe (2007) mentioned tea hospitality as one of the main areas for future tourism studies, thus the authors of this study were inspired to explore how hoteliers take tea to the next level.

Research design

A qualitative approach was followed to gather primary data using the multiple techniques of semi-structured interviews and observations. Semi-structured interviews were carried out to collect data on the areas of hotel tea products, tea-serving culture, aesthetic surroundings, occasions and tea ceremonies the hospitality industry promotes in bringing hospitality tea culture to the next level. Hotels in Nuwara Eliya are used as the study sites for observational research, and data obtained were integrated as auxiliary and confirmatory to the data collected through semi-structured interviews.

Sri Lanka has seven major tea production regions: Nuwara Eliya, Dimbula, Uva, Uda Pussellawa (hill country – high grown teas), Kandy (mid-country – mid-grown teas), Rathnapura and Galle (low country – low grown teas). A diverse climate has resulted in unique teas (in colour, taste, aroma and strength) in each climatic region in Sri Lanka. Among tea-growing regions, Nuwara Eliya is notable for being the best known of Sri Lanka's tea-growing districts, located in the central mountains and having the highest average elevation combined with lower temperatures. Therefore, the study is narrowed down to the Nuwara Eliya District. Purposive sampling was applied to develop this research on the discussion of hospitality tea culture. Within this context, the participants in the study were the hotel general managers, resident managers, and food and beverage executives of 20 selected accommodation establishments (composite city hotels, boutique hotels and villas, tea bungalows and heritage homes) in the Nuwara Eliya District. The data were collected between January and April 2021. Finally, content analysis was adopted to scrutinize data gathered from the semi-structured interviews and personal observations. The researchers used QDA Miner Lite qualitative data analysis software for better output.

Results and discussion

Nuwara Eliya was once the summer retreat of choice for British colonials, and now it is a popular tourist destination in Sri Lanka. It is blessed with a cooler climate than the lowlands, with pristine highlands, ancient colonial splendour, endless tea plantations and natural beauty. The designs of accommodation establishments in the Nuwara Eliya District are dominated by the earliest settlements in the colonial era. Most of the top-rated hotels are conversions of colonial housing structures modified without damaging their historical values. Hotels in converted tea bungalows, converted tea factories, and colonial-styled buildings amidst tea plantations give guests an authentic tea-related experience. Whatever the food and beverage trends indulged on in the hotels, Ceylon tea remains popular, iconizing Sri Lankan hospitality. Top ranking hotels are developing new offerings to appeal to the tea-drinking crowd.

Findings

Culinary experiences

Top-rated hotels integrate tea into multiple dining options such as tea lounges, tea bars, tea-infused fine dining, in-room dining, lobby bars and even in spas to enhance the rejuvenating and healthful aspects of travel.

High tea

The results show that high tea, with its own signature of finger foods served on an etagere (teatime serving stand) is quite common to most of the hotels in the area. The culture of serving high tea combines the best of Ceylon tea with colonial memories. Hoteliers offer high tea as a signature dining option arranged inside (tea lounges, hotel lobby or as in-room dining) or outside (al fresco high tea on hotel terraces, in gardens or in the surrounding tea plantations). High tea serves a limited variety of Ceylon premium teas or coffees with platters that include an array of fresh, organic and homemade finger foods like savouries (non-vegetarian, vegetarian and vegan), sandwiches, cookies, cakes and sweets with requisite trimmings (homemade jam, cream). The tea is served in elegant and fancy ceramic pots, and exquisite chinaware is used reflecting royalty in service. Boutique hotels patronizing luxury hospitality offer a tea butler service to educate guests on available tea offerings and blending options. The tea butler gives personalized attention to every guest's cup of tea brewed to its best taste. Immaculately dressed tea butlers add extra flavours to the guests' evening tea experience. High tea combined with ambience and personalized service denote the traditional royalty beyond just a cup of tea.

Tea lounges

Only 10% of the sampled hotels have on-premises tea lounges offering signature teas with various snacks. Tea lounges serve tea in multiple varieties which they have obtained from nearby tea factories or branded tea companies in the country. One sampled hotel enriched this offering with a full-fledged tea lounge as a well-executed branch under one of the leading tea companies (the t-Lounge chain). The rest serve a handful of tea varieties on premises. They have experienced that most of their guests prefer visiting tea factories that enable them to witness the tea production process that ends with tea tasting at factory tea shops. Moreover, some hoteliers do not think of having a separate tea lounge or shops inside due to the availability of inclusive tea shops in the surroundings. Their concerns are not aligned with having a dedicated tea lounge. Most of the time, hoteliers witness tourists visiting tea plantations, joining with tea factory guided tours, being involved in tea-tasting sessions and finally enjoying tea in the factory restaurant/tea shop.

Tea bars

Like tea lounges, tea bars are trendy dining options introduced by some of the hotels in Nuwara Eliya, making tea the spotlight of their menus. Some tea bars operate in the daytime as well as the night-time, offering tea-infused mocktails and cocktails that bring a new twist of flavours for tea enthusiasts who love to explore outside of the traditional cup of tea. Tea bartenders, dressed up in old-fashioned custom outfits, offer guests insights on how to enjoy their preferred teas and tea-related drinks from the tea bar's menu. Among the sampled tea-thematic hotels, a few promote traditional Sri Lankan–style teas in tea bars where guests can enjoy different aromas and tastes of tea varieties grown in plantations that surround the hotels.

Planters' lunch

In the early days of tea plantations, the planters covered the vast area of their tea estates by foot or horseback for inspection. At lunchtime, they took a rest under the shade of a tree. This old practice has been converted into a signature dining option by a hotel, naming it as Planters' Lunch, a lunch served in tiffin boxes in a picnic setting. This unique dining experience is hosted by the resident planter, who adds more colourful context into the dining experience by providing a fascinating account of life as a planter in the old days.

Tea-infused meals

Only one sampled hotel under the portfolio of a leading tea company offers a tea-infused dinner menu as its signature culinary experience for tea lovers. They introduce tea paired with meals that extends beyond simply drinking it. The hotel has a collection of tea-infused recipes and tea cuisine for dinner. As per the menus, the tea is incorporated with many of the meals to add extra taste. Moreover, boutique bungalows pair tea with fresh juices in offering welcome drinks to refresh guests on their arrival.

Non-culinary experiences

Walks and rides in the tea plantations

Tea bungalows located amidst the lush rolling tea estates inspire their guests to have a walk around the plantations in the mornings and evenings. Some hotels facilitate the walkers with guides called *PeriaDorai*: a formal tea estate superintendent who commentates on the history of tea, tea estates, tea growing and tea plucking, and accompanies them to close by tea factories and explains more of the process of the fascinating story of 'two leaves and a bud'. As the tea estates are also some of the best places to enjoy mountain cycle rides, hotels provide bikes and helmets inclusive to the hotel package for guests to use at their own pace. Some hotels embed guided birdwatching sessions into their cycling tours around the tea estates and surrounding jungle paths.

Tea plucking

Hotels offer their guests an authentic tea plucking experience with the estate workers. There, the guests can be tea pluckers (picker) for an hour or two. The guests are decked out in local attire – colourful sarees for the ladies, sarongs for the gentlemen – and provided with a basket in which to collect hand-picked tea leaves. The guests are shown which leaves to pluck and how to pluck by professional tea pluckers. Then the guests are given some time to practice on their own and finally the efforts are evaluated by the professionals. Old tea factory conversions into hotels exhibit this using a mini tea factory inside which guests can learn the whole process from picking, withering, rolling, fermenting, and sifting to produce different tea flavours. The guests can process the self-picked tea leaves in the mini-factory and take them home packed as a souvenir.

Tea factory visits

Nuwara Eliya is surrounded by multiple long-standing tea factories that provide visitors with an insight into the craft behind Ceylon tea leaves. Exploring the tea estates and visiting tea factories is one way of touching the tea culture that is promoted by most of the hoteliers in Nuwara Eliya. During the tea factory visits, the guests indulge in tea trekking experiences and learn how the tea leaves are picked, processed and packed. Guests who visit Nuwara Eliya never miss out on the tea factory visits during their stay.

Tea tasting

The final product may have substantially distinct flavours, aromas and aesthetics from different climatic factors, topography, manufacturing processes and clones of the tea plant (*Camellia sinensis*). The guests can get an experience in tasting these different flavours in tea-tasting sessions organized by the hoteliers (inside or outside) with a professional tea taster who gives insights on identifying the tea varieties and their qualities. Moreover, the guests can learn in detail the flavour parameters, as well as the leaf colour, size and shape that are categorized using a terminology established by the tea industry to define overall quality.

Explore the tea pluckers' life

Of the sampled hotels, 10% host traditional tea pluckers' line room experiences, where the guests can experience upcountry living. This gives a series of unique, immersive experiences designed around the life of the tea pluckers and their livelihoods. Moreover, the guests get a chance to taste authentic local meals that fuel the iconic tea pluckers; meals are prepared in firewood hearths and served in traditional pots.

Apart from the above categorized tea-related non-culinary experiences, the researchers could explore a few more techniques hoteliers initiate to infuse tea into the guests' stay. For example, one of the sampled hotels introduced tea-infused soaps and bath amenities in guest rooms, giving a herbal touch to guest room toiletries. Moreover, a few cases were recorded of hoteliers using tea in their spa treatments, such as green tea baths, tea face masks, massages with tea extracted oils and ointments to rejuvenate and nourish the guests' wellness.

As a concluding remark, the researchers here highlight the fact that hoteliers have thought of taking 'tea' beyond the traditional cup and introduce various culinary and non-culinary experiences to give their guests authentic tea-infused experiences during their visit to Nuwara Eliya. However, those efforts are limited only to a few hotels. The majority of the hoteliers are still framed into 'tea' serving as a 'beverage' and leave in-house guests to step outside to visit tea factories and enjoy a cup of tea at their tea shops enroute. Still, there are potentials in introducing tea gourmet experiences with tea-infused recipes into hotel menus.

Conclusion and recommendation

Tea production in Sri Lanka is pivotal to the Sri Lankan economy. The country is the world's fourth-largest tea producer, and the industry is one of the main sources of foreign exchange earnings and a significant source of rural income. Tea tourism has emerged as a recent niche in the world tourism scenario. Tea tourism is a contemporary concept researched and discussed since the beginning of the 21st century. As tea is the most widely consumed beverage globally after water, tea-drinking habits and tea cultures are transmitted worldwide. As contemporary tourists seek out unique and authentic experiences related to tea appreciation and consumption, tourism has the potential to improve the brand image and marketing of tea destinations.

This foundation motivated the researchers to explore how the hospitality industry takes tea beyond the traditional cup to its next level and examine their strategic techniques to do this. The findings were themed under culinary and non-culinary tea-infused experiences currently practised by the hoteliers in the Nuwara Eliya District. The majority of the hoteliers introduce tea as a value-added beverage through high tea services, but only a limited number of hoteliers, especially the boutique hotels, initiate introducing tea in signature dining experiences with on-premises tea lounges, tea bars and off-premises planters' lunches in tea estates in picnic settings. Hoteliers also promote non-culinary tea-infused experiences such as walks and rides in the tea plantations, tea plucking, tea-factory visits, tea tasting with a trained tea taster, and traditional line room experiences. A few hoteliers go beyond the traditional level of hospitality tea promotions by introducing more innovative experiences such as tea-infused herbal guestroom toiletries, tea baths, tea face masks, and the use of tea-extracted ointments in hotel spas. However, the majority of the hoteliers are promoting tea as a beverage and not using its higher potential in designing appealing tea-infused products, services and experiences, especially gourmet teas, tea ceremonies and themed events that would attract tea flirts. This research finding can help the hoteliers revisit their supply chain and implement strategies to bring tea to the highest level in the hospitality industry and strengthen the tea communities by directly involving them in shaping the tourists' tea experiences. Then Sri Lanka can bid to be the best tea tourism destination in the world with its great hospitality coloured by its tea culture.

References

Bohne, H. and Jolliffe, L., 2021. Embracing tea culture in hotel experiences. *Journal of Gastronomy and Tourism*, 6(1–2), 13–24.

Ceylon Tea Land, 2013. About Ceylon Tea. Available from: www.ceylontealand.com/about_tea.php [Accessed 24 January 2020].

Gunasekara, R.B. and Momsen, J.H., 2007. Amidst the misty mountains: the role of tea tourism in Sri Lanka's turbulent tourist industry. *In:* Jolliffe, L., ed., *Tea and tourism: tourists, traditions and transformations*. Clevedon: Channel View Publications, 84–97. https://doi.org/10.21832/9781845410582-008

Herath, H.M.U.N. and De Silva, S., 2011. Strategies for competitive advantage in value added tea marketing. *Tropical Agricultural Research*, 22(3), 251–262.

Jheng, L.Y., 1993. *The queen of England invites you.* Taipei: Linking Publishing.

Jolliffe, L., 2003. The lure of tea: history, traditions and attractions. *In:* Hall, D.M., Sharples, L., Mitchell, R., Macionis, N. and Cambourne, B., eds. *Food tourism around the world: development, management and markets*. Oxford: Butterworth-Heinemann, 121–136.

Jolliffe, L., 2006. Tea and hospitality: more than a cuppa. *International Journal of Contemporary Hospitality Management*, 18(2), 164–168. https://doi.org/10.1108/09596110610646718

Jolliffe, L., ed., 2007. Chapter 1. Connecting tea and tourism. *In:* Jolliffe, L., ed., *Tea and tourism: tourists, traditions and transformations*. Clevedon: Channel View Publications, 3–20. https://doi.org/10.21832/9781845410582-003

Jolliffe, L. and Aslam, M.S.M., 2009. Tea heritage tourism: evidence from Sri Lanka. *Journal of Heritage Tourism*, 4(4), 331–344. https://doi.org/10.1080/17438730903186607

Koththagoda, K.C. and Dissanayake, D.M.R., 2017. Potential of tea tourism in Sri Lanka: a review of managerial implications and research directions. *Equality and Management*, 51–68.

Martin, L., 2007. *Tea: the drink that changed the world.* Rutland, VT: Tuttle Publishing.

Pearce, M., 2016. A blend of tea and culture: tours and guides in Sri Lanka. Available from: http://theculturetrip.com/asia/sri-lanka/articles [Accessed 2 January 2021].

Sumuduni, M.C. and Piyumali, D.R., 2015. An analytical study on the tea culture of China and Sri Lanka. *Praba: Journal of Humanities*, 4, 142–153.

Walton, J.K., 2001. The hospitality trades: a social history. *In:* Lashley, C. and Morrison, A., eds. *In search of hospitality*. London: Routledge, 56–74.

Whitmore, J., 2000. Tea time is big business for hotels and resorts. Available from: www.prweb.com/releases/2000/01/prweb11192.htm [Accessed 11 January 2020].

World Travel & Tourism Council, 2020. Tourism satellite accounting. Available from: www.wttc.travel/eng/Research/Tourism_Satellite_Accounting_Tool/index.php [Accessed 2 January 2020].

Zhang, L.J., 2004. *Xian Dai Cha Wen Hua Yan Jiu.* Master's Dissertation, Zhejiang University.

13

HOMESTAY IN SMALL TEA GARDENS

The case of Meghalaya, India

Evarisa M. Nengnong and Saurabh Kumar Dixit

Introduction

In recent years, the homestay has been a popular choice for tourists. It is a sort of vacation that allows visitors to stay at someone's home outside of their own town and experience a different way of life, authentic culture, or even language. Homestay housing provides local communities with a direct, extra, or alternative source of income, assisting in community empowerment (Shyju *et al.* 2019). Heritage homes, farmhouses, bungalows, ancestral homes, and other homestay accommodations share an ambiance with traditional surrounds, an indigenous flavour, and a vernacular building where the host provides a comfortable stay experience. Homestays help preserve the community's cultural and ecological heritage while also providing a steady source of income (Vishwanath 2017). Through tea tourism, visitors can link themselves with tea culture, consumption, and legacy. Homestays can attract tourists to tea gardens, maintain the environment, and preserve traditional culture while also benefiting the region by creating jobs. Small tea garden homestays enhance tourist experiences while helping the local population socially, economically, naturally, and culturally (Venkatesh and Mukesh 2015). Rural tourism can be sustained if a comprehensive, inclusive planning strategy based on a multi-action, multi-stakeholder participatory approach is devised and implemented. Recommendations on tourism and rural development issued by the United Nations World Tourism Organization (UNWTO) aim to assist governments at all levels, as well as the private sector, in developing tourism in rural areas in a way that promotes inclusive, sustainable, and resilient development. This chapter illustrates two case studies of small tea gardens homestays in Meghalaya, India.

Homestay in the small tea garden

India is a diversified country that proudly displays its unique culture, heritage, flora, and fauna to the rest of the world. "The Indian civilization and culture inherit the hospitality tradition of 'Atithi Devo Bhava' (The Guest is like God) from time immemorial" (Dixit 2020b). Tea tourism helps preserve the region's legacy, culture, and environment, providing employment opportunities for residents and alleviating the region's socioeconomic concerns. The original colonial edifice has been re-established into beautiful homestays for tea-pickers who have opened their homes to visitors in India. It provides the tea-growing areas a unique avenue to visit, taste, and explore tea gardens, museums, and festivals. An increasing number of tourists feel that tea consumption can enhance mental and physical well-being (Zhang 2014). Therefore, the appreciation of local tea resources and tea

DOI: 10.4324/9781003197041-16

consumption establishes a substantial part of the social interface between tourists and hosts (Zhang *et al.* 2017) and is thus considered an effective tactic to attract prospective tourists (Cheng *et al.* 2012). Modern tourists have become very interested in feeling the pulse of any destination, for which they also engage closely with their hosts. A tourist's trip to India can be more memorable by staying in a family-run homestay unit. In India, various state tourism ministries have pushed homestays and granted homestay operators incentives (e.g., Kerala, Karnataka, Meghalaya, Himachal Pradesh, New Delhi, NCR, Goa, Uttarakhand, Sikkim, etc. Rajasthan, Gujarat; Ministry of Tourism 2018).

The product line of homestay in small tea gardens

Homestays focus equally on maximizing local resources and regeneration of tourism products. There are several implications of homestays, including accommodation, amenities, and facilities available to guests. Different units may decide on their offerings depending on cultural and physical differences.

Accommodation: The homestay is the house owner's accommodation unit run informally. These buildings are usually constructed and designed following the regional architectural and cultural legacies. India is exceptionally rich in its regional architectural and historical upbringing to present its social and religious traditions. Some examples of such accommodation are havelis, planter houses on tea plantations, colonial bungalows located at different locations – such properties provided with traditional furniture and moderate facilities to offer heritage ambiance and feel to its visitors.

Cuisines: Meals are an essential component of staying in a homestay. The majority of homestays provide guests with home-cooked meals, which helps to popularise tourists' ethnic cultures and cuisine.

Amenities: The guests usually expect essential amenities to make their stay comfortable in the homestay units. These may include amenities such as hot/cold water supply, clean and hygienic washrooms, parking space, television, hygienic foods and water, and cooler or air conditioning as per the property's location.

Participation in local activities: Local activities play a vital role in attracting visitors to the homestay units. Each unit has its distinct geographical location and social and cultural background, offering its guests specialized offerings. The homestay unit may decide the local activities as per their convenience and tourist preferences. Homestay provides a feel of the indigenous culture and closely makes visitors know these customs and traditions. Homestays offer a variety of local activities, including:

- Guided tea gardens tour
- Guided village tour
- Fishing, camping, campfire
- Trekking
- Tea garden/plantation visits
- Tea ceremonies

Sightseeing: The homestays, because of their unique natural, cultural, and historical locations, usually have enormous scope for sightseeing activities for their visitors. To make guests comfortable in such ventures, the family members/host usually serve as guides/escorts during these activities.

Review of literature

Concept of homestay

Homestay accommodation refers to offering "one's spare residential space to provide average boarding and lodgings facilities to the tourists. Homestays usually appeal to foreign tourists and budget class tourists willing to explore and experience local traditions, cultures, and architecture" (Kaushik 2021). Homestays offer boarding and lodging facilities at a relatively low cost and opportunities to intermingle with the host family and experience their lifestyle closely. Homestays are often related to other activities such as hiking, birdwatching, tea estate visitation, tea-tasting ceremonies, and participation in local fairs/festivals to remain appealing (Janjua *et al.* 2021). Different homestays offer varied activities to attract visitors (Chakraborty 2019). Some examples of such actions are homestays in Thailand are giving morning alarms to monks, giving insight on how to weave silk or cotton cloth, and engaging in agricultural activities. Organizing nature trips to tea estates to view the plants, to observe and practice village activities closely (i.e., making local handicrafts or cooking Thai food) of the local villagers are other examples (Kontogeorgopoulos *et al.* 2015). In Japan, one can learn doll-making art during their stay in a homestay. Similarly, in India, many homestays in Kerala offer backwater cruises in houseboats and canoe rides on the sea.

In Rajasthan, homestays offer jeep safaris or horse rides to visit tribal villages and nearby forests to spot wildlife (Guha 2020). The homestays are near tea plantations and hill stations offer trekking and guided walks through tea gardens and woodlands. Depending on the family, many hosts will offer to take guests around the sights of their town or village, sharing local insights and an experience far removed from the usual tourist trail (Ministry of Tourism 2018).

The core feature of homestay is its connection to nature and its ability to access an indigenous community's pure lifestyle and culture. These qualities and experiences can be packaged as part of a homestay package (Acharya and Halpenny 2013). Such tourism is centred on environmental sustainability to preserve the ecological ecosystem, biological variety, and biological resources. Furthermore, connected components include social and cultural sustainability to protect the destinations' social and cultural identities (Chakraborty 2019). According to Suvantola (2018), a place's experience is far more when the travellers find it "lived" rather than "seen."

Homestays serve as a starting point for a few days exploring the local area, but they may be considered overall holiday destinations. According to Bhan and Singh (2014), a homestay is a sheltered place visited by a tourist or a visiting foreign student whom a local family hosts. Also, homestays, due to their informal nature of operations, bring persons from varied backgrounds closer to each other and offer an ideal medium to develop long-lasting relationships.

The local community runs low-cost homestays in their own houses. It makes money from guests by promoting local culture and values. "Community involvement in homestay operations creates jobs and improves the local economy" (Bhuiyan *et al.* 2011). Homestay tourism aims to improve the quality of life of indigenous peoples while also allowing them to benefit from it.

Tea tourism is already popular in China, Sri Lanka, Thailand, and Japan. It is now spreading throughout India's tea-growing regions. Tourists want to spend time in nature's splendour at tea gardens, tea bungalows, trails, nature walks, and hiking. Interacting with tea workers, taking part in tea leaf plucking, and learning about local cultural customs and festivals are all possibilities (Jolliffe 2007). Guests will benefit from expert travel guides; cosy, homey lodging; and fabulous local cuisine during their tea journey. It's the ideal time for game enthusiasts to plan their vacation around their favourite indoor or outdoor game (Dixit 2020a).

Rural tourism can help the owners improve by planning and choosing the right marketing and, as a proper campaign, deciding on an investment. Rural tourism promotes rural life, art, culture, and heritage in rural areas while supporting the local community economically and socially and allowing tourists and locals to engage in a more fulfilling tourism experience. The term "rural tourism" refers to any activity in the countryside (Hossain 2007). Farm/agricultural tourism, cultural tourism, natural tourism, adventure tourism, and ecotourism are all examples of multifaceted tourism. "The enormous economic significance and scope of tourism in rural areas are largely unrecognized, as evidenced by national rural policy's continued tilt towards agriculture" (Su *et al.* 2019). Rural households have devised a survival strategy for small rural family houses and contribute significantly to rural destinations' social and cultural development (Sharpley and Craven 2001). Small, autonomous, and family-run rural households, on the other hand, frequently face a lack of funds for promotion and struggle to adapt to the present market scenario, which is controlled by today's technology and communication (Gannon 1994).

Gender influences people's opinions of tourism growth; tourism development in communities does not equally benefit men and women. While tourism may give jobs for young people and women, males in the community may believe it offers few sustainable, reputable, and acceptable economic opportunities (Kulshrestha 2019).

Meghalaya covers a wide range of appealing tourism resources and attractions, including streams and rivers, hillocks and mountains, cultures and indigenous festivals, and so on. These are enhanced by the lovely weather that prevails throughout the year. The expansion of tourism resources, attractions, circuits, and infrastructure and the proliferation of the tourism industry can be noticed, particularly in Meghalaya (Meghalaya Basin Development Authority 2012). Exploring tea estates and tea sale outlets has enormous tourism potential if promoted discreetly in Meghalaya. Many tea estate owners in Meghalaya (such as Anderson Tea Company and Meg Tea Estate in Ri Bhoi District) are developing cottages to lure tourists into experiencing the tea tourism resources of Meghalaya (Dixit 2020a).

Methodology

The chapter aims to examine tea tourism by highlighting the potential and development of homestay accommodations in a small tea garden in Meghalaya, India. Interviews were conducted with 40 respondents, 22 men and 18 women aged 25 years and older, which include homestay hosts, small tea garden owners, staff in the tea garden, and government officials. Researcher observation during field visits, and most importantly, interaction with the local people, also provided primary data. Secondary data was reviewed from various sources, including books, journals, and even helpful websites.

Small tea gardens in Meghalaya, India

The research area is in the East Khasi Hill District of Shillong, Meghalaya's capital. Shillong is located at a latitude of 25° 34′ N and a longitude of 91° 53′ E, at an average elevation of 1496 m above mean sea level, on the northern slopes of the Shillong peak. Shillong is a well-known tourist attraction in northeast India and is considered one of the country's most beautiful hill stations. It is approximately 100 km from Guwahati, Assam's capital. Shillong has a humid climate, as do the eastern sub-subtropical continent's regions. It is characterized by moderate warm wet summers and cool, dry winters. Compared to Assam or West Bengal, Meghalaya's tea gardens are smaller. As the tea-processing plant is within walking distance of the tea garden, this allows for the swift processing of high-quality green tea.

The homestay projects encourage cooperation between local communities in small tea gardens and native tour operators to engage indigenous people in different employment opportunities and income-generating tourism activities. Meghalaya is now taking steps towards tea tourism, with many private tea estates, such as the Anderson Tea Estate, venturing into the tourism trend and working towards offering cottages within its vast compounds. Government-run tea estates, such as Meg Tea in upper Shillong and the Tea Development Center in Umsning, also welcome visitors to observe and learn about the process of growing and tearing tea. As Khasi women, dressed in their traditional

Jainkyrshah (apron-like garment worn while doing domestic or other daily chores), pluck individual tea leaves, visitors take in the silence of the tea garden and its beauty (Marbaniang 2019).

Case study 13.1   Camellia Homestay in Sharawn Tea Garden, Sohryngkham

Camellia Homestay is situated within the Sharawn Tea Garden, located some 20 km from Shillong. It is a small homestay of three bedrooms located in the middle of the tea garden owned by Smt. Khamtimai Syiem (Figure 13.1). The homestay has been running for four years. The idea of a homestay in the middle of the tea garden contributes to developing tea tourism within the state of Meghalaya, India. Even with several natural and cultural attractions of Meghalaya, tourism potential, especially tea tourism, is yet to be explored by visitors. The homestay is by the tea gardens so that guests can explore and make their stay more memorable.

Figure 13.1   Camellia Homestay, Sharawn Tea Garden, Sohryngkham
Source: Authors

Guests at this homestay can take a guided tour of the tea garden and learn about the history while exploring the grounds. Breakfast featuring local cuisine is offered at the homestay and lunch and dinner can be arranged. Visitors can take a delightful walk down to the picturesque and colourful small village of Sohryngkham after their tea garden tour. Along with the natives, visitors enjoy the indigenous meals and cuisines available here, as well as the locally produced Sharawn tea.

Case study 13.2   Urlong Tea Estate Bungalow, Mawlyngot

Mawlyngot, located around 45 km from Shillong, is part of the Mawkynrew Development Block of the East Khasi Hills District. With the help of the Government of Meghalaya's Border Area Development Department and Horticulture Department along with the non-governmental organization World Vision, the Urlong Tea Integrated Village Cooperative Society was established in 2011 and commenced tea processing. They built a tea factory to process raw tea leaves grown and picked by the smallholder tea producers in Mawlyngot village and nearby villages such as Jatah Nonglyer and Mawsna. Under the name Urlong Tea, this factory produces black, green, and white tea, processed by both the CTC (cut, tear, curl methods which involved passing the leaves through a succession of cylindrical rollers with a large number of sharp teeth that crush, rip, and curl the tea into tiny, hard pellets) and Orthodox (the term "orthodox tea" describes loose-leaf tea made with conventional or orthodox techniques such as plucking, withering, rolling, fermentation, and drying) methods (Pathaw 2012). Urlong means "Dreams come true" in the local Khasi language.

Figure 13.2   Urlong Tea Estate Bungalow, Mawlyngot

Source: Authors

This cooperative society has offered livelihood and financial support to the residents of the entire village and neighbouring regions. In addition to producing tea, the society has lately expanded into tea tourism. With aid from the government of Meghalaya, the Urlong Tea Integrated Village Cooperative Society had also recently built a four-room guest house called Urlong Tea Estate Bungalow at Mawlyngot (Figure 13.2). The guest house was built with aid from the government of Meghalaya, provided in order to encourage the development of tourism in this part of the state (Government of India n.d.). Visitors can become involved in tea processing if they want to learn about different types of tea production.

Results and discussion

Small tea estates in tea-growing regions have benefited from the homestay mission. They made the host community aware that travellers visit their hamlet to learn about their rich cultural and natural sources. It makes the host community wealthy and encourages them to shield their particular biodiversity, lifestyle, and traditions. Homestays have empowered the nearby groups, each financially and culturally. The ladies' counterparts also can contribute to the family income, as they are the ones who usually operate the homestays.

The researcher conducted unstructured interviews with respondents in the area primarily based on open dialogue specific to themselves in their terms and analysed, more approximately, homestays in small tea farms and how they benefit the local financial system. Unstructured interviews are the first-class approach with the view that their natural spontaneity and conversational tone allow interviewees to tell their narratives in their very own tempo and style (Morse 2002). The interviews have been casual, with the small tea garden owner amassing records approximating the homestay, including its beginning date, a wide variety of rooms, the number of visitors hosted, and a fee per night. The focus of the interview questions then shifted to the current state of affairs and the functioning of a homestay. Interviews at Camellia Homestay in Sharawn Tea Garden, Sohryngkham villages, and Urlong Tea Society (Mawlyngot) are summarized below.

While asking about their primary incentive for starting a homestay, the owner of Sharawn Tea Garden (Case Study 13.1) stated it became a source of constant profits. The researchers found that homestay contributes to popularising tea brands' teas produced in the vicinity. Domestic tourists 50 years and younger journeyed to the homestay with the primary reason to revel in the village's herbal beauty by exploring an average tea lawn experience. The hosts make sure that visitors experience local tea culture and traditions, like witnessing and taking part in tea plucking, processing, tasting, forest treks, becoming a member of farming sports, and being part of the neighbourhood gala. In line with this remark, while travelling the tea estates, the researcher (one of the authors of this chapter) stayed inside the homestay to observe the services, as well as the tea production and processing. The neighbouring vacationers stayed in the homestay for leisure, to get peace of mind, and to revel in tea tourism. Even the travellers travelling the homestay need to explore the tourism resources and points of interest of the villages.

The owner and host in Case Study 13.1 reported, "Indian tourists say I preferred the balcony location with the small tea lawn in front . . . the area . . . calm and easy region . . . reached my expectation." "Tourism delivered infrastructural trends within the village," one local tea industry employee in this area explained. "We got roads, energy. The community is proud that tourists come to our sites, spend time with us, and take part in our cultural programs." "We're encouraging small tea plantations to spend money on tourism, and we're usually here to help them," the agricultural officer said, further indicating "we keep a close eye on the tea farms and their personnel, besides taking harsh measures when necessary. Since the inception of tea-tourism, there has been a noteworthy improvement within the social improvement of the place, with nearby inhabitants of the hills benefiting."

Observations and interviews were accomplished in relation to Case Study 13.2 of the Urlong Tea Estate Bungalow, Mawlyngot. The cooperation branch launched "Explore Exotic Meghalaya" (Travelers Nest) in 2012–2013 in collaboration with the tourism department; the authorities of Meghalaya, Shillong, and the Meghalaya village development; and Merchandising Tourism Cooperative Society Ltd. in Shillong.

The captivating landscape and beautiful hills and water and the availability of the unusual species referred to as the local blue worms have remained hidden from the human eye. This was noted through an interview with the homestay host and the cooperative member of Case Study 13.2, and statements from others throughout the sector within the region. The society is well known for tea

Table 13.1 Case study comparison of homestay concept and operation

	Case study 13.1 Camellia Homestay in Sharawn Tea Garden, Sohryngkham	Case study 13.2 Urlong Tea Estate Bungalow, Mawlyngot
Ownership	Tea garden owner	Cooperative society
Hosting	Workers on tea garden	Society workers
Capacity	Three bedrooms	Four bedrooms
Activities	Tea garden tour, village visit, local tea and indigenous cuisine	Tea factory visits, nature walks, local tea and cuisine, area excursions
Funding	Private	Society with government aid

cultivation and tea processing (manually). Tourists can experience the inside of the tea-processing area (organic tea) if they want to revel in one-of-a-kind styles of tea production like green tea, black tea, oolong tea, white tea, and so on. Travellers staying in this homestay may additionally experience the indigenous culture and delicacies. The maximum number of vacationers visiting there was observed to be domestic and international tourists under the age of 50. The proprietor of the small tea holdings are specifically worried about the development of social and economic benefits in the 40 villages in Mawlyngot to assist and generate process possibilities for the area people and maintain the standard of living in the society. Few vacationers touch on the benefits of homestay; however, one interviewee stated "that is an outstanding location to experience the treasures of nature far away from warmth, pollutants, and noise of metropolis life, the perspectives are breathtaking." The tourists stayed for two nights and had a notable experience. While the facilities don't correspond to those in five-star megaresorts, that is more than compensated for via the tremendous friendliness of the host and her crew, immaculate cleanliness, and eagerness to serve the site visitors. They put together and serve the delicacies in line with guest preferences. They may spend the night-time outside around the campfire lit by the host. Notably, the accommodation functions as a base camp for journeys to Dawki or other areas in the Smit valley.

Comparing the two case studies (Table 13.1), it is evident that in both instances the homestay concept has been adapted to different tea-producing situations. There have been similar results in terms of benefits for the local population – both direct, in terms of the diversification of employment related to the new accommodation offering, and indirect, such as the improvement of local infrastructure such as roads for tourism that also benefit locals. While the ownership and hosting arrangements differ between the two case study accommodations, the offerings are similar. Since they have differing locations, it is possible that in the future they might be able to work together to develop a package where visitors could stay for a few nights at each in order to experience different tea landscapes, growing and processing locations. As these types of accommodations develop in the province in support of tea tourism, it may be possible for operators to support and learn from each other.

Conclusion

This chapter took a look at contributing to cutting-edge homestay knowledge practices. It has found new information on the elements that impact the adoption of high-quality practice. These studies located significant variances inside the use and significance of strategies amongst Meghalaya's homestays. The findings show that tea tourism and homestay have a beneficial link. At this time, the homestay alternative is a new trend that is more appealing to tourists interested in learning about the nearby tradition and delicacies. Vacationers experience the natural and cultural splendour of the region because of the mild hospitality of the hosts creating a memorable vacation for them. Via travellers, homestay proprietors gain publicity to a global outdoor environment in their little surroundings, besides income through sales and impact on reputation. A small tea garden owner and a smallholder tea co-operative are extending their facilities to make room for travellers. Villagers were capable of complementing their earnings in this manner. It also facilitates the tea garden inhabitants to expand new opportunities. Homestay is a popular concept with potential, but tea tourism in Meghalaya is essentially unorganized.

Homestay in small tea estates in Meghalaya involves maintaining the environment, and maintaining lifestyle will benefit the country via growing employment possibilities, including for the positions of waiter, safety guards, helpers, cooks, gardeners, instructors, tour guides, chauffeurs, and so forth. It additionally allows for alleviating socioeconomic troubles inclusive of poverty growing financial development. As the tourism sector develops in small tea garden regions, it improves the infrastructure, instructional, and transportation facilities. As a result, the overall improvement of the tea garden location will appear. It additionally complements the community facilities and offerings. If tea tourism is a highlight with planning and suitable prospects, it can help the government earn sales and create a brand-new destination region on the map of the tourism area in the state.

The evolution of the homestay concept has the potential to boost the rural economic system. However, in the case of Meghalaya government policies for tea tourism need to be reviewed and the establishment of a central booking system for homestays implemented. The methods for implementing and adapting the homestay concept in Meghalaya offer real-life lessons for homestays on tea estates in other regions of India, as well as other tea-producing countries.

Discussion questions

1   What are the potential challenges of running a homestay in a small tea garden? Do you think the homestay concept will help in developing tea tourism?
2   What is the experience for the tourists and the homestay owner or host?
3   In what ways can the host or owner of the homestay in a small tea estate help tea tourism to boost the community's economy?
4   How will homestays contribute to both tea tourism and the community's livelihood?

References

Acharya, B.P. and Halpenny, E.A., 2013. Homestays as an alternative tourism product for sustainable community development: a case study of women-managed tourism product in rural Nepal. *Tourism Planning & Development*, 10(4), 367–387.

Bhan, S. and Singh, L., 2014. Homestay tourism in India: opportunities and challenges. *African Journal of Hospitality, Tourism and Leisure*, 3(2), 1–5.

Bhuiyan, M.A.H., Chamhuri, S., Islam, R. and Ismail, S.M., 2011. The role of home stays for ecotourism development in East Coast Economic Region. *Researchgate*, 8(6), 540–546.

Chakraborty, B., 2019. Homestay and Women Empowerment: a case study of women-managed tourism products in Kasar Devi, Uttarakhand, India. Delhi, Jamia Millia Islamia, India. *Tourism International Scientific Conference Vrnjačka Banja – TISC*, 4(1), 202–216.

Cheng, S., Hu, J., Fox, D. and Zang, Y., 2012. Tea tourism development in Xinyang, China: Stakeholders' view. *Tourism Management Prospectives*, 2–3, 28–34.

Dixit, S., 2020a. Tea tourism potential in Meghalaya (India). *In: World gastronomy institute global report 2020*. Madrid: World Gastronomy Institute, 173–175.

Dixit, S., 2020b. Editorial, special issue: tourism in India. *Anatolia: An International Journal of Tourism and Hospitality Research*, 31(2), 177–180.

Gannon, A., 1994. Rural tourism as a factor in rural community economic development for economies in transition. *Journal of Sustainable Tourism*, 2(1–2), 51–60.

Government of India, n.d. Urlong tea integrated village cooperative society. Available from: https://megcooperation.gov.in/success/coop-utivcs.pdf [Accessed 3 February 2022].

Guha, S., 2020. Worldnews features [Online]. Available from: www.worldteanews.com/Features/business-small-tea-growers-india [Accessed 20 March 2021].

Hossain, M.A., 2007. *Strategic promotion approaches to developing tourism in Bangladesh: an empirical study of some selected tour operators*, PhD Thesis, Dhaka University.

Janjua, Z.U.A., Krishnapillai, G. and Rahman, M., 2021. A systematic literature review of rural homestays and sustainability in tourism. *SAGE Open*, 11(2), 1–17.

Jolliffe, L., ed., 2007. *Tea and tourism: tourists, traditions and transformations*. Clevedon: Channel View Publications.

Kaushik, K., 2021. Newly Homestay: as homey as it gets. Available from: www.outlookindia.com/outlooktraveller/stay/story/68964/review-of-hewly-homestay-in-namsai [Accessed 6 February 2022].

Kontogeorgopoulos, N., Churyen, A. and Daungsaeng, V., 2015. Homestay tourism and the commercialization of the rural home in Thailand. *Asia Pacific Journal of Tourism Research*, 20(1), 29–50.

Kulshrestha, S. and Kulshrestha, R., 2019. The emerging importance of "homestays" in the Indian hospitality sector. *Worldwide Hospitality and Tourism Themes*, 11(4), 458–466.

Marbaniang, M., 2019. *The finest teas of Meghalaya*. Shillong: Tea Development Center, 8–51.

Meghalaya Basin Development Authority, 2012. Resource directory [Online]. Available from: https://mbda.gov.in/resource-directory [Accessed 30 August 2019].

Ministry of Tourism, 2018. Common national standards and guidelines for classification of bed & breakfast incredible India Establishments and Incredible India Homestay establishments [Online]. Available from: https://tourism.gov.in/sites/default/files/202002/Approved/Guidelines/Homestay/updated.pdf; https://tourism.gov.in/sites/default/files/2020-02/Approved%20Guidelines%20For%20BandB%20Homestay%202018%20update [Accessed 10 June 2021].

Morse, J., 2002. *Handbook of interview research context and method*. 2nd ed. London: SAGE.

Pathaw, S., 2012. Government of Meghalaya tourism department [Online]. Available from: https://megindustry.gov.in/policy/MIIPS_2016.pdf [Accessed 25 June 2021].

Sharpley, R. and Craven, B., 2001. Foot and mouth crisis – rural economy and tourism policy implications: a comment. *Current Issues in Tourism*, 4, 527–537.

Shyju, P., Rajeev, P. and Lama, R., 2019. Tourism sustainability through community empowerment and resource management: a case study of Sikkim. *International Journal of Agricultural Travel and Tourism*, 1, 84–93.

Su, M., Wall, G. and Wang, Y., 2019. Integrating tea and tourism: a sustainable livelihoods approach. *Journal of Sustainable Tourism*, 27, 1591–1608.

Suvantola, J., 2018. *Tourist's experience of place*. London: Routledge.

Venkatesh, R. and Mukesh, H.V., 2015. The role of Homestays in promoting rural tourism. *Global Journal for Research Analysis*, 4(4), 2277–8160.

Vishwanath, A., 2017. Unwind with the locals: eco-friendly homestays across the Northeast. *The Hindu*, 16 October. Available from: www.thehindu.com/life-and-style/homes-and-gardens/homestays-in-the-ne/article19853338.ece [Accessed 6 February 2022].

Zhang, J., 2014. *Puer tea: ancient caravans and urban chic*. Seattle, WA and London: University of Washington Press.

Zhang, J., Xu, H. and Xing, W., 2017. The host-guest interactions in ethnic tourism, Lijiang, China. *Current Issues in Tourism*, 20(7), 724–739.

14

LINE ROOMS

An authentic approach to heritage tea tourism

G.V.H. Dinusha, J.A.R.C. Sandaruwani and R.S.S.W. Arachchi

Introduction

The establishment of the tea sector was one of the results of the deadly coffee blight which first appeared in 1869. The British rulers introduced tea as an alternative to coffee. Though this colonial administration had initiated industries such as coffee, tea, and rubber, a lack of Sri Lankan labour participation forced them to bring Indian labourers to work in their commercial estates, which added a new ethnic identity to the country. Tourism was also administered initially by the British in 1937 through the formation of the Ceylon Tourist Bureau. With the passage of time, "tea" has become one of the largest foreign exchange earners of the country, bringing in 1.35 billion USD in 2019 (Central Bank of Sri Lanka 2020), while tourism earned 3.6 billion USD in 2019 (Sri Lanka Tourism Development Authority 2019). In the Sri Lankan tea industry, most tea leaves are plucked by the hands of Indian Tamils who are settled in and around the tea plantations. According to data from the Department of Census and Statistics (2012), 45% of the Indian Tamil population is scattered primarily around Nuwara-Eliya, which is predominantly a tea plantation district. Supported by the Indian Tamil population, Nuwara-Eliya District has the highest estate leaf production in the tea and green tea categories (Sri Lanka Tea Board 2017).

In the tourism context, this upcountry area in Sri Lanka, including Nuwara-Eliya, experienced a 57.5% hotel occupancy rate in 2019, a noticeable decline from 71.7% in 2018 (SLTDA 2019). What should cause such a decline in occupancy, however, is not an objective of this study. Nevertheless, this statistical information provides the framework of the tea plantations, the Indian Tamil communities in these estates, and tourism in general in the area. Therefore, tea and its Indian Tamil farming community in tea plantations cannot be separated, and that context is connected to tourism. Nuwara-Eliya is a very famous destination among foreign tourists as well as local tourists. As a destination, the city provides cooler weather and a natural setting, commercial accommodation, attractions like Gregory Lake, Victoria Park, Hakgala botanical garden, and other places related to tea heritage that are fundamental facets of the supply-side view of tourism in this area. However, in this bucket of tourism attractions, little attention is paid to the exploration of upcountry tea pickers' cultures and their lifestyles, which are so socially and culturally vibrant. Among these lifestyles, line rooms are vital. Tourism entrepreneurs in the hills have paid little attention to incorporating this unique cultural element into the tourism product, but neither has academia. This chapter compares the few line rooms used by the tourism industry in Nuwara Eliya, Sri Lanka, through the lens of authenticity. The potential repercussions on sustainability and the romanticizing of the culture of Indian Tamils in the hill country when maintaining the authenticity of the line rooms for tourism is discussed.

DOI: 10.4324/9781003197041-17

Literature review

The authenticity of the tourism product

"Since modern society is inauthentic, those modern seekers who desire to overcome the opposition between their authenticity-seeking self and society have to look elsewhere for authentic life" (Cohen 1988, 373). Tourism has a share of this quest based upon the notion that authentic experience resides away from the borders of day-to-day life in present-day society (MacCannell 1976). Therefore, the typical western conceptualization of the tourist gaze is the quest for authenticity (MacCannell 2011; Urry 2011; Ratnayake and Hapugoda 2014; Hapugoda and Ratnayake 2021). When former western colonial ruler populations come back to Nuwara-Eliya as tourists, they would prefer to witness the authentic experience of their own history, as well as the "natives". As this chapter focuses on the authenticity of Indian Tamils living in line rooms on tea estates, its applicability to tourism depends on the commodification of their cultural originality. As Cohen (1988, 372) stated, "local culture generally serves as the principal example of such commoditization".

In particular, colourful local costumes and customs, rituals and feasts, and folk and ethnic arts become touristic services or commodities, as they come to be performed or produced for touristic consumption. Cohen (1979) modified MacCannell's six approaches to staging authenticity into two: "real" and "staged". This was further elaborated into four: authentic; staged authentic (covert tourist space); denial of authenticity (staging suspicion); and contrived (overt tourist space) using the angles of "nature of scene" and "tourist impressions of the scene". There is a widespread agreement present about the meaning of authenticity, which is "real", "genuine", or "true" (Lehman *et al.* 2019). However, Reisinger and Steiner (2006) noted that authenticity in tourism is not about the consumption of genuineness, nevertheless experiences that are subjective to tourists may construct a sense of identity and their link with the world. As Cohen's approach is much more illustrative than MacCannell's approach to authenticity, it is used to interpret the authenticity of two prominent cases used in this study from a touristic, experiential perspective.

The Indian Tamil community

As far as tourism is significant in the economy of Sri Lanka, the Indian Tamil community settled mainly upcountry, especially in Nuwara-Eliya, is also a concern as they are the primary workers in Ceylon tea production. Employment in a lush tea estate and plucking tea leaves is their role. Thus, for any tourists who have gone through any promotional material about Sri Lanka tourism, these Indian Tamil tea pluckers are in it, although in reality their life is misery. And this fact is not reflected by the golden Ceylon cup of tea offered to the tourist. According to Sabaratnam (1990), Ceylon had a free flow of labour from Tamil Nadu, India, via an arrangement called the indenture system, especially from the Ramnad, Tinnavelly, Tiruchirapally, Salem, and Arcot Districts of India due to two reasons: the location of Ceylon and India in close proximity and the development of the *kankani* system.

As further explained by Sabaratnam (1990), a kankani was an enterprising migrant labourer with organizational flair sufficient to collect a gang of 40 or 50 labourers under him. He was a father figure, a leader of his group, who looked after their welfare. He normally recruited his gang from among his relatives and friends in his village. This gave the gang unity and a common interest. This system enabled British planters to tap an easy source of labour and freed them from the worrisome burden of conforming to the British Indian government's labour ordinances. The Madras officials and the government of Ceylon did not interfere (Sabaratnam 1990).

In the multi-ethnic context of Sri Lanka, 80% of the tea estate workers are Hindu Tamils who are socially marginalised and economically deprived (Philips 2003).

Often eclipsed within representations of the country's civil and political conflict, Malaiyaha or Hill Country Tamils, who primarily reside and work on Sri Lanka's tea plantations, have experienced protracted forms of discrimination that directly result from the social and economic matrices of escalating civil violence, legal and affective exclusion, and neoliberal policies of worker dispossession.

(Jegathesan 2015, 255)

The religious beliefs of Indian Tamils vary estate to estate. However, all of them are commonly devoted to the goddess Mariamman in droughts, viruses like chickenpox, praying of women for successful pregnancies, and many other diseases. Every Indian Tamil tea estate has a *devala* for the Goddess Mariamman, and this is often filled with banana leaves and margosa (*Azadirachta indica*) leaves, which are scientifically believed to be antibacterial. The housing condition of these communities is described as follows:

About two-thirds of estate residents live in "line rooms", which are sets of four to twelve one-story homes with shared walls, like row houses, often with another set of line rooms sharing the rear walls. The size and condition of line rooms vary, but they can be as small as seventeen square meters, and most are at least several decades old and suffering from significant official neglect. Another third of estate residents live in newly built detached housing, which has been a priority of upcountry Tamil political parties in recent years. Estate workers receive housing as a condition of employment. While upcountry Tamils increasingly try to find better-paying, higher-status work off the estates, often one member of the family continues to work on the estate, to retain this benefit.

(Bass 2018, 27)

Heritage tea communities and tourism

Heritage is often associated with inheritance (Nuryanti 1996; Fyall and Garrod 1998). Since Indian Tamil communities have served for more than 160 years, especially by settling in the upcountry tea plantations in Sri Lanka, with their unique culture that has been taken forward, they can be easily identified as heritage. Nuryanti (1996) shows that people's work, the combined work of nature and people, and the areas of archaeological sites, have distinct exceptional assessments from ancient, artistic, ethnological, or anthropological views. As explained by Fyall and Garrod (1998, 213), "heritage tourism is, as an economic activity, predicated on the use of inherited environmental and socio-cultural assets to attract visitors". Fyall and Garrod (1998) further state the need for sustainability in the careful management of those inherent assets to save enough resources to cater to the needs and wants of the generations. However, this should pay attention to the accumulation of resources rather than depending upon what is available. Otherwise, the right for a better life for the communities who live in those heritage areas would be a problem.

In the Asian context, tea serves to entertain the visitors who come to houses; hence Jolliffe (2007) named tea as an instrument of hospitality, which is, however, inherited by the region as a result of western cultural colonialism. For marginalised Indian Tamil communities in upcountry plantations, tea is merely a fundamental resource to battle hunger and tiredness when working as labourers in unfavourable weather conditions on mountain slopes, while tourists are tasting tea at a luxury heritage-themed hotel. For the western tourists and the other rich edge of society, tea is a fashionable drink, but Jolliffe (2007) argued it is a threat to the authenticity of tea traditions. Most of the time, the tea-related experience is limited to tea tasting and the surrounding physical and natural environment. However, sporadic attention is paid to experience the culture of the traditional tea pluckers' community or indigenous community in the tea value chain initiating phases with few benefits. Jolliffe and Aslam (2009) developed a typology of tea tourism supply in Sri Lanka, where they identified tea estates as attractions and tourist experience as part of the livelihood of the tea country.

Methodology

The research area can broadly be defined as being located in the Nuwara-Eliya District in hill country Sri Lanka. As mentioned, the researchers intended to find Indian Tamil community-related ongoing tourism ventures that are practised in these communities' living houses (the line rooms). There, the researchers identified two main attractions; Meena Amma's Line Room experience operated by a prominent local hotel chain in Sri Lanka, Jetwing Hotels; and Kandapola Village House, operated by a young Indian Tamil who was originally born and raised by tea plucker parents. These two cases from the hill country Indian Tamil communities are used in a multiple case study method. A case study is "an exploration of a bounded system or a case (or multiple cases) overtime through detailed, in-depth data collection involving multiple sources of information rich in context" (Ratnayake 2018, 28). As such, the researchers draw a line under both authenticity and sustainability in these two offerings by utilising the case study design to take a qualitative approach within an interpretivism paradigm. Here, the researchers used site observations, interviews with owners, and web content relating to the digital marketing appearances of the product and the literature to ensure the rigour and trustworthiness of the discussion.

Case Study 14.1   Meena Amma's tea experience

Meena Amma's tea experience is a value-added experience offered to tourists by Jetwing Warwick Gardens, one of the luxury bungalows owned by the Jetwing Hotel chain and located in Ambewela. Meena Amma's experience is offered using a line room in the lower Ambewela division. According to the Jetwing Hotels (n.d.), Meena worked in a regional estate until she joined Warwick Garden Bungalow as a gardener. According to Hiran Cooray, the chairman of the Jetwing group, they have modified two traditional line rooms to accommodate guests. Also, this effort would help earn some more income for Meena since she needed to raise two daughters. However, Jetwing is still struggling to promote and market this product since it is such a novel experience (Tourism in Paradise 2020, 1:35:04). Even so, this new tourism experiential product has shown the possibility of employing Indian Tamil tea pickers in alternative industrial sectors. Jegathesan (2019), however, viewed this movement as a trans-sector tactic of the state to retain these women workers remaining on the estates, and as noted in the very beginning, there was a backlash, especially via social media, to this initiative. Responding to this criticism, Hiran Cooray, the chairman of Jetwing, had published a letter on Twitter that included the following information:

> Meena has been a loyal employee of Jetwing for 12 years, a former tea plucker who we recognized as a talented individual and welcomed into the Jetwing family. She is the housekeeper at Jetwing Warwick Gardens, and it was on her request that we refurbished two-line rooms within the grounds, as she herself wanted a property to manage in the style and life she knows best. . . . We do not seek to romanticize their lives, which have historically been very difficult. . . . Rather, it's an environment that Meena herself welcomes visitors from all over the world, speaks plainly about her life and experiences and accompanies them on their excursions.
>
> (Jetwing Hotels 2018)

The Jetwing chairman's statement shows that Meena Amma's Line Room experience has been standardized to suit tourist's needs. As explained by Cohen (1988), such standardizations lead to commoditization of the culture and livelihood for tourist consumption; and commoditization destroys the authenticity of culture in what MacCannell (1973) called "staged authenticity". Such staged authenticity is said to prevent tourists' quest for the authentic tourist experience, especially for MacCannell's western tourist gaze. As further explained by Cohen (1979), Meena Amma's line room experience is in the tourist establishment phase, where tourists do not become aware of its staged authenticity and thus accept it as reality. Cohen suggested this phenomenon is the situation of a "covert tourist experience". That may be the reason Meena Amma's Line Room experience has a lot of satisfied customers.

Figure 14.1   Refurbished bedroom at Meena Amma's Line Room
Source: Jetwing Symphony PLC

Figure 14.2   Bathroom at Meena Amma's Line Room
Source: Jetwing Symphony PLC

Figure 14.3 A photo of a bed in newly built house for the tea community at Hethersette Estate
Source: Authors

Figures 14.1 and 14.2 depict the layout of the refurbished line room, which has been provided with facilities similar to a standard hotel room. Figure 14.2 shows the attached bathroom with standard bathroom fittings, towels, mirror, finished walls, and tiled floor pleasantly serving the visitors. Figure 14.1 also provides shreds of evidence about standardization and commercialization of this unique cultural product by adding bedside tables with lamps, comfortable pillows, and expensive bed covers. However, in offering an authentic experience for visitors, the authors consider this has moved up to a luxury level in which the stay does not provide any unique experience other than the feeling of staying in another hotel room in a different place. This argument can further be understood by observing Figure 14.3. The researchers photographed a bedroom of a newly built house of the Indian Tamil communities. It has a basic bed only with a cupboard. Hence, Meena Amma's Line Room experience could have provided such basic facilities for the tourist, but the marketing and promoting issue will remain, as the chairman mentioned.

If the appropriate market segments, the authentic experience seekers, are promoted, they will be staying with minimum facilities without hesitation since they are searching for real experience. When an original experience seeker exposes these communities' authentic culture, as Cohen (1979) explained, there will be a threat of "denial of authenticity". However, for this location food is different, as this is situated in a line room area, and Meena Amma serves tourists. Furthermore, as explained by the chairman of Jetwing, in fact, they do not intend to romanticize social life and their living conditions by this practice. From that, they expect Meena Amma to have a better life. However, when thinking of heritage tea tourism products, there is a question about sustainability and authenticity, so romanticizing their culture in the context of tourism in which these concepts seems to be a conflicting goal.

Case study 14.2   Kandapola Village House

Kandapola Village House is located at the Hethersette Estate, Kandapola, Sri Lanka. It is a tiny line room offered as a place to experience the Indian Tamil life in the upcountry. Unlike Meena Amma's place, this line room exclusively provides an experience to visitors. Although Jolliffe and Aslam (2009, 338) mentioned, "many are accidental visitors willing to experience tea and tea culture", a requirement by European tour operators led to the start-up Kandapola Village House. Surprisingly, the owner of this place is also a young Indian Tamil whose parents used to be tea pickers. His name is Suresh Sathyanathan. After his secondary education, he went overseas (Germany) and then came back to Sri Lanka. After joining a reputed destination management company in Sri Lanka, he is now the revenue manager for all source markets of a reputable European tour operator for Sri Lanka. Of course, as Jegathesan (2019) noted of others, Sathyanathan is also disconnected from his inherited job of picking tea leaves.

Sathyanathan's Kandapola Village House offers a snapshot experience about the culture within two and a half hours. It also can be interpreted as staged authenticity, since Sathyanathan packed the experience with simplified cultural experiences of a standardized duration. Firstly, the experience starts with a welcome for the visitors from Indian Tamil ladies who are cousins of Sathyanathan. They paste *pottu* on the foreheads of visitors and invite them to light the home's traditional oil lamp. Then, they dress visitors with sarees and veshti (*dhoti*). Sarees are for women, and veshti are for men. Once visitors are dressed, they accompany them to the small garden to colour *Kolam*. The outline of the Kolam is already drawn, but visitors are invited to colour it. During these activities, Sathyanathan or a member of his family gives a simple demonstration of these practices. The next stop in this tiny house is the kitchen, which is also visible in very close proximity. The kitchen is attached to the living area, as the space is limited. Demonstrations of dry grinding and wet grinding, firewood cooking, their traditional kettles, and so forth are very interesting, as their traditional food is prepared while they watch. Food that can be prepared by steaming or frying is in the majority. Once the food is cooked, the visitors have to serve themselves on banana leaves provided by the house. As there are no tables available, visitors enjoy the food by holding the banana leaves in their hands. The kitchen environment of their ancestors, older than a hundred years, is presented in this tiny house. These moments are authentic. As Cohen (1979) explained, such items can be said to be objectively real, hence authentic.

The most valuable element of the Kandapola Village House experience is the meaningful discussion that can be initiated with Suresh Sathyanathan. As he has travelled around the world, he can deliver his culture comparative to that of the tourists. Though the experience is over quickly, the essence of this meaningful intercultural discussion remains in the mind of the visitors. However, visitors still have no overnight staying facility at Kandapola Village House, unlike at Meena Amma's place.

Discussion

Authenticity, sustainability, and romanticizing the local culture

Both Meena Amma's Line Room and Kandapola Village House provide unique experiences for visitors. Defining these two products using the idea of an authentic framework is a difficult task, as the term "authenticity" is a fluid concept. As Cohen (1979) mentioned, once the customer perceives the experience as real, it is authentic (Figure 14.4). Moreover, this judgment of the customer is subjective. When looking at the surface level, Kandapola Village House provides a comparatively authentic experience in two and a half hours. As Hiran Cooray mentioned, Meena Amma's Line Rooms serve the customers on a commercial basis. In defining this product as authentic, it has been commercialised a lot in terms of dining, washrooms, bedrooms, and so on. However, giving a final judgement is subjective, as the term authenticity is overdefined in the tourism literature. Aronsson (1994) explained that authenticity is analogous to social changes that happen over time. According to his research, the boundary line of genuineness with artificiality is unclear. Aronsson (1994, 86) further linked the concept of authenticity and sustainability as follows:

> Encounters in authentic meeting places generally take place on the residents' terms. The authentic meeting-place can be characterised as a sustainable tourism product in itself as long as it is not changed by tourism.

As per Mr. Sathyanathan, Figure 14.4 shows the original kitchen that was used for the preparation of food by his mother. Brass and copper containers are used in the kitchen. Most importantly, their kitchen is situated on the floor. The person who cooks must sit on a short bench. Visitors may enjoy food by sitting on the clay parapet (the right bottom corner in Figure 14.4).

Unlike with the Kandapola Village House, the location of the firewood cooker in Figure 14.5 is situated on the top of the bench. Food is also served on a table with fancy utensils. As such, compared to Kandapola Village House this shows a multistaged authenticity due to standardization for tourists' needs. But saying that is also questionable, as some tea families are living a better life.

Hence, as explained by Aronsson (1994), a product should be authentic to be sustainable, and tourism should not then change the authentic features of the product. This argument suggests that local communities remain unchanged in terms of their lifestyle, dress, housing conditions, and so forth to search for modern tourists, which does not seem fair. Therefore, maintaining the authenticity of a local community, as suggested by Aronsson (1994), is not sustainable because that concept

Figure 14.4   Traditional kitchen in the Kandapola Village House

Source: Suresh Sathyanathan

Figure 14.5   The kitchen in Meena Amma's Line Room

Source: Jetwing Symphony PLC

of authenticity restricts people from upgrading for a better life level. They must earn money from tourism and should remain culturally unchanged or pretend to be culturally unchanged for touristic needs. Phillips *et al.* (2021) claim this kind of practice questionably encourages the romanticization of poverty, and it vindicates continuous labour exploitation in tourism by strengthening the intentional economic inequalities. Therefore, romanticizing the authenticity of Indian Tamil culture and their living conditions in the upcountry for heritage tea tourism contradicts the real meaning of sustainable tourism. When there is an opportunity to use a washroom, as shown in Figure 14.2, why would they still have to manage with less sanitized common toilets, as shown in Figure 14.6. If this

Figure 14.6 Common toilets used by the line room community in Kandapola Hethersette Estate

Source: Suresh Sathyanathan

society is ready to change, this should happen rather than make them prisoners of the postcolonial ideology. On the other hand, how far such changes satisfy the modern tourist quest for authenticity is questionable. If tourism is contributing to inclusive growth, the occurrence of such cultural changes is inevitable.

As seen in Figure 14.6, one toilet is commonly used by two families, which leads to hygienic problems for these people. In the context of authenticity, these are the toilets used by the Indian Tamil community. For a real authenticity seeker, this may be attractive. However, from the point of tourism, as explained by Phillips *et al.* (2021), this could indirectly motivate these communities to remain at the same level of life to market their heritage tea tourism experience, which is inhumane. Since tourism takes these communities to higher levels, for instance, as per Figure 14.2, it is irrational for these communities to remain unchanged. However, if they changed, as explained by Cohen, the authenticity of the product would be questioned by authenticity seekers, while another customer from another segment (e.g., a comfort seeker) will be satisfied in the meantime.

Conclusion

As argued in the discussion, 100% pure authenticity cannot be achieved, especially in the traditional line rooms and the lifestyle in upcountry Nuwara-Eliya, when practising heritage tea tourism. Authentic or genuine tourism products problematize sustainability or heritage forms of tourism when it connects to the culture. Therefore, suppliers of these line room–related tourism products have to depend upon staged authenticity as far as helping visitors to believe that the product as genuine, which signals the requirement of cultural manifestos. Sometimes, staged authentic experiences may be rejected by extreme authentic seekers, but they may be suitable for another customer from a different market segment. Further, realising line room experience as original or a duplicate is subjective depending on the visitor profile. Meena Amma's Line Room and Kandapola Village House experiences are heritage tea tourism products offering the livelihood experience of upcountry Indian Tamils who migrated from India years ago. Moreover, these experiences are not romanticizing the real stories of these communities. Rather than that, these two unique heritage tea tourism products contribute to inclusive growth for these marginalized communities.

References

Aronsson, L., 1994. Sustainable tourism systems: the example of sustainable rural tourism in Sweden. *Journal of Sustainable Tourism*, 2(1–2), 77–92.
Bass, D.M., 2018. The goddess of the tea estates. *The South Asianist Journal*, 6(1), 23–23.
Central Bank of Sri Lanka, 2020. Annual report of Central Bank of Sri Lanka 2020. Available from: www.cbsl.gov.lk/sites/default/files/cbslweb_documents/publications/annual_report/2020/en/9_Chapter_05.pdf [Accessed 3 July 2021].
Cohen, E., 1979. Rethinking the sociology of tourism. *Annals of Tourism Research*, 6(1), 18–35.
Cohen, E., 1988. Authenticity and commoditization in tourism. *Annals of Tourism Research*, 15(3), 371–386.
Department of Census and Statistics, 2012. Census of population and housing 2012. Available from: www.statistics.gov.lk/pophousat/cph2011/pages/activities/reports/finalreport/finalreporte.pdf [Accessed 3 July 2021].
Fyall, A. and Garrod, B., 1998. Heritage tourism: at what price? *Managing Leisure*, 3(4), 213–228.
Hapugoda, M. and Ratnayake, I., 2021. Can the Western Tourist Gaze be deconstructed through Buddhist Ontology? *South Asian Journal of Tourism and Hospitality*, 1(1), 102–116.
Jegathesan, M., 2015. Deficient realities: expertise and uncertainty among tea plantation workers in Sri Lanka. *Dialectical Anthropology*, 39(3), 255–272.
Jegathesan, M., 2019. State-industrial entanglements in women's reproductive capacity and labor in Sri Lanka. *South Asia Multidisciplinary Academic Journal*, (20), 1–21.
Jetwing Hotels, n.d. Meena Amma's tea experience. Available from: www.jetwinghotels.com/meena-ammas-tea-experience/things-to-do-in-ambawella/ [Accessed 3 July 2021].

Jetwing Hotels, 2018. Regarding Meena Amma's line rooms. Available from: https://twitter.com/JetwingHotels/status/1008398564588969985/photo/1 [Accessed 3 July 2021].

Jolliffe, L., ed., 2007. *Tea and tourism: tourists, traditions and transformations*. Clevedon: Channel View Publications.

Jolliffe, L. and Aslam, M.S., 2009. Tea heritage tourism: evidence from Sri Lanka. *Journal of Heritage Tourism*, 4(4), 331–344.

Lehman, D.W., O'Connor, K., Kovács, B. and Newman, G.E., 2019. Authenticity. *Academy of Management Annals*, 13(1), 1–42.

MacCannell, D., 1973. Staged authenticity: arrangements of social space in tourist settings. *American Journal of Sociology*, 79(3), 589–603.

MacCannell, D., 1976. *The tourist: a new theory of the leisure class*. New York: Schocken Books.

MacCannell, D., 2011. *The ethics of sightseeing*. Berkeley, CA: University of California Press.

Nuryanti, W., 1996. Heritage and postmodern tourism. *Annals of Tourism Research*, 23(2), 249–260.

Philips, A., 2003. Rethinking culture and development: marriage and gender among the tea plantation workers in Sri Lanka. *Gender & Development*, 11(2), 20–29.

Phillips, T., Taylor, J., Narain, E. and Chandler, P., 2021. Selling authentic happiness: Indigenous wellbeing and romanticised inequality in tourism advertising. *Annals of Tourism Research*, 87, 103–115.

Ratnayake, I., 2018. *Research process simplified*. Sri Lanka: Stamford Lake Publication.

Ratnayake, I. and Hapugoda, M., 2014. Diverged meaning of heritage: a critique on visual authenticity of the Golden Rock Temple of Dambulla. *In:* Cooper, M.J.M., Aslam, M.S.M., Othman, N. and Lew, A.A., eds. *Sustainable tourism in the global south: communities, environments and management*. Cambridge: Cambridge Scholar Publishing.

Reisinger, Y. and Steiner, C., 2006. Reconceptualising interpretation: the role of tour guides in authentic tourism. *Current Issues in Tourism*, 9(6), 481–498.

Sabaratnam, T., 1990. *Out of bondage: the Thondaman story*. Colombo: The Sri Lanka Indian Community Council.

Sri Lanka Tea Board, 2017. *Annual report*. Colombo: Sri Lanka Tea Board.

Sri Lanka Tourism Development Authority, 2019. *Annual statistic report*. Colombo: Sri Lankan Tourism Development Authority.

Tourism in Paradise, 2020. Emerging dynamic tourism paths towards a sustainable new normal [Video file]. Available from: www.youtube.com/watch?v=RmLozN4_SQg [Accessed 3 July 2021].

Urry, J., 2011. *The tourist gaze*. London: SAGE.

15

EMPLOYMENT ISSUES IN TEA TOURISM

A way forward

P. Gayathri, D.A.C. Suranga Silva, Krishantha Ganeshan, Baghva Erathna, K.M.B.S.Y. Kulasekara and Teeshakya Weerakotuwa

Introduction

The history of Ceylon tea goes back to when tea was brought from China to Sri Lanka and planted in the Royal Botanical Garden for non-commercial purposes (Sri Lanka Export Development Board 2015). In 1867, James Taylor began commercial cultivation, this soon spread to other regions, and Sri Lanka managed to send its first shipment to the London tea auction in 1875. Later, Governor Barnes took the initial steps in developing the right infrastructure, and the first Colombo public auction of tea was held in 1883, making the city the oldest and largest tea auction center in the world (Sri Lanka Export Development Board 2015). Currently, the tea industry plays a major role in bringing the foreign exchange to the country and increasing GDP. However, the contribution from the tea industry has dropped along with declining tea production. This downward trend has been observed since 2017, and the main reasons for it are the labour issues that emerged during wage negotiations in the sector and the recent dry weather conditions in the country (Central Bank of Sri Lanka 2019b).

Tea tourism can be identified as a peripheral industry associated with tea production, and it mainly focuses on the goods and services that target tea-loving travellers. The traveller is motivated to learn more about the heritage, culture, and traditions associated with tea production and consumption (Wipulasena 2020). This is a novel concept that emerged with the niche tourism segment of a responsible form of tourism focusing on information about tea, tea accessories, and the tea-themed items available as souvenirs. With the boost in traveller numbers from this concept, industry development alongside employment development is expected. This change can be seen in the growing number of gift shops, the tourism Small and Medium Enterprise sector development, tea factories, and the tea experience that provides jobs in hotels and other accommodation sector facilities and can be expected to grow in the future (Table 15.1).

To further develop this connectivity, Sri Lanka Tourism has partnered with the Tea Board to create a connection with its target audience (Ministry of Tourism Development and Christian Religious Affairs 2017). But the authorities have also identified the need to reduce the overcrowding effect in tourism and spread visitors travelling to identified key locations such as Nuwara Eliya, which is the best known of Sri Lanka's tea-growing districts (Sri Lanka Tea Board 2014). As a result, a higher number of both local and foreign travellers are being relocated to tourism getaway towns such as Bandarawela to avoid congestion during peak season. Furthermore, to bring out a diversified product experience, the authorities are conducting feasibility studies on interactive museums,

DOI: 10.4324/9781003197041-18

Table 15.1 Nature of tourism industry in Sri Lanka with other sectors

2017				2018			
Sector	FE Earnings (Rs. million)	Share of Total FE Earning (%)	Rank	Sector	FE Earnings (Rs. million)	Share of Total FE Earning (%)	Rank
Workers' remittances	1,091,972	27.1	1	Workers' remittances	1,138,124	25.4	1
Textiles and garments	767,254	19.0	2	Textiles and garments	865,975	19.4	2
Tourism	**598,143**	**14.8**	**3**	**Tourism**	**711,961**	**15.9**	**3**
Transport	356,741	8.9	4	Transport	402,806	9.0	4
Tea	233,338	5.8	5	Tea	231,750	5.2	5

Note: FE, foreign exchange.

Source: Adapted from Foreign Exchange Earnings by Industry 2017 and 2018, Sri Lanka Tourism Development Authority 2018

including the Tea Museum in Colombo and many more, to showcase tea heritage and to encourage visitor learning. All these sectors help in increasing employability in both the tourism and tea sectors (Ministry of Tourism Development and Christian Religious Affairs 2017). According to statistics, the total of direct and indirect employment in the tourism sector in 2019 was estimated at 402,607 (Sri Lanka Tourism Development Authority 2019), and in 2018 the tea sector contributed 11% of all employment in Sri Lanka (International Labour Organization 2018).

The present study is based on primary and secondary data that focuses mainly on three areas: a resort; a tea factory café in Ella, Sri Lanka; and interviews about tourist guides in that area. The study looks to reveal how the issues in employment can be overcome with the promotion of heritage cultural tea tourism. It examines the existing problems and challenges faced by employees in the industry and measures that can be taken to overcome them.

Literature review

Tea tourism is associated with cultural tourism characteristics and follows most of the core concepts of food-related tourism, such as wine tourism in European nations (e.g., France), that many scholars who are interested in concept-based tourism are investigating. Jolliffe (2007), for example, has explored the possibilities for adopting tea tourism in the Sri Lankan plantation business using several concept-based tourist models.

Even though both men and women engage in tea plucking, it is mostly done by women (Chandrabose 2015). According to that research, 52% are engaged in full-time work, 35% are partly involved, and the rest are engaged in non-estate work. The non-estate workers who study up to secondary level leave the rural area looking for jobs in the cities, and estate owners then have an issue with finding labour. This highlights the fact that higher literacy leads to higher labour turnover in the sector through rural-urban drift, and something must be done to reverse this situation.

With the major contribution from the tea and tourism industries to the economy, there is an interconnection between tea and tourism as well. This has been there for a very long time; however, the economic advantages have only been seen very recently. The combination of these unique features will help tea become more attractive in the potential guest's mindset given the stronger relationships with those customers who are in love with Ceylon tea and Sri Lanka beauty. This is very similar to wine tourism, where travel is also known as wine escapes and includes wine-motivated travel, the wine itself, and food and wine festivals as recreation.

In the tea plantation industry, most of the workers are low-wage-earning laborers on a basic salary plus allowances. These workers have participated in strikes for higher wages. Recent strikes have reduced the labour available in the country by a total of 54,919 person-days, and out of this, 51.4% occurred in the plantation sector according to the Department of Labor (Central Bank of Sri Lanka 2019a). The Central Bank report discusses the measures the government has taken to minimize this impact and to maintain a harmonious relationship in the tea industry, with a six-year strategic plan for the sector to motivate the planters to improve sector productivity and competitiveness and many other favourable laws.

Even though Sri Lanka has great potential in the tea tourism industry, there are certain issues among the employers and employees that have not been taken into consideration by researchers. Previous studies show that employees in this sector are dealing with issues such as lack of recognition, low or no salary, seasonality, and many more.

Research methods

Primary data refers to the information and evidence gathered from direct sources, which include discussions and the webinar on Tea Cultural Heritage Tourism: A Way Forward. This webinar was held on 23 March 2021 and was organized by students in the postgraduate program in tourism economics and hotel management, Department of Economics, University of Colombo. In-depth interviews were conducted with employees, including five each with tour guides, employees in the tea tourism accommodation sector, employees in the tea café sector, and management/ownership stakeholders. The on-site observation was also used as a data source. The group discussions conducted with the employees are the main source of primary data in this research. Secondary data collection was based on published materials such as reports, books, and existing research studies.

Case study context

The tea tour that is provided by a hotel can teach tourists everything they need to know about tea production, from start to finish. Since the hotel may also play a major role in tourism, the employee problems in the tea industry can be highlighted by group tour guides, accommodation workers, and the café employees in tea tourism hotel properties.

Case study 15.1   Employee issues in tea tourism: Ella City, Sri Lanka

Our selected tea estate maintains that tea-related tourism services can be recognized as the combination of a tea resort and restaurant located in a former tea factory. The resort is an eco-friendly, romantic hideaway located at a picturesque tea plantation overlooking a scenic hillside. It presents a restaurant with a fusion of Sri Lankan and international cuisine, snacks, factory-fresh tea mocktails, tea bar, and a tea boutique facility, architecturally designed to incorporate heritage elements of the former tea factory. Tea was brought to Sri Lanka for commercial purposes by the British during the colonial era, and the everyday practice of drinking tea has since become a firmly established institution in Sri Lankan culture. Serving tea to friends and family in Sri Lanka is a true sign of hospitality.

When considering the employment issues tour guides face, the guest directly contacts the resort for a tea tour without any third-party involvement. However, only 20% of travellers are known to participate in tea tourism. In this context, tea tourism involvement is at a neutral level for a hotel. Hence, the opportunity for tour guides in tea tourism is unstable. The guides themselves do not have enough facilities to build a proper tea tour, and they do not receive other benefits, allowances, compensation, or acceptance from government or private societies to succeed in that service.

Most of the staff involved in tea tourism does not get accommodation and allowances by comparison to staff in other industries. Recreation officers, café managers, and stewards are active employees in tea tourism in this resort and restaurant. Many other sectors in the tourism industry provide a more beneficial remuneration package that is enough for the employee's family. For example, the tea tourism staff in the Labookellie Tea Center, Nuwara Eliya, are given accommodation with land for cultivation. So, the unavailability of a sound welfare system for staff supporting tea tourism is recognized as a significant problem in this sector.

Also, employers typically only provide personal development training per year for selected heads of departments. However, it would be an advantage if they could provide special training programs for all staff, including those who participate in tea tourism. In this situation, the employees in this hotel's tea tourist attractions provide their services based on language skills, guest handling, communication, tea selling and purchasing, tea process explanation, and so on. But they do not receive proper training and development programs to improve their skills.

However, the employees in the resort restaurant are also facing numerous problems. Although the staff in the tea café must work according to an eight-hour shift, they often work more hours during the peak season. However, they do not report receiving overtime payments, allowances, or incentives other than the basic salary and service charge for their extra hours. Most of the staff in the tea tourism industry get leave according to a proper system including annual leave, medical leave, public holidays, and so on. However, the café staff only get leave for about five days per month. Importantly, though, in most cases their workplace is where people with various personalities, communication styles, and worldviews interact. Therefore, management should provide more opportunities for staff to develop their personalities and communication styles if they want the destination to succeed.

Having said this, reviews and guest comments show that the resort has done an excellent service for the tea tourism industry, but its employee services and performance are not much appreciated, unlike in other organizations that select the best employees of the month or the best seller of the month and appreciate their kind cooperation for the tea tourism industry.

During the current pandemic, the executive-level employees in other departments are still in their positions and are working according to the guidelines given by local health authorities. However, at the time of this research the cafés, restaurants, and factories have been closed because of the COVID-19 situation. Therefore, they are not receiving benefits or compensation to survive further in the industry. As a result, most of the employees have left their positions and have selected other ways to gain income. Some of them express disappointment regarding their current situation.

Results and discussion

The data collection approach for this study is based on thematic analysis. It is an effort to investigate employee issues in tea tourism based on tour guides' comments and the experience of tea-related accommodation and the cafés. The sample was based on five employees in each field with whom the structured interviews were conducted. According to the employment issues faced by tour guides, there are numerous problems. An unorganized tourist handling procedure is one of the problems in such a situation. According to the case study, almost 75% of the guides reported that there is no proper appreciation for their service by the resort or the guest.

Tour Guide B: "Since we are here to conduct tea tourism some guests directly contact the tea factory and complete their tea tour without any third party."

If the guest contacts tourist guides directly, they will get sufficient payment. According to the guest satisfaction, they will provide gratification with a positive review.

Low-level employee facilitation is another considerable issue among tour guides in this sector. During their interviews, most participants highlighted that they do not receive proper facilitation and licensing and do not have sufficient infrastructure and superstructure.

Tour Guide C: "Most of the tea trails are conducted in hill countries with cold and misty climates. Therefore, we need proper uniforms favorable for climate and equipment, first aid facilities. If we have any insurance recovery for tea tourism guides and licensing procedures, we can build trust among the travelers. If the guest requests a family tea tour, we don't have appropriate transportation and other facilities. So that we must arrange outside vehicle facilities."

According to their responses, if they have insurance recovery for tea tourism guides and licensing procedures, they can build trust among the travellers. Also, since they do not have sufficient transportation facilities, they must hire them to fulfil family tours. On the other hand, the lack of a global promotional campaign is an arising problem. Participants reflected that guest involvement in tea tourism is relatively low due to the lack of a promotional campaign. During their period of working, different kinds of activities are promoted.

Tour Guide D: "I've been doing tea trails for ten years. During this period, different kinds of activities are promoted. Recently around the Ella area, 20% of were interested in hiking, 40% of travelers looking for Flying Rawana Zipline, 20% of travelers preferred yoga and bird watching. Only 20% of travelers participate in tea tourism."

As with the data provided by participants, the accommodation sector is suffering from a lack of employees for different services.

Employee C: "The resort is enriched with different services but not enough employees in tea tourism such as recreation officers, cafe managers, and stewards. Limited employees are providing different kinds of services such as arranging interesting tea trails, tea products selling and promotion, other recreational activities, tea tasting activities, arranging tea fiesta, and so on. In the accommodation premises, we are providing facilities to buy tea products and outlets consisting of other eco-friendly products such as souvenirs, sculptures, paintings, etc."

Employee D: "Compared to other tea-related accommodation, we are not receiving sufficient facilities to provide the best service for the travelers. We are not receiving a proper transportation system, uniform facilities, infrastructure development, proper communication system such as app development, payment procedures."

Training is essential for employee development. The employees are looking for proper training to enhance their skills and knowledge at least once a year. The pandemic situation has affected the employees and the entire industry a lot. When employees were getting tourists during the peak season, they received enough income to support their families. Now they don't have any response from the public or the private sector to overcome these problems.

Employee E: "As a result of the COVID-19 pandemic tea tourism sector was affected on a large scale. Since we were getting tourists during the peak season, we received capable income to survive our families. Now we don't have any response from the public or the private sector to overcome these problems."

According to the view of participants working in the cafés, a lack of selling procedure to attract the guests has been observed. If any guest visits the café after visiting the factory, they will try the food and beverages, but most of the time the café does not get a chance to sell tea products as they have already purchased them from the factory. On the other hand, seasonality also considerably affects the cafés.

Employee A: "Most travelers arrive during the winter season but in the Ella area, there is a big competition to attract guests during peak season as more restaurants are in the heart of the city. If we have many travelers in peak season, we don't receive many travelers the rest of the year. It will affect café income and the staff will get low service charges according to the revenue."

Like the participants, cafés need product differentiation due to the competitive market. If they provide a unique experience for the guests, guests will be satisfied, and this will result in a smooth operation for their outlets. Employees in cafés are facing several drawbacks in their working environment. Many employees have experienced awful employers, and unjustified layoffs. Therefore, the employees have to constantly depend on ungrateful environments, without any compensation for their hard work. This dependence leads to disadvantages in being an employee. In the present day, the most affected problem is COVID-19. This situation has directly influenced the employees in cafés. Cafe employees are faced with the unenviable task of dealing with the challenges. Most of the employees have lost their jobs, faced salary deductions, or moved to self-employed jobs. They are facing economic difficulties to fulfil their basic needs and wants because of this unemployment.

As discussed above, the study identified employee issues related to touring guides, the accommodation sector, and Cafés. In the light of these, the discussion proceeded to identify specific issues in the tea tourism sector based on five major themes: training and development, welfare system, competitive market, employee satisfaction, and employee situation after COVID-19. Each theme elaborates on different issues in the working environment. The identified themes can be defined and labelled as follows.

The first theme can be identified as training and development. Training is an essential element for employees to gain knowledge, develop skills, and help to improve corporate performance. It brings long-term and short-term benefits. When there is a gap in this process, it will affect the productivity of the organization. When looking at the tour guides, employees in tea-related accommodation, and in cafés, it can be observed that they too are not experiencing proper training at least once a year.

The study identified employee issues based on the second issue, the welfare system. The outcomes of welfare dissatisfaction harm and affect the quality of employee work life. Without adequate welfare, the current employees of the tea tourism industry are likely to leave, and replacements will be difficult to find.

The tourism industry consists of different competitive markets, our third issue. This is also a contemporary issue in the tea tourism sector. If tea tourism can organize a global promotional campaign to position tourists, it will bring more benefits to the employees in tea tourism.

For an effective product or service the fourth area, employee satisfaction, is very important. The employees in this industry are influenced by lack of compensation, employee benefits, and other dissatisfactions within the facilities. If the employees do not face these drawbacks, it will be hard for them to give the best service or retain talented staff. Due to these drawbacks, job satisfaction is negatively affected.

Fifth and last is the most recent problem that affects the employees in the tea tourism industry: COVID-19. During this pandemic, the employees are unable to secure their livelihoods. Our data based on interviews provides more clarification on this point.

Conclusions

One of the key solutions to the issues facing tea tourism and its employees is to correctly position and brand the product. Currently, most tourists only consider tea tourism as a minor component of their Sri Lanka itinerary. This damages industry personnel, especially those vulnerable groups like tea pluckers and tour guides. Increasing the industry's reputation would improve employee conditions. Of course, marketing Sri Lanka only as a tea destination isn't feasible. However, Sri Lanka has the advantage of being a destination for adventure travel. This is because, despite its small size, the island offers many tourist attractions including beaches, animals, ancient sites, tea plantations, and train rides. Tea tourism should be promoted as part of this package.

But changes must be made to achieve these objectives. Several historic factories, homes, and clubs are abandoned, especially in the hill country. If refurbished, they may become luxurious experiences. As a result of the increased cash flow, the working conditions of employees would immediately improve. Upselling is an advantage of tea tourism. In addition to experiencing the authentic flavour of Ceylon tea, travellers staying at a bungalow can learn about the agricultural elements of tea, wellness tea, and other activities like trekking and mountain climbing, which creates new job prospects in the country. Rather than relying on large numbers of tourists, this model should be branded and positioned as a highly valued small tour luxury business.

An accessible destination is a major tourist draw. Due to their remote location, even locals struggle to reach the tea estates. Most of the tea regions lack access to the railway network, and driving is difficult. As a result of this, the proprietors of these otherwise tourist-friendly territories may not appreciate the concept of guests visiting their lands. Improving accessibility would benefit both locals and tourists. For the people in the industry, this means increased travel to these places to participate in activities.

The infrastructure and living conditions of workers in this industry should also be improved. It is difficult to sell tea as a luxury commodity in the global market without improving the living conditions of plantation employees. In addition to fundamental needs like shelter and education, these excluded communities face many challenges. To earn a living, most affected families have dedicated their lives to activities like tea picking, even if the compensation or job stability is not always satisfactory. Their environment and conditions do not satisfy the tourists visiting these regions either.

Reducing congestion in destinations and other issues requires realistic answers. The government should introduce fresh combinations of attractions to ensure that the tea-related sites visited are varied, as congestion frequently happens in a well-known tourist destination. Promoting identified regions for various additional market segments might help introduce fresh attractions to the market. For the staff, this would be a solution where everyone benefits.

In the tea tourism locations, not only the guests but also the personnel must be taught the right technique for seamless operation. Digitalization can help with this. For the guests that visit the locations, new apps, online payment systems, and so on would be advantageous. Workers in this business should be properly trained and licensed, as part of their job involves directly engaging with and serving guests. A specialist who can manage challenges and find answers would be created. A solid welfare system is also required for industry personnel, as most of them rely on tourism and any changes in the industry will affect them directly.

Finally, we have several questions that should be discussed in-depth with tea tourism industry employers and employees based on the results of our study. These questions could be adapted for research on tea tourism in plantation settings beyond Sri Lanka.

Discussion questions

1   As a tourist, are you satisfied with the tea tour guide with which you are currently engaged?
2   As an employee, do you experience a good welfare system in Sri Lanka, and what is your opinion on the current welfare system? What are your suggestions for change, if any?
3   To all respondents, what is your view on the impact of the pandemic situation in Sri Lanka and the world, and how has the COVID-19 pandemic affected you?

References

Central Bank of Sri Lanka, 2019a. *Annual report 2019: prices, wages, employment and productivity*. Colombo: Central Bank of Sri Lanka.

Central Bank of Sri Lanka, 2019b. *National output, expenditure, and income* (Annual Report 2019). Colombo: Central Bank of Sri Lanka, 64–65.

Chandrabose, A.S., 2015. Outgoing labour and its impact on the Tea Plantation sector in Sri Lanka. *5th International Symposium 2015 – IntSym 2015*, SEUSL, 301–303.

International Labour Organization, 2018. International Labour Organization (ILO) [Online]. Available from: www.ilo.org/colombo/info/pub/pr/WCMS_632466/lang-en/index.html [Accessed 9 October 2021].

Jolliffe, L., ed., 2007. *Tea and tourism: tourists, traditions and transformations*. Clevedon: Channel View Publications.

Ministry of Tourism Development and Christian Religious Affairs, 2017. *Sri Lanka tourism strategic plan 2017–2020*. Colombo: Ministry of Tourism Development and Christian Religious Affairs.

Sri Lanka Export Development Board, 2015. Sri Lanka export development board [Online]. Available from: www.srilankabusiness.com/blog/wonderful-history-of-ceylon-tea.html [Accessed 9 October 2021].

Sri Lanka Tea Board, 2014. Sri Lanka tea board [Online]. Available from: www.srilankateaboard.lk/index.php/nuwara-eliya [Accessed 10 October 2021].

Sri Lanka Tourism Development Authority, 2018. *Statistical report 2018*. Colombo: Research & International Relations Division, Sri Lanka Tourism Development Authority.

Sri Lanka Tourism Development Authority, 2019. *Annual statistical report 2019*. Colombo: Research & International Relations Division.

Wipulasena, A., 2020. Tea tourism: beyond just a cup of tea. *Features, Daily News*, 25 August [online]. Available from: www.dailynews.lk/2020/08/25/features/226821/tea-tourism-beyond-just-cup-tea [Accessed 27 January 2022].

PART III

Management and marketing of tea tourism

16

SERVICE QUALITY IN AN ENGLISH TEA ROOM

A picture is worth a thousand words

Belinda Davenport

Introduction

The UK is one of the largest per capita consumers of tea in the world, below only Turkey and Ireland (Statista 2016). Afternoon tea in the UK has emerged as a major tourism attraction (Boniface 1998), and the increase in tea rooms has further driven this popularity. As such, a UK-based tea room specialising in afternoon tea is worthy of further research. Davenports Tea Room in Cheshire was selected for the present research, as it is a fine example of a UK heritage tea-tourism attraction that gives the visitor "an historical and cultural experience of a traditional English afternoon tea which is central to the consumption of tea" (Jolliffe 2007, 9). The UK Tea Guild inspectors scored Davenports as almost perfect across 16 different categories and awarded it the UK Tea Guild's last-ever Top Tea Place (BBC News 2013).

Situated in a rural location in the Northwest of England, Davenports gives the impression of being in the middle of nowhere, but it is, in fact, located on routes to many places. It has good links to the motorway, meaning that several large towns and cities are within a 30-minute drive. It is an established destination for afternoon tea; though most of its guests are from the Northwest of England, others travel from all over the world to visit this quintessential English tea room. Visitors regularly make an effort to return, with social media describing Davenports as a high-quality experience, with staff who provide excellent service and have good product knowledge.

The owner, and author of this chapter, has a background in tourism management, many years of experience in that field, and has developed the business since 2007 with her husband. Qualified with the UK Tea Academy, she has fine-tuned techniques for encouraging guests to try different teas and to give them an experience they feel is worth repeating. She has trained the Davenports staff to do the same and to deliver a unique, high-quality experience that increases revisit and word-of-mouth intentions.

During the 2020–21 COVID-19 pandemic, Davenports' good reputation as an experience with high product standards enabled it to adapt and create a successful new product. Rather than disappointing the 90 guests booked in for Mothers' Day weekend by simply closing when England's lockdown was announced, Davenports created "Afternoon Tea @ Home". More than 70 of the existing guests – mostly those living within a 10-mile radius – converted their in-person reservations to this product. Between March 2020 and September 2021, Davenports sold more than a thousand "Afternoon Tea @ Home" boxes for two, with approximately 10% of customers reordering, and several ordering more than five times in a 12-month period. This revealed Davenports Tea Room's

DOI: 10.4324/9781003197041-20

clear understanding of their customer base and the importance of visitor satisfaction in developing high-quality experiences, especially given that their marketing strategy largely consists of encouraging repeat business and word of mouth.

To understand the influence of service quality on tea tourists' satisfaction, this study collected longitudinal qualitative data from participant observation and interviews over a 14-year period (2007–2021). Additionally, the researcher analysed three years' worth of Davenports Tea Room's most recent online reviews (2019–2021) to triangulate the findings. The Meal Eating Establishment Experience Instrument (MEEEI) adapted from Hansen (2014) was utilised as the conceptual framework for online review analysis. This model was chosen because it accounts not only for food but also for the environment in which it is served and customers' intention to return. It is expected that the findings of this study will help tea-room managers, owners, and stakeholders in the field of tea tourism to understand the importance of service quality as the basis for improving the marketing and management strategies of tea-tourism destinations.

Literature review

Chen *et al.* (2021, 306) pointed out that "tea tourism is a very young phenomenon, with some of the earliest recorded literature less than twenty years old and with the majority of work only emerging in the past three years." Unlike tea-tourism research, which remains in its infancy, studies related to service quality are well established. For example, the assessment tool SERVQUAL (Parasuraman *et al.* 1988) has been widely used in various contexts to measure guests' perceptions of service quality. Nevertheless, Seth *et al.*'s (2005, 34) review of service-quality models found that "none . . . currently satisfies the set framework". Similarly, Oh and Kim's (2017) review of studies related to customer satisfaction, service quality, and customer value published from 2000 to 2015 found little consensus among their theoretical models, hypotheses, research instruments, or analytical methods, and recommended that scholars in this field "rethink and reinvigorate our research efforts" (23). Therefore, it would seem that further research aimed at expanding the theoretical lens of SERVQUAL is warranted.

As research interest in dining experiences has increased, many scholars have developed tools to assess not only service environments but also meal experiences. Among these models, the MEEEI is one of the most comprehensive, in part because it combines elements of other models, including DINESERV, the Five Aspects Meal Model (FAMM), and the Customer Meal Experience Model (CMEM), to understand the complexity of dining experience (Hansen 2014). In addition to service dimensions, the MEEEI covers restaurants' atmosphere and the company and, therefore, is likely to yield a more comprehensive view of the dining experience.

In recent years, posting images of one's meal or the decoration of the restaurant on social media has become an important part of the dining experience. As Edlinger (2015) pointed out, moments captured in pictures can facilitate the creation of memorable tourism experiences. Although this phenomenon has been investigated in the field of service management, no prior study is known to have explored the relationship between photo-taking and customers' perceptions of service quality in the context of tea tourism. Because the staff of Davenports Tea Room are trained to use customers' own cameras and smartphones to capture picture-perfect afternoon-tea stands and table displays, and this practice is often mentioned positively in customer reviews, one of this study's uses of MEEEI is to better understand photo-taking as a marketing strategy and its influence on tea-room service quality.

Methods

This study primarily used qualitative research methods, including participant observation and interviews, to achieve its research purposes. Kawulich (2005) pointed out that, despite some limitations, participant observation has several advantages in social science inquiry. It allows for richer detailed descriptions and facilitates the development of new research questions or hypotheses. Therefore, the author conducted participant observation from 2007 to 2021 to collect qualitative data related to visitors' preferences (including tea strength) and their satisfaction. Additionally, the author observed Davenports visitors' travel characteristics and behaviours, such as where they had come from, what motivated them to visit, what key elements influenced their travel decisions, and whether they had previously visited. Data were collected from more than 129,000 visitors during the 14-year data collection window.

In terms of interviews, more time was allocated to discussing tea options, preferences, and other service-relevant topics during the COVID-19 pandemic when no menus were given out to customers. Using verbal and non-verbal techniques – analysing what people said, what "wowed" them, and body language – a clear, comprehensive picture of the products and services making Davenports a high-quality experience, sufficient for customers to repeat and/or recommend, was established.

To deepen the researcher's understanding of customers' thoughts, TripAdvisor and Google reviews posted between 2018 to 2021 were also analysed using the MEEEI to triangulate the findings derived from participant observation and interviews. Ratings without any written comments and 24 negative reviews in which the reviewer was unable to visit because the tea room was fully booked or closed were omitted. In total, 250 reviews were analysed. Using multiple data sources increases the credibility of the analysis and avoids biases imposed by the use of a single research approach.

Case study context

Davenports Tea Room is a quintessential English tea room situated in the Cheshire countryside, close to the birthplace of author Lewis Carroll. Visits to Davenports, with its Victorian décor, Victorian staff uniforms, William Morris wallpaper, antique chandeliers and period furniture, are sometimes likened to stepping back in time. Hand-painted *Alice in Wonderland* murals adorn the walls, other *Alice* memorabilia is displayed around the room, and the tables are laid to maximise visual impact. Music is carefully chosen to create the right ambience, and scones are cooked just prior to the guests' arrival to enhance sensory delight.

The root of Davenports Tea Room's success as a special afternoon tea experience is that its staff have consistently sought to understand what its customers consider a "proper afternoon tea" and ensured that its offerings evolved accordingly. Today, Davenports classifies its afternoon teas as Traditional, Combination (similar to a Traditional, but with more savoury), Champagne, and Tea Tasters (a unique tailor-made tea-tasting experience including a full afternoon tea). These offerings are supplemented by further themed events, including an annual Mad Hatter's Tea Party and Victorian Christmas (mid-November through the end of December) and a seasonal afternoon tea cruise aboard a traditional canal boat.

All guests are guided through the range of 50 loose-leaf speciality teas and given background information about them. Staff check each individual's unique tea preferences before brewing for each guest according to the type of tea selected. The staff encourage guests to use this phase of the visit as a tea-tasting experience – an opportunity to try different teas, including at least one tea that complements the selected food. Each type of cake is also lovingly described, with wording and tone of voice carefully developed over time to set the scene for a mouth-watering culinary experience. In contrast to the standard industry practice of providing the same range of cakes to each diner, Davenports provides a range of different cakes to encourage discussion and interaction. Afternoon teas are also set up on the assumption that each one is likely to be photographed, so it should therefore be a delight to the eye, including at its moment of delivery to the table.

As the owners of Davenports live on-site, all scones are home-baked just prior to the first afternoon tea being served. All scones are served warm, generally towards the end of the meals, but this

Figure 16.1   A server of Davenports
Source: Belinda Janette Davenport

is down to customer preference. Scones are presented on silverware with silver tongs and served with locally made jam and clotted cream as part of a theatrical afternoon tea experience. Attention to detail and choreographed precision all form a part of this quintessentially English cultural experience. The moment is captured by staff on guests' own cameras or smartphones to ensure that they have a good-quality image, not only to facilitate their own remembering but also to share via social media or among friends.

Davenports' business is mostly driven by repeat visits and word-of-mouth recommendations that it is a high-quality experience. Even during the COVID pandemic, Davenports was able to offer a new product, the delivered 'Afternoon Tea @ Home', with equal success.

Analysis

Demographic profile of customers

The demographic profile of Davenports customers changed markedly during the 14-year data-collection window. Initially, local people from within a 10-mile radius used it as a refreshment stop when visiting the on-site farm shop or when passing by on foot or by bicycle. Other visitors included those en route from Scotland to more southerly parts of England using the A49 as an alternative to the motorway. After television coverage in 2013, however, customers increasingly visited from within a 25-mile radius, as well as from further afield in the UK and a few from other countries.

Customers' ages varied greatly in all periods, from young children through to centenarians, but the core visitor group consisted of females aged 35 to 60, with those outside of that demographic often accompanying them. Generally, at a given table, all the customers were on a return visit, or one was a returner who also brought new visitors.

The main ways people said they had discovered Davenports Tea Room were word of mouth, followed by internet searches and website visits, and social media. The tea room's television appearances have been numerous and various, ranging from short news interviews during coverage of a local canal breach to antiques programmes. However, the most-mentioned television appearance was the *BBC Breakfast* news programme in 2013, when Davenports won the last-ever Tea Guilds "Top Tea Place in the UK" award. This was linked to a 100% increase in footfall, as well as to the first marked numbers of visitors from foreign countries, including Japan, the United States and France. Interestingly, one local couple discovered the tea room after seeing it on the news while on holiday in Barbados.

Core product: the English afternoon tea

When Davenports Tea Room first opened, its tea menu was limited and included a well-known British brand, Yorkshire Tea, considered by many to be a typical English tea. However, customers requested other brands of UK breakfast tea and less strong teas. To introduce the tea room's offerings that are not readily available at home, customers were encouraged to choose alternatives based on their strength preferences: for example, Taylor's of Harrogate Tea Room Blend for strong, KwaZulu-Natal Breakfast Tea for medium, and Atkinson's Assam for weak or weak-to-medium. The maltiness of the Assam tea was often cited as a reason for its popularity at weaker strengths. KwaZulu was also popular, in part, due to its rarity or novelty. At the time of writing, Davenports has increased its tea menu to include 50 different loose-leaf teas.

The afternoon tea market built up over time, and although the tea room was open from 10 a.m. to 4 p.m., it became the predominant reason to visit Davenports prior to the COVID-19 pandemic. Due to limited capacity and a commercial decision to only offer pre-booked afternoon teas, it later became the sole reason. As part of the afternoon tea experience, customers were always encouraged to try different teas. They tended to be more receptive to this if they were having a formal afternoon tea, rather than tea as a refreshment, especially if the table consisted entirely of adults. Customers were asked what they typically drank, which was generally breakfast tea brewed from tea bags throughout the day. This is consistent with information published by the UK Tea and Infusions Association (2021). An interesting observation in the tea room was that generations of the same family often had similar tea-strength preferences. For some visitors, even trying an unknown brand of breakfast tea was considered daring. Some also considered it traditional to take breakfast tea with raisin scones and strawberry jam, which is not technically true but seems to have become embedded in British culture.

As the owner's skills improved through both experience and training, customers who agreed to do a spot of tea tasting as part of their afternoon-tea experience would generally take a Darjeeling afternoon tea (or other less familiar varieties, depending on their tastes and appetite for novelty) instead of their normal breakfast fare. Guests would judge the strength of tea based on the colour of the liquor; however, staff explained that, for example, a breakfast tea is a certain colour due to the blend, while a Darjeeling afternoon tea is a pure tea, so its clearer liquor should still have strength of taste. Customers were then more willing to try the Darjeeling. In the majority of cases, they enjoyed Darjeeling, often without milk or sugar, when normally they would take both. Customers were familiar with green tea, and some had tried it because they had heard it was healthy. Almost all considered it to be bitter, which staff explained was due to incorrect brewing. Interestingly, all customers were aware that pouring boiling water on coffee would scald it, but very few knew about the same effect on some types of tea. Hardly any Davenports customers had heard of white tea, and few understood that white, green, red, and black tea were from the same plant. None had thought about the fact that, because tea is a plant, it can be affected by growing conditions and, as a result, they underappreciated the skills of the master blenders.

Encouraging customers to first try teas as tasters without milk led to numerous individuals discovering that they liked tea without milk. Jing Darjeeling, Cheshire Tea's, Chocolate/Mint Rooibos, and Quintessential Tea's White Elixir all proved particularly popular in this context. Generally, non-tea drinkers gave Davenports staff one of two reasons for not liking tea (i.e., the chemical aftertaste of tannin or not liking any hot drinks). In almost all such cases, when encouraged to try a suitable tea, such as a cold infusion green tea, they liked it. Customers whose only preferred hot drink was coffee, in almost all cases, liked Jing Tea's Pineapple and Chamomile, or smoky varieties such as Lapsang Souchong, Russian Caravan or Pu-erh.

Atmosphere and restaurant interior

The analysis of online reviews indicated that the tea room's strong positive attributes included staff knowledge, its core product, and "delicious" food. However, when it came to customers' clear intention to return or recommend, almost equal importance was assigned to the atmosphere. The quality of the experience was not just based on the quality of the tea but on how it was presented as part of the afternoon tea experience. Much as supermarkets use fresh produce at their entrances to send a message about freshness and quality, Davenports ensured that entering customers were greeted by the aroma of freshly baked scones. Moreover, just as high-street shops can manipulate their visitors' shopping speed via music, especially at certain times of the year, Davenports carefully chose music to complement the setting and to enhance the ambience, on the basis that sound "is the forgotten flavour sense", according to Prof. Charles Spence of the University of Oxford Crossmodal Research Laboratory (personal communication, 24 March 2017). Understanding the importance of keywords (e.g., homemade, locally made, free range, freshly baked) and relaying them to customers were also part of the Davenports business strategy to encourage repeat visitation.

The room, its décor and ornaments, background music, the table layout, the food presentation, how the tea was delivered (e.g., warming the cups), cleanliness, staff presentation, and staff knowledge all contributed to the overall experience. These attributes featured strongly in observations and the analysis of online reviews and are consistent with the findings of prior research, such as Lin and Chang's (2020, 1) conclusion that "atmosphere and service performance influence customer well-being, which can positively affect customers' repurchase intentions".

The unique selling point: photography

Davenports' emphasis on photography ensured that kitchen staff and servers prepared every afternoon tea to be as picture-perfect as possible in a real-world situation. The importance of this approach was confirmed during the pandemic when, during a three-week period minimising contact, photographs were taken on Davenports' own camera and forwarded to afternoon tea customers via WhatsApp. If photographs were delayed or not received, Davenports was chased for them. One online reviewer noted that the photos "have prompted many others to promise a visit".

Conclusion

Some features of British culture have become so embedded over generations that breaking through them can be quite difficult – one such barrier is tea meaning only breakfast tea. This case study, however, has demonstrated that it was possible to break through this barrier by utilising another British cultural event, afternoon tea. Tea rooms and afternoon tea are an essential tourism product for a tea-tourism destination. Davenports established that afternoon teas can generate local tourism, both as an add-on to a wide variety of accommodation stays or to non-food activities such as Bird of Prey experiences. At the other end of the spectrum, these events can become international destinations, with some Japanese visitors coming to Cheshire solely for Davenports Tea Room.

The present study's findings make it clear that investment of time by customers and suppliers can allow both these groups to enter into the "world of tea" and the fantastic range of tastes that it offers. Proverbially, however, time is money. To deliver a high-quality experience requires a time investment in training, in understanding the customer through listening and observation, and in ensuring that what is delivered is picture-perfect. In other words, delivering a high-quality experience goes beyond what is on the plate or in the cup, to encompass the whole afternoon tea experience as a product. Whilst a tea room can be considered a mere refreshment stop, this case study has demonstrated that tea-tourism destinations and attractions can bring much more to the tea industry through investment in high-quality customer experiences, and perhaps even encourage some customers to visit other tea-themed tourist attractions by sparking their interest in knowing more about tea.

This study has also shown how Davenports invested in understanding their customers' needs and committed themselves to delivering a high-quality experience that encourages customers to tell others about it and to visit repeatedly. Consistent with previous research (e.g., Haghkhah *et al.* 2011; Kim *et al.* 2012; Lemon and Verhoef 2016; Kuhn *et al.* 2018; Huang *et al.* 2019), this holistic approach is the key to their success. The quality of food and drink, staff knowledge, atmosphere, cleanliness, and empathy for the needs of the customer all play important roles in delivering their afternoon tea experience. Whilst attractions serving tea have been established in the UK since the late 1800s, the concept of tea tourism is in its infancy in the UK (Chen *et al.* 2021). As such, this study of Davenports Tea Room, and especially its tests of visitors' tea-palate preferences, should be invaluable to the ongoing development of tea tourism in the UK and beyond.

Discussion questions

1   How could the findings from this case study be used to improve tea-tourism attractions around the world?
2   Using the MEEEI model, compare the findings from this study with a heritage tea tourism attraction from another culture.
3   Consider the pros and cons of working in a tea room like Davenports.

References

BBC News, 2013. Davenports in Northwich wins Tea Guild best afternoon tea award [online]. Available from: www.bbc.com/news/av/uk-england-merseyside-22123150 [Accessed 3 September 2021].

Boniface, P., 1998. Tourism culture. *Annals of Tourism Research*, 25, 748–750.

Chen, S.H., Huang, J. and Tham, A., 2021. A systematic literature review of coffee and tea tourism. *International Journal of Culture, Tourism and Hospitality Research*, 15(3), 290–311.

Edlinger, P., 2015. *Tourism and photography: an analysis of the impact of photographing on the visitors' emotional experience at cultural spaces and their post-visit behavioural intentions.* Unpublished Master's Thesis, University of Girona.

Haghkhah, A., Nosratpour, M., Ebrahimpour, A. and Hamid, A.B.A., 2011. The impact of service quality on tourism industry. *In: 2nd International Conference on Business and Economic Research (2nd ICBER 2011) Proceeding*, 13–16 March, Langkawi, n.p.

Hansen, K.V., 2014. Development of SERVQUAL and DINESERV for measuring meal experiences in eating establishments. *Scandinavian Journal of Hospitality and Tourism*, 14(2), 116–134.

Huang, F., Huang, J. and Wan, X., 2019. Influence of virtual color on taste: multisensory integration between virtual and real worlds. *Computers in Human Behavior*, 95, 168–174.

Jolliffe, L., ed., 2007. *Tea and tourism: tourists, traditions and transformations*. Clevedon: Channel View Publications.

Kawulich, B.B., 2005. Participant observation as a data collection method [online]. Available from: www.qualitative-research.net/index.php/fqs/article/view/466/996 [Accessed 3 September 2021].

Kim, J.H., Ritchie, J.R.B. and McCormick, B., 2012. Development of a scale to measure tourism experiences. *Journal of Travel Research*, 51(1), 12–25.

Kuhn, V.R., Benetti, A.C., dos Anjos, S.J.G. and Limberger, P.F., 2018. Food services and customer loyalty in the hospitality industry. *Tourism Management Studies*, 14(2), 26–35.

Lemon, K.N. and Verhoef, P.C., 2016. Understanding customer experience throughout the customer journey. *Journal of Marketing*, 80(6), 69–96.

Lin, S.Y. and Chang, C.C., 2020. Tea for well-being: restaurant atmosphere and repurchase intention for hotel afternoon tea services. *Sustainability*, 12(3), 778.

Oh, H. and Kim, K., 2017. Customer satisfaction, service quality, and customer value: years 2000–2015. *International Journal of Contemporary Hospitality Management*, 29(1), 2–29.

Parasuraman, A., Zeithaml, V.A. and Berry, L.L., 1988. SERVQUAL: a multiple-item scale for measuring customers perceptions of service quality. *Journal of Retailing*, 64(1), 12–40.

Seth, N., Deshmukh, S.G. and Vrat, P., 2005. Service quality models: a review. *International Journal of Quality & Reliability Management*, 22(9), 913–949.

Statista, 2016. *Annual per capita consumption worldwide as of 2016, by leading countries* [online]. Hamburg: Statista Research Publication. Available from: www.statista.com/statistics/507950/global-per-capita-tea-consumption-by-country/ [Accessed 3 September 2021].

UK Tea and Infusions Association, 2021. *Bag it up* [online]. London: UK Tea and Infusions Association. Available from: www.tea.co.uk/tea-facts [Accessed 3 September 2021].

17

TEA FACTORY TOURISM EXPERIENCES

Pearl milk tea in Taiwan

Nikki Wu and Li-Hsin Chen

Introduction

The history of tea in Taiwan spans more than 200 years and is deeply rooted in local customs and traditions. Many varieties of tea are produced throughout Taiwan, including oolong tea, Pouchong tea, green tea, and black tea. The popular drink is associated with entertaining guests and gift-giving moments. Taiwan's tea culture is often referred to as "tea art," and many tea art houses have been established across the country (Huaxia News 2021). In these establishments, people learn about the tea-making process and then experience the culture of Taiwanese tea. The development of tea culture has also prompted the emergence of chain tea shops throughout Taiwan. These tea shops offer a variety of popular specialty tea drinks, particularly pearl milk tea, which originated in Taiwan. Pearl milk tea is made by combining tea (black, green, or oolong), milk, sugar, and tapioca balls. It is also known as *bubble tea, boba milk tea*, and *tapioca tea*. The popularity of pearl milk tea has also caused a dispute between two beverage industries in Taiwan over who created the drink. After ten years of litigation, the court offered a peaceful result, ruling that the drink would not be patented.

In addition to Taiwan, pearl milk tea or bubble tea has recently become popular around the world. Thus, Taiwanese chain tea shops selling this drink, such as Hanlin Teahouse, Chatime, and Gong Cha, have sprung up in many countries. Apart from its economic value, pearl milk tea is also part of the Taiwanese identity. People's love for the drink goes beyond taste and texture; the combination of traditional tea and tapioca balls evokes nostalgic Taiwanese culture. The success of pearl milk tea on the international stage has showcased Taiwan's tea culture, and the drink has become an international symbol of Taiwan (Wong 2020).

Existing studies on pearl milk tea generally explore the factors affecting consumers' purchasing decisions (Iswara and Rahadi 2021; Ong *et al.* 2021), the reasons for consumers' repurchasing habits (Lei and Lei 2020), the health problems associated with the beverage (Min *et al.* 2017), and the meaning of it (Lin and Tzeng 2010). However, few studies have examined pearl milk tea within the context of tourism, despite the drink's role in shaping Taiwan's reputation as a tourist destination and enhancing the local economy (Gupta *et al.* 2020; Bohne 2021). This study fills the gap by investigating the tourism experience at a pearl milk tea tourism factory in Taiwan.

Furthermore, unlike other analytic models of tourism that only focus on the emotional or cognitive aspects of the experience (Scott and Le 2017), this study uses the orchestra model of tourism experience (Pearce 2011) to explore tourists' sensory, affective, cognitive, behavioural, and relational experiences in tea tourism. It is expected that the findings will offer valuable suggestions for the

DOI: 10.4324/9781003197041-21

management and planning of tea tourism. Additionally, this study notes the need for future research on the links between tea and tourism in Taiwan.

Literature

Tourism experience and experiential marketing

Tourism experience refers to "an individual's subjective evaluation and undergoing of events related to his/her tourist activities" (Tung and Ritchie 2011, 1369). Typically, the tourism experience can be divided into three stages: before, during, and after travel (Scott and Le 2017). Furthermore, according to Godovykh and Tasci (2020), it includes cognitive, affective, sensory, and behavioural components (i.e., people's behaviour), and is affected by consumers, brands, and situations. Therefore, using brands and places to create a unique tourism experience and market it to consumers becomes essential for tourism planners. As such, experiential marketing has recently become a popular research topic in tourism.

Experiential marketing is a strategy to attract consumers to obtain a unique experience and be loyal to the brand and destination. Schmitt's (1999) strategic experiential model is the most widely used framework among all relevant theories and models. It illustrates the five aspects of experiential marketing – sense, feel, think, act, and relate – as well as how managers can use marketing techniques to stimulate consumers' five senses, inner emotions, problem-solving and cognition, behaviour and interaction, and connection with the brand community. Regarding the time dimension of experiential marketing, Le *et al.* (2019) found that the marketing messages consumed by visitors before travelling significantly influence their travel behaviour at the destination. Although the influence of experiential marketing for visitors during and after travel has been established, Kim and Jang (2016) pointed out that the relationships among tourism experiences, experiential marketing, and tourists' willingness to revisit and word-of-mouth recommendations were not clear. Therefore, this study aims to fill the research gap and investigate the links between the critical factors of tourism experiences and experiential marketing and their impacts on tourists' behaviours in the context of tea tourism.

The orchestra model of tourism experience

Many theories conceptualising the tourism experience often fail to fully capture the interrelated dimensions of tourism (Aho 2001; Cutler and Carmichael 2010; Scott and Le 2017). However, the orchestra model of tourism experience illustrates how sensory, affective, cognitive, behavioural and relational elements are interrelated in the context of tourism (Pearce 2011), which is valuable for designing and evaluating tourism experiences (Pearce and Zare 2017). It has been successfully used in previous empirical research (Pharino *et al.* 2018). Furthermore, it parallels Schmitt's (1999) strategic experiential model regarding the diverse aspects of human experiences. Therefore, it is selected as the guiding theory of this study to provide a thorough analysis of tea tourists' experiences at a tourism factory. A detailed overview of the orchestra model and its components is shown in Figure 17.1.

Tourism factories

According to Taiwan's Ministry of Economic Affairs (2003), a tourism factory is a place that provides tourism products while also having educational and cultural value. Previous research has explored different aspects of Taiwan's tourism factories, such as tourist satisfaction (Lee 2015), willingness to

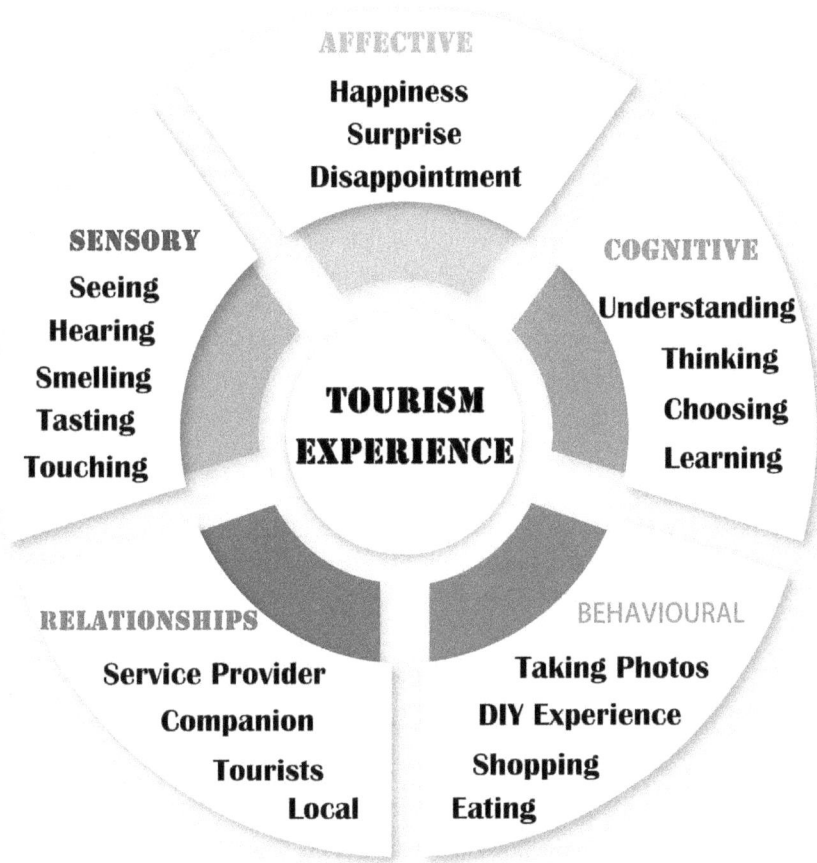

Figure 17.1 The orchestra model of tourism experience

Source: Authors

revisit (Lin *et al.* 2018), cultural protection and regeneration of tourist factories (Weng and Yang 2016), and customer relationship management (Chung and Chen 2016). Tourism factories' relationships with experiential marketing, innovative services, and satisfaction have also been discussed (Yeh *et al.* 2019). However, existing research on Taiwan's tourism factories has focused on health, paper, and textile factories, not tea. Since tea tourism can offer experiences that are educational and authentic, unique and unforgettable (Seyitoğlu and Alphan 2021), tea tourism factories should also be investigated to better understand their value to Taiwan and other tourism destinations. Therefore, this study explores the tourism experience at the Kili Bay Pearl Milk Tea Tourism Factory in Taiwan and hopes to provide suggestions for future improvement.

Methods

This research uses a qualitative approach to explore consumers' experiences in the tourism factory and improve current marketing strategies. The research design included on-site observations and an analysis of online reviews of Kili Bay Pearl Milk Tea Tourism Factory. While online reviews have long been a useful data source in the field of hospitality and tourism (Xiang and Gretzel 2010; Guo *et al.* 2017), we used observational data to compare and verify the data obtained online (Jamshed 2014). The on-site observation was conducted during August 2021, the peak season for the tourism attraction. We used field notes and photos to record the physical environment, interactions among visitors and service providers, and the tourists' behaviours. In short, the five major aspects of tourism experiences were documented: sensory, affective, cognitive, behavioural, and relationships. Online reviews of the tourism factory were collected from Google Reviews and TripAdvisor to triangulate the analysis of observational data. A total of 450 reviews, posted between 2017 and 2021, were obtained. Using the orchestra model as the conceptual framework, the data were thematically analysed by generating codes, determining the themes, rechecking the themes, and presenting the results (Braun and Clarke 2006).

Results

Kili Bay Pearl Milk Tea Tourism Factory includes shopping areas, western restaurants, a pearl milk tea cultural centre, DIY classrooms, and playing areas. Consumers' tourism experiences at the tourism factory were divided into five major elements: sensory, affective, cognitive, behavioural, and relationship. The following sections outline each of these in turn.

Sensory

This category includes seeing, hearing, smelling, tasting, and touching in the factory.

Seeing

Most consumers commented on the novel appearance of the building, which was designed using an environmentally friendly concept. Both the venue and the attached restaurant are spacious, comfortable, and clean. Visitors can pose for photos with the cutely designed, pearl-shaped mascot and a huge light bulb milk tea at the entrance. Pearl-shaped dolls are also available for purchase. In the cultural centre of the factory, the organiser introduces the history, raw materials, and production process of pearl milk tea through text billboards and displays showcasing the physical machinery. Some online comments describe the "seeing" aspect of sensory experience, for example, "There are roselle planting outside the restaurant area, it's really beautiful!" and "There is a light bulb pearl milk tea as a special product at the entrance. It is so interesting and eye-catching."

Hearing

Consumers learn about the history and manufacturing process of pearl milk tea during a self-guided tour. Visitors can scan QR codes to listen to audio guides and commentary. Overall, visitors can hear "many people talk and laugh. This place has a good vibe and energy!" (Online review). However, lack of interaction between tour guides and visitors may influence the tourism experience, as one online comment revealed: "although there is no guide inside, tourists can use the mobile phone to scan the QR codes to listen to the audio guide."

Smelling

In the cultural centre of the factory, the visitors get some smell-based experiences, such as smelling the raw materials for pearl milk tea in jars. Tourists said, "it is great that some of the spices and syrups in jars can be opened to smell". Consumers also have a good experience in the gift shop because "there are many tea bags of different flavours in the commodity area, and they smell very good".

Tasting

Taste is obviously an important part of the sensory experience in this tourism factory. Consumers can make their own pearl milk tea to drink. Visitors say the milk tea sold here is moderately sweet and the pearls are delicious. Consumers who participated in the DIY experience especially enjoyed the pearl milk tea they made themselves. One of the online comments disclosed, "I love the DIY activities. I feel this is the best pearl milk tea because I make it by myself." However, the restaurant's meals are described as ordinary, with low-quality food. One visitor commented, "The taste of the meals in the restaurant are average."

Touching

Consumers' touch senses are stimulated by the hands-on activities, including the experience of making pearls and pearl milk tea. Most comments related to the touching aspect of tourism experiences are positive, such as "Come here to make the DIY of pearl milk tea. It is the first time that I touch the raw materials of the milk tea. It has so much fun playing with it"; "The pearl milk tea made today is very successful to create an unforgettable experience for me"; and "It's a good experience to actually make a delicious cup of tea and DIY hand-made pearls."

Affective

This element analyses the consumers' emotions during their trips. Consumers experienced an array of emotions, such as happiness, accomplishment, disappointment, surprise, and boredom. Typically, families with children said that the children's play area generates a lot of fun. One comment mentioned: "The DIY activity is very fulfilling. I would love to do it again." Furthermore, the DIY activities commonly stimulated happiness and surprise. Tourists were surprised by the variety of pearls; the cultural centre introduces pearls made from fruits, and the sales department offers pearl drinks in a variety of colours. This arrangement triggers a feeling of surprise. One comment revealed, "Unexpectedly, there are so many kinds of pearls for beverages!"

However, the DIY experience also disappointed some consumers. Employees over-promoted the products for sale and there were not enough ingredients for everyone to make the milk tea. Some adult consumers were also disappointed by the simplicity of the procedure. They felt slightly bored because the theme of the exhibition was not clear and it only took a short time to visit. Some visitors gave negative comments about the depth of this DIY experience, such as the following:

> The course only teaches you how to make black tea with creamer and sugar. The whole course is promoting their products. The DIY activity is only suitable for children under 5 years old. Our children are already junior high school students, and they are very disappointed and feel a waste of time.

Others get bored because of the exhibition's poor design and leave comments like, "It's boring, the theme of the exhibition is not clear enough."

Cognitive

The cognitive component analyses consumers' perceptions of the on-site experiences, the moments that made them stop and think about their choices, and the moments of learning.

Understanding

Consumers describe the exterior of the tourism factory as fresh and environmentally friendly, and the tour route through the pearl milk tea cultural centre to be clear and easy for self-guiding. Consumers also thought the diagrams were clearly illustrated and that the pearl milk tea culture's nostalgically decorated scenes were interesting. One comment indicated, "Interior milk culture and nostalgic decoration are interesting!" However, some felt the exhibitions in the cultural centre were too simple and mundane. Some visitors said, "The knowledge of the exhibition in the hall is not in-depth enough." Furthermore, although some visitors were interested in the DIY activity procedure and the uniqueness of making their own drink, they thought the experiential activities were too expensive and did not produce value.

Regarding other facilities, consumers found the restaurant affordable and well-planned and the gift shop selling pearl milk tea products to be special. The products for sale include pearls, various teas, and coffees. A guest mentioned, "The small farm restaurant is attentive, and the price is very affordable. It is well worthy to be visited!" The tourist factory sells pearl milk tea in a creative light bulb–shaped glass which consumers say adds to the novelty. The place has a joyful atmosphere. One visitor commented, "I think the light bulb milk tea is cool and novel." However, some comments also revealed that the shop sometimes felt incoherent, as classic Taiwanese sweets and local specialities were displayed next to the pearl milk tea products. This part should be further improved.

Thinking and learning

Thinking represents consumers' evaluations and opinions of the on-site experience. Consumers thought about the tourism factory experience, mainly when in visits to the cultural centre and during DIY experiences. The experience also let most consumers believe the tourism factory was suitable for family travel. Furthermore, the on-site observations found that the cultural centre displays its corporate products, information on cross-industry cooperation, and provides information for people who want to start a pearl milk tea, handmade coffee, or snow ice business. However, the cultural centre is overly commercial and the information is generally presented in Chinese, with only a few English translations located in the corners of the exhibition hall. The entire DIY activity is also conducted in Chinese, which is not ideal for foreign consumers, as tour guides cannot keep up with the translation. One visitor mentioned: "The cultural hall is almost all in Chinese. It seems that this place is only suitable for locals. They can at least have English in all descriptions." Another tourist said: "The entire DIY activity is conducted in the local dialect. Although we have a Taiwanese tour guide to help with the translation, there is no pause during the course for the tour guide to translate."

Regarding the dining experience, the meals in the restaurants are sufficient and the prices are reasonable. However, the restaurant menu does not offer novel meals related to pearls and tea. There is also a lack of staff in the tourism factory, so tourists are unsatisfied with the long waiting time.

Behavioural

Many behavioural components of tourism experiences can be observed in this factory. For example, consumers take many photos of the mascot and bulb-shaped pearl milk tea, participate in hands-on experiences, purchase souvenirs, dine, and visit the pearl milk tea cultural centre. Some visitors mentioned: "it's a place where you can take pictures crazy." Visitors also said that they "come here to make DIY of pearl milk tea. It's a very interesting experience."

Relationship

This factory offers the consumers many opportunities to build relationships with service providers, companions, other tourists, and locals. From both the observational data and online reviews, we found that the service staff were seen as kind, enthusiastic, patient problem-solvers. Many consumers mentioned the "very warm hospitality by the service provider." The staff of the DIY activity was professional, and consumers considered it a "professional bubble tea teaching." The overall interaction between consumers and service providers was decent; however, there were not enough service staff members at the snack bar, so the service was slow. As some visitors mentioned, "There are really few service staff inside, but the service attitude is good."

Discussion of the results

Among the five components of the orchestra model, the most salient dimension of the tourism experience at Kili Bay Pearl Milk Tea Tourism Factory is the sensory experience. In addition to the unique, environmentally friendly architectural appearance that generates an enjoyable "seeing" experience, the self-service audio guide and jars of pearl milk tea ingredients stimulate consumers' hearing and smelling experience. The visitors are also offered many opportunities to stimulate their tasting and touching experiences during the DIY milk tea activity. Furthermore, based on the analysis, we found that the interactive activities at this factory can generate various types of tourism experiences and are the most important element to enhance brand association and brand loyalty (Chow *et al.* 2017). Some research results are worthy of discussion.

Firstly, regarding the cognitive component of the experience, the text boards and exhibits that introduce the culture of pearl milk tea clearly convey their messages. However, consumers believed the exhibition lacked sufficient depth and interactivity. Consumers also negatively perceived the lack of rich and diverse introductions to tea in the cultural centre. Additionally, the Chinese-language DIY activities and exhibitions inconvenienced foreign consumers. These weaknesses should be addressed in the future. Furthermore, we suggest that the cultural centre should add more interactive experiences for consumers, such as a virtual game where one makes milk tea. Introductory films about tea culture can also be used to enrich the transfer of knowledge and enhance the cognitive experience. The literature on cultural attractions has also confirmed that the atmosphere, layout, and design related to consumers' word of mouth are important to managers. This includes colour, lighting and sound effects to enhance the atmosphere of the environment, interactive and multilingual displays, and appropriate displays and passage spaces (Bonn *et al.* 2007).

Secondly, visitors experience a wealth of emotions at this tea tourism factory, mainly through DIY experience and the cultural centre. The organiser conveys its uniqueness through the DIY activity, which stimulates the tourists' positive emotions and emphasises the importance of the brand (Lee and Chen 2017). The DIY experience was considered easy to understand and interesting. However, some consumers believed that there were not enough ingredients available and that the activity was not suitable for adults. This part of the activity may produce negative feelings and should be improved to maintain the service quality. Furthermore, since most people visit tourism factories with their families, this factory should create more parent-child interaction activities to increase the satisfaction level of tourists with families (Yeh *et al.* 2019).

Besides the interactive activities, the tourism factory also connects consumers with a service provider, which produces a positive relationship experience. Consumers think highly of the staff but note that many venues were understaffed. Since customer satisfaction is reliant on service providers (Tsai *et al.* 2012), more should be done to correct the problems with labour assignments. In addition, Bonn *et al.* (2007) found that although staff on-site have little effect on consumers' word-of-mouth recommendations, their attitude towards consumers may significantly influence customers' willingness to revisit the place. This study also found that consumers write positive comments about professional and courteous employees.

Finally, the factory has little connection with the local area and local culture. While local products and desserts are sold at the gift shop, the factory does not introduce the host community well.

According to Chien *et al.* (2018), a consistent image and co-production experience with tourists and the local area is required for the successful development of a tourist destination. The combination of souvenirs and local culture helps tourists identify with the business and, ultimately, spend more at the venue. Therefore, the factory should design distinctive souvenirs and related products highlighting local tea culture to improve their brand image (Yeh *et al.* 2019).

Conclusion

Despite pearl milk tea's international popularity, few studies have investigated the drink's role in tourism. This research aims to fill the gap and investigates the tourism experience of a Taiwan pearl milk tea tourism factory. The orchestra model was used to analyse the five dimensions of tourism experiences. The results indicated that tourists' five senses, emotions, and cognition can be stimulated by the cultural centre and DIY experiences. However, the content and service quality of the interactive activities did not satisfy all age groups. Additionally, the tourism factory is not well connected to the local area. Kili Bay Pearl Milk Tea Tourism Factory has the potential to promote both the development of local tourism and Taiwan's tea culture. Site managers should include more Taiwanese tea culture in the factory. They should also employ diversified marketing methods, supplemented by interactivity and multi-language services to enrich the consumer experience. This could include virtual games and videos or incorporating pearl milk tea into regional cuisines such as ox tongue–shaped pastry with pearls, or pearl milk tea blancmange roll. Scholars should continue to explore the intersection of tea and tourism, including how to attract more tourists to tea tourism factories by improving the on-site tourism experiences. Furthermore, tourism managers should design experiences that address the five major aspects of tourism to help tourists have memorable experiences. Although this study provides the industry with an understanding of experiences at a tea tourism factory and recommends improvements, it only focuses on one site. Therefore, for the purpose of comparison, future researchers should continue to use the orchestra model to study more tea tourism sites.

Discussion questions

1 Do you know of any other tea tourism factories? Can you describe the tourism experience offered in these factories? What aspect of the factory interests you the most?
2 Could you apply the orchestra model to another tourism factory-type facility and compare it with this case? What are the similarities and differences?
3 What experiential marketing strategies do you think the Kili Bay Pearl Milk Tea Tourism Factory should pursue in the future?

References

Aho, S.K., 2001. Towards a general theory of touristic experiences: modelling experience process in tourism. *Tourism Review*, 56(1/4), 33–37.
Bohne, H., 2021. Uniqueness of tea traditions and impacts on tourism: the East Frisian tea culture. *International Journal of Culture, Tourism and Hospitality Research*, 15(3), 371–383.
Bonn, M.A., Joseph-Mathews, S.M., Dai, M., Hayes, S. and Cave, J., 2007. Heritage/cultural attraction atmospherics: creating the right environment for the heritage/cultural visitor. *Journal of Travel Research*, 45(3), 345–354.
Braun, V. and Clarke, V., 2006. Using thematic analysis in psychology. *Qualitative Research in Psychology*, 3(2), 77–101.
Chien, S.H., Wu, J.J. and Huang, C.Y., 2018. "We made, we trust": coproduction and image congruence in the food-tourism factories. *Asia Pacific Management Review*, 23(4), 310–317.

Chow, H.W., Ling, G.J., Yen, I.Y. and Hwang, K.P., 2017. Building brand equity through industrial tourism. *Asia Pacific Management Review*, 22(2), 70–79.

Chung, Y.C. and Chen, S.J., 2016. Study on customer relationship management activities in Taiwan tourism factories. *Total Quality Management & Business Excellence*, 27(5–6), 581–594.

Cutler, S.Q. and Carmichael, B.A., 2010. The dimensions of the tourist experience. *In*: Morgan, M., Lugosi, P. and Ritchie, J.R.B., eds. *The tourism and leisure experience: consumer and managerial perspectives*. Bristol: Channel View Publications, 3–26.

Godovykh, M. and Tasci, A.D., 2020. Customer experience in tourism: a review of definitions, components, and measurements. *Tourism Management Perspectives*, 35, 100694.

Guo, Y., Barnes, S.J. and Jia, Q., 2017. Mining meaning from online ratings and reviews: tourist satisfaction analysis using latent Dirichlet allocation. *Tourism Management*, 59, 467–483.

Gupta, V., Sajnani, M., Dixit, S.K. and Khanna, K., 2020. Foreign tourist's tea preferences and relevance to destination attraction in India. *Tourism Recreation Research*, 1–15.

Huaxia News, 2021. Taiwan tea culture history [online]. *Huaxia News*. Available from: www.a14913.com/archives/352186 [Accessed 7 September 2021].

Iswara, C. and Rahadi, R.A., 2021. Consumer preferences and pricing model on bubble tea purchase in Bandung, Indonesia. *Advanced International Journal of Business, Entrepreneurship and SMEs*, 3(7), 111–119.

Jamshed, S., 2014. Qualitative research method-interviewing and observation. *Journal of Basic and Clinical Pharmacy*, 5(4), 87.

Kim, J.H. and Jang, S., 2016. Memory retrieval of cultural event experiences: examining internal and external influences. *Journal of Travel Research*, 55(3), 322–339.

Le, D., Scott, N. and Lohmann, G., 2019. Applying experiential marketing in selling tourism dreams. *Journal of Travel & Tourism Marketing*, 36(2), 220–235.

Lee, C.F., 2015. Tourist satisfaction with factory tour experience. *International Journal of Culture, Tourism and Hospitality Research*, 9(3), 261–277.

Lee, Y.J. and Chen, C.I., 2017. Application of analytic hierarchy process to the tourism factory DIY experiential value evaluation. *In*: *Proceedings of the 2017 international conference on organizational innovation (ICOI 2017)*. Weihai: Atlantis Press, 213–220.

Lei, S. and Lei, S., 2020. Repurchase behavior of college students in boba tea shops: a review of literature. *College Student Journal*, 53(4), 465–473.

Lin, C.Y. and Tzeng, H.Y., 2010. Research on local food globalization in Taiwan: the meaning of pearl milk tea among college students. *Cross-cultural Studies*, 1(4), 37–61.

Lin, Y.C., Lee, Y.H., Su, J.M. and Hsieh, L.Y., 2018. Relationships among service quality, experiential marketing, and the revisit intention of visitors to tourism factories. *International Journal of Economics and Research*, 9(2), 22–37.

Min, J.E., Green, D.B. and Kim, L., 2017. Calories and sugars in boba milk tea: implications for obesity risk in Asian Pacific Islanders. *Food Science & Nutrition*, 5(1), 38–45.

Ministry of Economic Affairs, 2003. *Passing local characteristics, local industry transformation*. Taipei: The Industrial Development Bureau of the Ministry of Economic Affairs.

Ong, A.K.S., Prasetyo, Y.T., Libiran, M., Lontoc, Y.M.A., Lunaria, J.A.V., Manalo, A.M., Miraja, B.A., Young, M.N., Chuenyindee, T., Persada, S.F. and Perwira Redi, A.A.N., 2021. Consumer preference analysis on attributes of milk tea: a conjoint analysis approach. *Foods*, 10(6), 1382.

Pearce, P.L., 2011. *Tourist behaviour and the contemporary world*. Bristol: Channel View Publications.

Pearce, P.L. and Zare, S., 2017. The orchestra model as the basis for teaching tourism experience design. *Journal of Hospitality and Tourism Management*, 30, 55–64.

Pharino, C., Pearce, P. and Pryce, J., 2018. Paranormal tourism: assessing tourists' onsite experiences. *Tourism Management Perspectives*, 28, 20–28.

Schmitt, B., 1999. Experiential marketing. *Journal of Marketing Management*, 15(1–3), 53–67.

Scott, N. and Le, D., 2017. Tourism experience: a review. *In*: Scott, N., Gao, J. and JianYu, M., eds. *Visitor experience design*. Wallingford: CABI, 30–49.

Seyitoğlu, F. and Alphan, E., 2021. Gastronomy tourism through tea and coffee: travellers' museum experience. *International Journal of Culture, Tourism and Hospitality Research*, 15(3), 413–427.

Tsai, C.H., Peng, Y.J. and Wu, H.H., 2012. Evaluating service process satisfaction of a tourism factory-using brands' health museum as an example. *In*: *2012 6th international conference on new trends in information science, service science and data mining (ISSDM2012)*. Taipei: Institute of Electrical and Electronics Engineers, 244–247.

Tung, V.W.S. and Ritchie, J.B., 2011. Exploring the essence of memorable tourism experiences. *Annals of Tourism Research*, 38(4), 1367–1386.

Weng, H.Y. and Yang, C.H., 2016. Culture conservation and regeneration of traditional industries derived by tourism factory – case study of Kwong Xi paper factory in Taiwan. *International Journal of Humanities, Arts and Social Sciences*, 2(5), 172–180.

Wong, M.H., 2020. The rise of bubble tea, one of Taiwan's most beloved beverages [online]. *CNN*. Available from: https://edition.cnn.com/travel/article/taiwan-bubble-tea-origins/index.html?utm_term=link&utm_medium=social&utm_content=2020-04-30T03%3A15%3A08&utm_source=twCNNi [Accessed 8 September 2021].

Xiang, Z. and Gretzel, U., 2010. Role of social media in online travel information search. *Tourism Management*, 31(2), 179–188.

Yeh, T.M., Chen, S.H. and Chen, T.F., 2019. The relationships among experiential marketing, service innovation, and customer satisfaction – a case study of tourism factories in Taiwan. *Sustainability*, 11(4), 1041.

18

FACILITATING TEA STORIES ON INSTAGRAM DURING THE COVID-19 PANDEMIC

Joan Pan and Wayne Buente

Introduction

In 2018, the co-founder and CEO of Instagram, Kevin Systrom, announced that the popular social media application, 'a place to connect with the people who inspire, educate and entertain you every day', reached one billion monthly active users (Systrom 2018). The hospitality and tourism industry has noticed Instagram's growing prominence. In recent years, scholars have explored consumer behaviour in travel and tourism and their use of the internet and social media platforms. Such studies show how consumer use of social media and mobile devices to search for information on travel destinations, seek recommendations, and share experiences reshapes the tourism experience (Choe *et al.* 2017; Wong *et al.* 2020). However, there is little academic research on destination and brand use of social media and even less research on tea businesses.

Tea tourism is a journey through which tourists can experience tea-related activities at a destination or host community. A widely accepted definition of tea tourism is travel 'motivated by an interest in the history, traditions and consumption of tea' (Jolliffe 2007, 9). Activities at tea destinations include visits to sites of tea production, processing, and exhibition (Jolliffe 2007). The public health guidelines around COVID-19, beginning in March 2020, reduced in-person contact and travel. The pandemic's effects increased the use of digital media in all aspects of society, altering the way tea retailers and tour operators conducted business. Given the pandemic's impacts on travel and global supply chains, this chapter explores how two tea brands adapted to online business models during the pandemic. This chapter investigates how two tea businesses used Instagram to facilitate storytelling and marketing for pre-, during, and post-tour experiences between April 2020 and April 2021. In doing so, it contributes to conversations about how tea tourism is mediated by technology and the potential for the development and marketing of experiential tea tourism using social media.

Literature review

Instagram is best defined as a social media platform with a visual focus serving as a conduit of communication to visual social media cultures (Leaver *et al.* 2020). In this regard, this chapter examines how tea brands develop their marketing and cultural practices on Instagram. The first and second sections of the literature review examine Instagram as a platform, its visual aesthetics, and how it serves as a marketplace for attention and commerce. The last section provides an overview of social media during the three phases of the travel planning process and tourism experience

DOI: 10.4324/9781003197041-22

(i.e., pre-, during, and post-tour) from the perspective of travellers and considers how businesses may collaborate with consumers.

Instagram

Instagram began as a free iPhone application that allowed instant photography within a square frame with the option to apply filters for stylistic effect. In addition, Instagram allowed networked connection through photography, as other users could like and comment on the uploaded images. As a result, the mobile app could improve the appearance of mobile phone pictures, facilitate sharing across multiple platforms, and allow quick and easy upload of images (Laestadius 2016).

The visual social media platform has developed over the years, adding more features including video, live streaming, and ephemeral content. For example, Instagram introduced the Stories feature in 2016, as social media was becoming more ephemeral. Stories posts expired in 24 hours, limiting the number of times content could be viewed. Highlights were added a year later, which allowed users to curate their Stories and group them in a section that resides on the profile, under the account's bio (Vázquez-Herrero *et al.* 2019).

Instagram, aesthetics, and attention

The visual nature of Instagram, centred around the productive activity of producing, circulating, and attending to images, encourages affective experiences that help to channel attention (Carah and Shaul 2016). Scholars have noted how Instagram is a unique platform for storytelling and imagination as it uses creative filters, captions, and emojis (Hurley 2019). In addition, aesthetics and communicative styles can signal membership within an Instagram community of like-minded individuals. For social media influencers who seek to monetise their content and build authenticity, their social connections can range from the performative to the fantastical. Thus, brands seek out creative influencers to act as both production and distribution channels – to produce compelling visual content and also distribute this content to their own network of followers (Carter 2016).

Instagram functions as a marketplace for attention and commerce. Pointing, tapping, and scrolling on a never-ending image feed encourages intense competition for limited attention resources (Cotter 2019). Opaque algorithms hide key elements of visibility and engagement for Instagram users. Accordingly, brands partner with key influencers and cultural intermediaries to calibrate attention and affect others, often relying on agreed brand templates (Carah and Shaul 2016).

Social media and the phases of tourism experience

Some scholars approach the tourism experience in a multistage process, consisting of pre-trip (anticipatory), during trip (experiential), and post-trip (reflective) consumption activities (Gretzel *et al.* 2006; Minazzi 2015). During the travel planning phase, consumers use social media to search for information and recommendations from their friends and social network (Buhalis and Foerste 2015; Minazzi 2015). In this pre-trip phase, the dreaming and collecting of information for a future trip can influence the meanings and interpretations of a tourism experience. During the trip, travellers use their social media to share their activities, which contributes positively to the tourism experience (Kim and Tussyadiah 2013). Finally, social media functions as a space for sharing post-trip experiences (Kim and Fesenmaier 2017), with storytelling being an important part of the process of deriving meaning from one's travel and tour experiences (Gretzel *et al.* 2006).

Improved wireless connectivity, mobile devices, and the co-creation of value and meanings by tourists have changed the marketing practices of businesses and brands. Companies and DMOs have adapted their branding strategy to integrate consumers' experiences and informal conversations into their social media narratives (Yoo and Lee 2017). Besides establishing and fostering personal connections with their social media viewers, Lund *et al.* (2018) recommend that brands collaborate with consumers in the telling of stories and narratives.

Methods

Data collection

Consistent with the preferred methodology in tourism research and teaching in recent decades, the authors used a case study approach in this study. The case study methodology is well suited for exploring 'how' research questions and for topics that are underexplored (Beeton 2005). Multiple methods inform the case study analyses, including (1) content analysis of Instagram posts and (2) semi-structured interviews. Cases were selected through purposive and convenience sampling. The criteria for inclusion are cases which exhibit specific characteristics, allow study of the businesses' Instagram use during the pandemic; being convenient through one author's online participation.

Using an application program interface, the authors collected Instagram posts (image and caption) and metadata from the two tea businesses' Instagram accounts between April 2020 and April 2021, during the COVID-19 pandemic ($n = 211$). Instagram videos over 5 minutes in length and IGTV posts were excluded ($nr = 29$) due to time constraints. The authors also conducted online semi-structured interviews ($n = 5$) on Instagram marketing practices with the business owners and staff from March to July 2021.

Data analysis

Content analysis involves the discovery of patterns and categories in data (Luborsky 1994) and has been increasingly applied by tourism research on advertising and images (Camprubí and Coromina 2016). This study's codebook was adapted from Yoo and Lee's (2017) study on hotel groups' Facebook communication and marketing. The following four topics were included in the codebook: (1) Brand News and Information; (2) Marketing and Promotion; (3) Education; and (4) Conversation and Community Outreach. Interviews with the founders and staff of the two case studies further informed and clarified the codes for this study. After codebook creation, the authors determined the parameters for data set inclusion. Given the sample size, the authors relied on a complete consensus model for coding, established first through an analysis of a random set of Instagram posts. The authors independently coded the sample, and when there was disagreement, the responses were discussed until a consensus was reached. Through an iterative process of coding and discussion, the authors reached a consensus on all descriptions and codes for each Instagram post. We then calculated engagement for each Instagram post using the industry standard of likes + comments. Engagement is important 'because each interaction with an image generates data that makes the image available in wider flows of content on the platform' (Carah and Shaul 2016, 75).

Case study context

The first case study is Grass People Tree, a retail and wholesale tea brand based in London, UK, and Guizhou, China. The founder and CEO of Grass People Tree was born and raised in Guizhou in southwest China, where ancient and indigenous tea trees grow freely. Prior to the pandemic, tea-related travel to Guizhou with Grass People Tree was typically arranged for tea buyers who had an interest in direct sourcing, not consumer tours. Their consumer services were held in-person at worksites and various venues in the UK and through online retail. When the national lockdown was imposed in late March 2020, the CEO of Grass People Tree was forced to adapt her business model by exploring and strengthening online avenues to reach current and potential customers. As part of their 2020 online adaptation, Grass People Tree hired a tea consultant and writer to help with their social media and community engagement.

The second case study is Nagasaki Ikedoki Tea (henceforth Ikedoki), a small tea tour operation in Nagasaki, Japan, that launched in February 2020 (Nagasaki Ikedoki Tea 2000). Nagasaki is located on the western side of the island of Kyushu in southern Japan. Its geographic location, topography, and several islands makes this prefecture an important site of contact for the movement of people, goods, and ideas. Ikedoki's founder has many years of experience with a Japanese Tea Association and in managing another tea brand that provides guided tours, mostly to foreign visitors, on tea farming and production in another prefecture. Ikedoki was intended to attract foreign travellers based on Nagasaki's historical significance with tea and the atomic bombing in more recent history. The founder hired a foreign national to assist with the project, but this individual could not complete their move due to pandemic-related restrictions. Shortly after the project launched, the Japanese government enforced strict travel regulations that barred most foreign nationals from entering Japan. This had an immediate effect on Ikedoki. The original plan was for the new staff member to work on-site, with part of her work being customer communications (e.g., newsletter, website, social media content), as well as assisting with the tours and tea tastings. Although she could not move to Japan during this time, she worked remotely for Ikedoki from the Netherlands. She is a serious tea enthusiast who has a personal Instagram account of tea-focused content, attends tea study programmes, has visited tea areas, and has volunteered at tea farms.

Both businesses incorporate experiential aspects of tea tourism into their Instagram posts to ensure that the tour experience begins prior to physical travel. In both cases, their use of Instagram shifted due to the COVID-19 pandemic and its consequences on in-person contact and travel.

Case study 18.1   Grass People Tree

Central to Grass People Tree's work is sharing the history and story of the mountain region in Guizhou through local tea, tea culture, and associated practices. Their Instagram feed contains visuals and stories on the culture, people, tea trees, landscape, and natural views of the destination, such as the winding mountain roads that connect the Yi People and the yellow tea they prepare for the world. Grass People Tree also attempts to build community conversation on their Instagram by asking invitational and reflective questions, such as 'Join me with tea and let's share our stories. Because without sharing, how do we understand ourselves truly and where we will be?' (Grass People Tree 2020a). They also used Instagram's long-form, immersive video feature, IGTV. As part of their business adaptation during the COVID-19 pandemic, they hired a community engagement staff member in August 2020. This individual had already been working in the tea industry in various capacities and has a dedicated tea blog. In his first months with Grass People Tree, he experimented with different visuals and formats on Instagram. He also coordinated a posting schedule for the brand's social media accounts. Their Instagram practices evolved as the brand built a routine and momentum through their virtual tea house in fall 2020. This was reflected by an increase in post frequency and engagement in October 2020, peaking in March 2021 (see Figure 18.1). Although the IGTV posts were excluded from the study due to time constraints, this was their preferred mode of communicating and sharing participant reflections. IGTV allows for longer live video streams and replay, which is well suited for live tea-tasting sessions and conversations with their community members.

Of the eligible Instagram posts analysed (N = 119), Marketing and Promotional content was the most frequently (47.90%) posted during the study period; however, it generated the lowest amount of engagement (see Table 18.1). The next most frequent category was Conversation and Community Outreach (28.57%), followed by Education (15.13%) and Brand News and Information (8.40%). They received the most engagement when talking about their brand and educating their audience about Guizhou and its culture, perhaps due to their emphasis on storytelling and transparency.

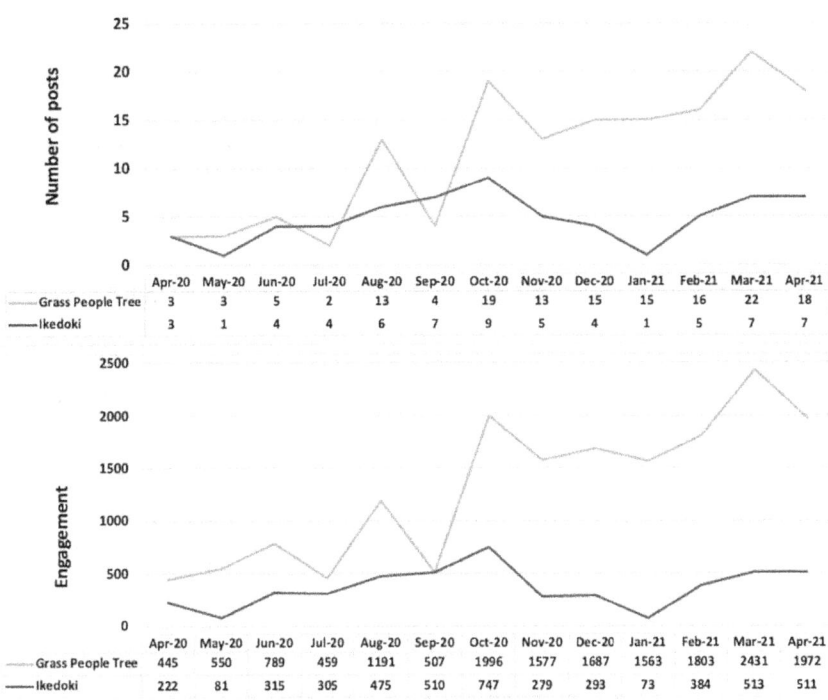

	Apr-20	May-20	Jun-20	Jul-20	Aug-20	Sep-20	Oct-20	Nov-20	Dec-20	Jan-21	Feb-21	Mar-21	Apr-21
Grass People Tree	3	3	5	2	13	4	19	13	15	15	16	22	18
Ikedoki	3	1	4	4	6	7	9	5	4	1	5	7	7

	Apr-20	May-20	Jun-20	Jul-20	Aug-20	Sep-20	Oct-20	Nov-20	Dec-20	Jan-21	Feb-21	Mar-21	Apr-21
Grass People Tree	445	550	789	459	1191	507	1996	1577	1687	1563	1803	2431	1972
Ikedoki	222	81	315	305	475	510	747	279	293	73	384	513	511

Figure 18.1 Frequency and average engagement per post for both case studies

Source: Authors

Table 18.1 Frequency and average engagement per post for both case studies

	Grass People Tree (n = 119)				*Nagasaki Ikedoki Tea* (n = 63)			
	Frequency	*% of* n	*Frequency Rank*	*Avg. Engagement per Category*	*Frequency*	*% of* n	*Frequency Rank*	*Avg. Engagement per Category*
Brand News and Information	10	8.40%	4	136.2	7	11.11%	4	74.6
Marketing and Promotion	57	47.90%	1	119.7	31	49.21%	1	74.2

(Continued)

179

Table 18.1 (Continued)

	Grass People Tree (n = 119)				Nagasaki Ikedoki Tea (n = 63)			
	Frequency	% of n	Frequency Rank	Avg. Engagement per Category	Frequency	% of n	Frequency Rank	Avg. Engagement per Category
Education	18	15.13%	3	134.9	16	25.40%	2	69.3
Conversation and Community Outreach	34	28.57%	2	125.3	9	14.29%	3	86.2
	119				63			

Note: Counts by categories, percentage of *n*, frequency rank, and average engagement per post for both case studies.

The reflections and takeaways of their (virtual) tea house participants were important to Grass People Tree's use of Instagram. They viewed the social media platform as a site to foster community through conversations, ask reflective questions, and engage in live tastings. When asked about the entry point of someone's Grass People Tree experience, both key informants agreed it was through the brand's Instagram page. The founder likened their Instagram bio to a poster that you see on the motorway and their Instagram feed as a ticket window 'to step into the tea tour or the tea experience, the Master's Tea House Guizhou experience' (Interviewee A 2021).

Case study 18.2   Nagasaki Ikedoki Tea

Nagasaki Ikedoki Tea is a small operation resembling a standard tea tour, including visits to the tea fields, local processing sites, and the tasting of teas in the Higashi-Sonogi tea region. Since Nagasaki is not yet widely known for its tea production and long history of growing tea, the founder's goal is to share Nagasaki's connection to tea with the world. In the early months of the pandemic, Instagram was a space to keep Ikedoki's name and presence alive. With uncertainties in global travel, their staffer, a tea enthusiast and student of formal tea study programmes said, 'we've been holding it [Ikedoki] warm until the world opens up and we can start tours again,' and described Instagram as 'an entrance hole of Ikedoki, where the main interactions occurred' (Interviewee B 2021). She posted images and descriptions that felt hospitable and welcoming, such as views from the tea fields, overlooking the bay, of the new tea shoots, and images of their recent tours. The images also offered potential visitors an idea of the tour prior to booking. This was because Ikedoki's founder understood the pre-tour process and booking period as a highlight of visitors' total tour experiences.

In August 2020, Ikedoki hosted their first virtual tea tour. These tours, which were free to attend, were conducted in partnership with the local government and tourist agency. The tours presented a live view of the founder and the local staff in the tea fields, brewing tea, and switched camera views to provide a tour of the tea factory. In the pre-tour phase, Instagram was used to announce and promote the events. These posts were shared by those who were interested in and/or already registered for the tours. During the tour, which was held on another platform, some attendees took photos or videos of their view screens, capturing their experience while on the virtual tour. Screenshots were posted to their Instagram Stories, a feature that allows a quick share of content to one's profile and disappears after 24 hours.

Ikedoki reciprocated by reposting participants' Instagram Stories to their own Stories and later saved them as Highlights. The motivation for reposting their guests' Stories and using Highlights was to remind past participants of their experience (a post-tour experience); for potential partic-ipants to get a taste of what the events are like (to imagine for a future tour); and for the brand to showcase participants' feedback (Interviewee B 2021). As shown in Table 18.1, most of Ikedoki Tea's Instagram content (n = 63) during this study period were categorised as Marketing and Pro-motion (49.21%) and Education-Related (25.40%). About a quarter of the posts were considered Conversation and Community Outreach (14.29%), and Brand News and Information (11.11%). Despite their low frequency, the Conversation and Community Outreach category generated the highest average engagement of the four categories. For example, one of their higher-engagement conversational posts was a short video of a fox spotted in the tea fields in September 2020. The key informant expressed enthusiastically, 'That was fantastic, so many people loved it. It was right at that moment and people can feel like they are here' (Interviewee B 2021).

Discussion

The COVID-19 pandemic had a significant impact on tourism and travel in 2020–2021. As a result, tea retailers and tours adapted to online business models and experiences. Our study highlights two tea companies and their use of Instagram to facilitate storytelling and marketing activities throughout the tourism experience. Through Instagram, tea businesses provide brand information, educate users about tea production processes and culture, market products and events, and convene community conversation and concerns. These findings differ from Yoo and Lee's (2017) work on social media marketing in the hotel industry. Hotels emphasise corporate information over marketing, while our two tea companies valued marketing, education, and community conversation over brand informa-tion. Storytelling and marketing are vital in the tea industry, where tea brands focus on the experi-ential nature of tea drinking.

Instagram engagement in tea tourism

In both case studies, influencers made a significant contribution toward improving engagement with tea brands. Though both brands placed a primary emphasis on marketing and promotion posts dur-ing the pandemic, engagement was driven by influencers who helped create compelling content and distribute it to their own networks of followers (Carter 2016). For Grass People Tree, a tea enthu-siast and Instagram tea influencer helped to provide an inside look at the tea brand and some of its key influential staff and stakeholders. The founder of Grass People Tree also had a wide network of followers on her personal Instagram and was closely involved in their social media process and sharing. Both directed attention to cultural and geographical aspects of Grass People Tree's production site, allowing content related to the destination to trigger Instagram tea users' interests. Likewise, Ikedoki's tea influencer assisted with conversational and community content, which had the most engagement.

Between April and August 2020, the two brands' engagement was relatively similar (see Figure 18.1). However, in fall 2020, engagement with Grass People Tree's Instagram content surged far ahead of Ikedoki. This coincided with the development of Grass People Tree's online curriculum and the promotion of their virtual tea house offerings. This business model shift to move classes and workshops online was accompanied by a schedule of set types of Instagram posts each day of the week. This strategy led to increased Instagram posts, which may explain why engagement for April 2021 was over three times higher than Ikedoki's. Part of this explanation is that they simply posted 2–3 times more than Ikedoki since fall 2020. Yet, previous research has shown that cultural intermediaries or influencers produce the majority of high engagement images on Instagram, oftentimes exceeding engagement produced by the brand itself (Carah and Shaul 2016).

Both brands hired a tea influencer to help produce and coordinate their brand's social media content and community outreach. Grass People Tree was far more effective at harnessing the efforts of their influencer to gain attention directly from the customers. This demonstrates the importance of planning and coordination between brand and influencer to communicate not only information about the product but also affective elements that inspire Instagram users to become part of the tea community and its destination culture. This is best accomplished by utilising tea influencers to apply their creativity on posts that discuss inside information about the tea brand, educating consumers about the tea production destination and its culture, and encouraging conversation about relevant social concerns. However, Grass People Tree's effectiveness with Instagram engagement was not solely dependent on Instagram or tied to the tea influencer/brand community liaison. Rather, it could be argued that their pandemic-forced business model adaptation and the CEO's close involvement with their Instagram account leveraged the role of the tea influencer for Grass People Tree.

Instagram and travel experiences

In the interviews, the founders of the brands both considered Instagram to be a gateway to their brands, raising viewers' awareness of their work. In the multistage process of the tourism experience, indirect experience of travel can influence consumer planning and meaning creation (Gretzel *et al.* 2006; Kim and Tussyadiah 2013; Minazzi 2015). Instagram can be the site of these 'indirect experiences.' In the pre-tour phase, both Grass People Tree and Ikedoki posted announcements about their offerings and raised awareness of their mission and work on their Instagram accounts. The during-tour experience of both brands would sometimes include direct tactile and sensory experiences like preparing and sampling teas. These experiences, along with their social aspects, are qualities of tea appreciation and drinking. Grass People Tree also offered classes on Guizhou traditions, while Ikedoki facilitated live interactions with tea producers. Sometimes participants took photos of their experience with the brand's tea during the session, which they posted on their Instagram after the live session ended. They would usually tag and/or mention the brand in their posts. Then, to recap the sessions, the brands summarised and thanked their participants for joining by posting a screenshot of the event to their Instagram. These post-tour shares served as reflections for those who attended the session and as a promotion to potential future participants. Interestingly, the temporal aspect of the post-tour experience sharing to Instagram can happen at any point along the travel process. Tea drinkers on Instagram continue to post about and tag the teas of both brands after an event took place, which may help recall their memory of the 'tour' and re-present the story of the product and its heritage.

Conclusion

This chapter explored how two tea brands used Instagram to facilitate storytelling and marketing experiences across the three tour experience phases during the COVID-19 pandemic. It reviewed the literature on Instagram as a platform, Instagram's functions, social media, and the phases of tourism (pre-tour, during, and post-tour). This work contributes to the existing literature on tea tourism and the role of social media in facilitating tea experiences. Insights from this chapter may open doors to future research into technology-mediated tea tourism and the phases of the tourism experience, as well as the role of tea influencers in tea tourism and marketing research and practices in the wake of COVID-19.

Discussion questions

1 How have COVID-19 adaptations changed the traditional model of the phases of travel planning and tour experiences?
2 Discuss possible differences in the role of cultural intermediaries or influencers of niche tourism (e.g., tea tourism).
3 Content about the brand was an important way to build engagement on Instagram for the two case studies. How do other tea tourism companies (or companies in other tourism industries) build engagement on Instagram? How do the mobile, visual, and network characteristics of Instagram lend themselves to creative storytelling for brands?

References

Beeton, S., 2005. The case study in tourism research: a multi-method case study approach. *In*: Ritchie, B.W., Burns, P. and Palmer, C., eds. *Tourism research methods: integrating theory with practice*. Wallingford: CABI, 37–48.

Buhalis, D. and Foerste, M., 2015. SoCoMo marketing for travel and tourism: empowering co-Creation of value. *Journal of Destination Marketing & Management*, 4(3), 151–161.

Camprubí, R. and Coromina, L., 2016. Content analysis in tourism research. *Tourism Management Perspectives*, 18, 134–140.

Carah, N. and Shaul, M., 2016. Brands and Instagram: point, tap, swipe, glance. *Mobile Media & Communication*, 4(1), 69–84.

Carter, D., 2016. Hustle and brand: the sociotechnical shaping of influence. *Social Media + Society*, 2(3), 1–12.

Choe, Y., Kim, J. and Fesenmaier, D.R., 2017. Use of social media across the trip experience: an application of latent transition analysis. *Journal of Travel & Tourism Marketing*, 34(4), 431–443.

Cotter, K., 2019. Playing the visibility game: how digital influencers and algorithms negotiate influence on Instagram. *New Media & Society*, 21(4), 895–913.

Grass People Tree, 2020a. [Instagram], 28 April. Available from: www.instagram.com/p/B_iMy8Og2Ha/ [Accessed 26 September 2021].

Grass People Tree, 2020b. [Instagram], 9 October. Available from: www.instagram.com/p/CGIGDezAG87/ [Accessed 2 September 2021].

Gretzel, U., Fesenmaier, D.R. and O'Leary, J.T., 2006. The transformation of consumer behavior. *In*: Buhalis, D. and Costa, C., eds. *Tourism business frontiers: consumers, products and industry*. London: Routledge, 9–18.

Hurley, Z., 2019. Imagined affordances of Instagram and the fantastical authenticity of female Gulf-Arab social media influencers. *Social Media+ Society*, 5(1), 1–16.

Jolliffe, L., ed., 2007. *Tea and tourism: tourists, traditions and transformations*. Clevedon: Channel View Publications.

Kim, J. and Fesenmaier, D.R., 2017. Sharing tourism experiences: the posttrip experience. *Journal of Travel Research*, 56(1), 28–40.

Kim, J. and Tussyadiah, I.P., 2013. Social networking and social support in tourism experience: the moderating role of online self-presentation strategies. *Journal of Travel & Tourism Marketing*, 30(1–2), 78–92.

Laestadius, L., 2016. Instagram. *In*: Sloan, L. and Quan-Haase, A., eds. *The Sage handbook of social media research methods*. Thousand Oaks, CA: SAGE, 573–592.

Leaver, T., Highfield, T. and Abidin, C., 2020. *Instagram: visual social media cultures*. Cambridge: Polity Press.

Luborsky, M.R., 1994. The identification and analysis of themes and patterns. *In*: Gubrium, J. and Sankar, A., eds. *Qualitative methods in aging research*. Thousand Oaks, CA: SAGE, 189–210.

Lund, N.F., Cohen, S.A. and Scarles, C., 2018. The power of social media storytelling in destination branding. *Journal of Destination Marketing & Management*, 8, 271–280.

Minazzi, R., 2015. *Social media marketing in tourism and hospitality*. New York: Springer.

Nagasaki Ikedoki Tea, 2020. [Instagram], 18 October. Available from: www.instagram.com/p/CGhSuSNgetS/ [Accessed 2 September 2021].

Systrom, K., 2018. Welcome to IGTV, our new video app. *Instagram Blog*, 20 June [Blog]. Available from: https://about.instagram.com/blog/announcements/welcome-to-igtv [Accessed 25 September 2021].

Vázquez-Herrero, J., Direito-Rebollal, S. and López-García, X., 2019. Ephemeral journalism: news distribution through Instagram stories. *Social Media+ Society*, 5(4), 1–13.

Wong, J.W.C., Lai, I.K.W. and Tao, Z., 2020. Sharing memorable tourism experiences on mobile social media and how it influences further travel decisions. *Current Issues in Tourism*, 23(14), 1773–1787.

Yoo, K.H. and Lee, W., 2017. Facebook marketing by hotel groups: impacts of post content and media type on fan engagement. *In*: Sigala, M. and Gretzel, U., eds. *Advances in social media for travel, tourism and hospitality: new perspectives, practice and cases*. London: Routledge, 131–146.

19

CULTIVATING SENSE OF PLACE

Sabah Tea experience in Malaysian Borneo

Balvinder Kler and Paulin Wong

Introduction

Across the world, tea cultivation contributes to the agricultural sector and supports local communities' livelihoods. Tea plantations, estates or gardens embody a rich, unique, location-specific cultural heritage that is suitable for tourism. Tea gardens, as visitor attractions, connect beautiful landscapes with international tea consumption culture for aficionados. A systematic review of the literature confirms that studies on tea tourism are on the rise (Su and Zhang 2020; Chen *et al.* 2021; Mondal and Samaddar 2021). Yet, gaps remain, specifically a dearth of theoretical frameworks to examine or explain the expansion of both coffee and tea tourism (Chen *et al.* 2021). Tea tourism is categorised as a special interest or niche tourism market segment. Niche markets attract smaller groups of dedicated tourists who are willing to pay more for non-exploitative and authentic experiences (Smith *et al.* 2010). Tea tourists seek specialised experiences related to a specific tea brand, its heritage, and related activities. According to Mondal and Samaddar (2021), tea attractions can be naturally based (tea gardens, plantations, factories), artificially created (museums, exhibits), special events (tours, festivals), or social and cultural interactions (exploring village life or cultural programmes).

However, relying singularly on tea tourists limits alternative revenue streams from other kinds of tourists. A diversification strategy to attract other suitable niche markets to the estate, such as a different on-site activity, might entice other visitors. Benefits would include additional revenue and an opportunity to foster an interest in tea tourism among non-aficionados. However, the success of this idea depends on how well the tea garden qualifies and projects its sense of place (Christou 2020). Sense of place (SoP) refers to the specific qualities making a place unique and fostering emotional attachments, or 'topophilia' (Christou 2020). SoP can guide the sustainable planning and development of tea gardens as visitor attractions. This chapter conceptualises a relationship between SoP and niche tourism to produce a framework that delineates a novel way for tea estates to connect supply to demand.

Literature review

Sense of place

SoP embraces the tangible and intangible characteristics of a geographic location that make it distinctive and memorable, including the physical and cultural environment, the products, and the

Table 19.1 Deeper meanings of *genius loci*

Deeper meanings of genius loci *for a site*	Elements
The physical, inner environment: traditional architecture, the natural environment (endemic flora) and objects and items in the static environment (wooden seats and icons)	Physiological
The sensory environment: the creation of an experience linked to the site which addresses all human senses	
The human interrelationship factor: for example, allowing social interactions between key persons (staff) and guests to establish personal connections with the site	Social
The aura, spirit, and soul of the site: the deepest and most challenging aspect to maintain and promote; for example, enabling and encouraging guests to experience the most genuine and deepest aspects of the site, such as taking part in an authentic event	Psychospiritual

Source: Christou *et al.*, 2019

people associated with place (Jarratt *et al.* 2018; Christou 2020). SoP arises from the interaction between a person and place that leads to an emotional bond. The attachment is supported by the qualities or attributes inherent to the destination, the experiences available on-site, and the myriad conceptions emanating from this interaction. SoP is a useful framework for identifying and communicating the uniqueness and emotional tone of a destination (Jarratt *et al.* 2018).

When SoP is strong, it is felt by all visitors. Christou (2020) explains that SoP, in its original interpretation, can be denoted as the *genius loci*. This ancient concept, worthy of revitalisation, captures a location's uniqueness and recognises the soul and spirit of a place. Visitors to places with a strong SoP experience it so clearly that they foster an emotional attachment. As Christou *et al.* (2019) suggest, three components make up a site's *genius loci* – physiological, social, and psychospiritual elements. Table 19.1 displays the deeper meanings of *genius loci* for a site.

SoP has not been directly utilised within tea tourism (Chen *et al.* 2021; Mondal and Samaddar 2021). Tea plantations can cultivate SoP due to their location, climate, environment, history, and status. These factors offer an element of authenticity to the tourist experience that comes from utilising the unique characteristics of place (Jarratt *et al.* 2018). Recent literature has introduced the notion of the 'therapeutic landscape', a place for physical, mental, and spiritual healing. A study of tea drinking as a catalyst to promote well-being concluded that tastescapes of well-being rely on spatial, temporal, and social aspects of tea tourism (Su and Zhang 2020). Well-being was achieved by building closeness to nature through absorption into place, slowing down to taste tea on-site, rural settings (different rhythms of life), and the opportunity to build connections with hosts and other tourists and create meaningful experiences over tea conversations (Su and Zhang 2020). We propose that such activities also nurture SoP for all visitors. In sum, SoP identifies a location's unique elements, which could be utilised for branding and marketing visitor attractions or destinations (Jarratt *et al.* 2018). The SoP, or the *genius loci*, of a tea estate should be identified, preserved, and used to attract other niche tourism markets that are suitably related to the main niche, tea tourism.

Niche tourism

Niche tourism is defined as 'travelling with the primary motivation of practising or enjoying a special interest. This can include unusual hobbies, activities, themes, or destinations, which tend to attract niche markets' (Smith *et al.* 2010, 160). Tea tourism is quintessentially a niche market 'motivated by an interest in the history, traditions, and consumption of tea' (Jolliffe 2007, 9). It focuses on tea gardens (plantation or estate), tea factories (production), tea shops (consumption), and cultural institutions which preserve and interpret tea culture (Jolliffe 2007). The picturesque landscapes of

Table 19.2 REAL model

Experience	Definition of Component
Rewarding	achievement/precious to the psyche/once in a lifetime/deep satisfaction
Enriching	improve awareness/something unexpected/express oneself/adds value (emotional or intellectual)/personal growth
Adventuresome	unusual/exciting/daring activity (for that individual)/hard and soft adventure activities
Learning	positive engagement with site or host community/traditional crafts/way of life/homestays/ farm stays/skills development/knowledge acquisition

Source: Authors, 2021

historic plantations, coupled with the brand image of specific teas, have elevated some tea gardens into well-known tourist attractions in countries like China, India, Kenya, and Sri Lanka. Such luxury, niche experiences are high yield and good for business.

However, for destinations in countries that are 'off the beaten tea tourism track', it is not cost-effective to only attract specialist tea tourists. A recent study suggested that tea tourism should also target more mainstream budget and mid-budget travellers to address the socioeconomic inequality between service takers and providers (Mondal and Samaddar 2021). Diversification to capture new markets is often necessary. Moreover, non-specialist tea tourists may also visit tea gardens as a stopover in their itinerary or to experience an overnight stay at a well-established tea resort. Yet, the literature has not conceptualised the joining of tea tourism and other related niche markets in one estate.

Discerning tourists undertake niche tourism for distinct and specific purposes, such as a desire for authentic and meaningful experiences (Smith *et al.* 2010). However, according to Read (1980), special interest travel is characterised by the tourist's search for novel, authentic, and quality tourist experiences; travel should be a *rewarding, enriching, adventuresome,* and *learning* (REAL) experience. This concept of niche tourism explains what drives demand and motivates people to pursue their hobbies. For suppliers, it offers a model around which to design products for niche markets. Though REAL is a pithy concept, there has been limited research on how to achieve a REAL experience. This study uses Read's REAL framework to explore niche tourism. A critical analysis operationalised each component and produced the definitions listed in Table 19.2. Niche tourists are discerning and, as this model suggests, are seeking REAL experiences to satisfy their expectations.

Niche tourism thrives on its unique supply and demand to attract visitors. This chapter explores the connection between place and experiences using two conceptual models, SoP and REAL. The case study presents a tea garden that caters to a range of interrelated niche markets that complement tea tourism.

Methods

This study explores the potential of two frameworks (SoP and REAL) to explore the value of enriching supply and demand for a visitor attraction. The Sabah Tea Resort (STR) in Malaysian Borneo is used as a case study to understand visitors' experiences and relationships with the tea garden. This study used an interpretative inquiry paradigm and Q-Methodology (QM), which systematically focuses on exploration and subjectivity (Stergiou and Airey 2011). The process involved categorising labels and statements (the 'Q-set'). In QM, participants ('P-set') conduct a Q-sort of the Q-set into relevant categories. The QM was adapted for this study. The 'concourse' (materials on the topic of interest) consisted of social media material about STR and Tables 19.1, 19.2, and 19.3. Table 19.1 summarises four deep meanings of place which enhance *genius loci*. Table 19.2 defines the components of a REAL experience. Table 19.3 depicts 28 product categories and activities at

Table 19.3 Sabah Tea Resort product categories and activities

No.	Product Categories	Activities (28)
1	A Tea Adventure (4)	Tea harvesting in traditional costume with local woven basket (*wakid*) and hat (*sirung*)
		Tea making (Journey of Tea)
		Tea sampling
		Sabah Tea shop: variety of Sabah Teas (organic black tea, floral and fruit teas); Borneo herbs; T-shirts and local souvenirs
2	Sporty and Motivational (8)	Adventure races
		Team building
		Obstacle crossing/Low rope and commander course
		Jungle treks; river treks
		Swimming (river)
		Mountain biking
		Self-guided walks through the plantation
		Misty morning sunrise guided walks
3	Socially Responsible (2)	Tea planting
		Volunteer vacations
4	Educational (4)	Guided Tea factory tour
		Batik painting
		Insect night walk (identification activity)
		World War II history (Quailey's Hill, Garden of Remembrance)
5	Fun time with Animals (3)	Rabbit farm
		Fish pond
		Desa deer 'gated safari' (opened 2021)
6	Great Photo Shoots (Insta 'Famous' Sites) (3)	Oversized outdoor frames 'Mt. Kinabalu and Plantation' backdrop
		Super-sized Love Locked Tea Pot and 'I ♡ Sabah Tea' signage
		Wedding photography
7	Treehouse Experience (2)	Canopy walk and viewpoint
		Photo opportunity
8	Sabah Tea Restaurant (1)	Sabah *Tealicious* signature dishes menu (tea-infused pancakes, scones with cream and jam, pudding, ice-cream, waffles)
9	Accommodation (1)	Tea estate stays: colonial cottages, traditional longhouses and camping

Source: Sabah Tea Resort website and Facebook, 2021

the STR. A purposive sample of 60 final-year tourism management students enrolled in the Special Interest Tourism module at Universiti Malaysia Sabah participated in this study. Twelve groups (five members each) were asked to complete a theoretically grounded Q-sort of statements in two phases (Huang *et al.* 2017).

In phase one, groups conducted a Q-sort matching product categories (for each activity) to niche markets. Each group had 28 yellow labels ('activities') which they categorised into niche markets using the concourse as a guide. This exercise uncovered their perceptions about the types of niche markets at STR. In phase two, the groups conducted a Q-sort to match niche markets at STR to REAL experiences (e.g., 'We feel *adventure tourism* at STR is an *enriching* experience'). This Q-sort clarified their perceptions of niche markets in relation to REAL experiences. The Q-sort results were reviewed by both authors to ensure credibility. A supplementary task asked the groups to use Table 19.1 to identify the special features of STR, according to the four deeper meanings of *genius loci*. The Q-sort results, the supplementary task, and the STR case all contributed to the proposed conceptual framework for diversifying tea tourism.

Case study context: Sabah Tea Resort

Tea plantations in Malaysia remain understudied within the literature on tea tourism. The Desa Tea Pte Ltd plantations are located in the foothills of Mount Kinabalu (13,455 feet) in Malaysian Borneo. Kinabalu Park is Malaysia's first designated World Heritage Site and the mountain is a spiritual site for the local people. The plantation itself is at an elevation of 2272 feet and spans 6200 acres, surrounded by tropical rainforests 130 million years old. The view of majestic Mount Kinabalu dominates the landscape of this quaint tea garden in the lush green hills (see Figure 19.1).

Established in 1984, Sabah Tea is the only organic tea farm in Borneo certified by SKAL International B.V. of the Netherlands to produce organic tea. Desa Tea focuses on the planting, manufacturing, and distribution of tea, while STR focuses on tourism activities. The STR is located within the Kundasang-Ranau tourism corridor and has become a must-visit attraction. There is a thriving domestic market consisting of urban residents who make the three-hour drive from Kota Kinabalu, the capital of Sabah, to escape the hustle, bustle, and heat of the coast. The crisp mountain air and temperate climate are a welcome relief. An international market consists of Western tourists, mainly from Australia and the United Kingdom. STR has evolved into an important stopover for tourists who choose land transfers between the west and east coasts of Sabah. The management of STR is constantly working to maintain a regular flow of visitors. Table 19.3 depicts the product categories and related activities available at STR. Notably, not all are tea related, yet all were built on the strengths of the geographical landscape of the tea gardens.

As a visitor attraction, STR upholds its tagline of 'Always Something New' by offering new products each year. In a recent interview, the general manager, Ismail Martin Kong, identified three key

Figure 19.1   View of Mount Kinabalu from Sabah Tea Resort

Source: Ismail Martin Kong, Sabah Tea Resort

selling points: the environment, the food, and the genuine staff (Barnes 2019). He also spoke of plans to increase the number of cottages on-site and to introduce a zip line to complement the existing adventure tourism activities. History and the pursuit of an enriching experience plays an important role, even in the design of the cottages:

> the cottages have white picket fences. A tea plantation is colonial . . . it's English. So, let's put some English elements in it. We try to tie everything into the colonial history of Borneo. When you come here, it's educational. For example, there's war history, with the most important memory being the infamous Death March. They passed through this very place. People also learn about the rainforest and its flora and fauna.
>
> *(Barnes 2019)*

Although tourism initially supported the tea plantation, by continuously developing niche markets on-site STR has become a popular destination and is now less dependent on the tea itself (Barnes 2019). STR meets the demands of tea and other niche tourists who collectively savour the Sabah Tea experience.

Case study 19.1   Quailey's Hill Memorial, Malaysia

Sabah's offerings include dark tourism to a World War II–era prisoner-of-war (POW) camp in the East Coast town of Sandakan, where British and Australian POWs were interned in 1942. In 1945, the Japanese moved the POWs to the West Coast by foot, 260 km away (Silver 2007a). Two additional marches took place later. The rainforests of North Borneo (Sabah) proved inhospitable for men who were severely malnourished, tortured, and ill-equipped for the hike. Many POWs perished en route or were killed in cold blood if they could not keep up (Silver 2007a). Sandakan is considered Australia's worst wartime tragedy (see Silver 2007a for a more in-depth account).

STR has developed an array of niche tourism markets to cater to those seeking adventure, gastronomy and photography. The 'Quailey Booklet' for sale in the souvenir shop lists 'Ten Reasons to Visit Sabah Tea Garden'. The eighth reason is 'Visit Quailey's Hill to pay respects to a fallen Australian soldier and learn more about the Sandakan-Ranau Death Marches (Silver 2007b).

In 2007, a local trekking expert and an Australian historian mapped the route of the Sandakan-Ranau Death Marches. The route went through Kampung Nalapak, the present-day site of Desa Tea plantation (Silver 2007b). They identified the location where, in 1945, Private Allan Quailey, an Australian prisoner of war, was killed by Japanese soldiers. The site was renamed Quailey's Hill (Silver 2007b), and the STR management dedicated the area as a 'Garden of Remembrance' with a granite plaque on a memorial stone explaining the circumstances of Quailey's death. Visitors are encouraged to visit the memorial to learn about wartime history from the informative display.

A visit to the garden of Remembrance is an opportunity to experience the preservation of a site's SoP. Standing in front of the plaque, in the spot where a young soldier drew his last breath, awakens one's emotions. The interpretation of this site is aided by a display board with maps, photographs, and explanations. The story tugs at your emotions. The white picket fences and

Australian and Malaysian flags fluttering in the wind emphasise the site's importance to national identity. Families with children can be seen explaining the war that now seems like distant history but is brought to life by the garden. The memorial is a special place, a precious and authentic nod to history, leading some students to visit other World War II memorials in Sabah. This REAL experience was made possible by the preserved *genius loci* that boosted the sense of place.

Those who painstakingly mapped the Death March route, and others who supported this quest, including the owners and management of STR, accomplished something meaningful. Here, in the tea gardens of Nalapak, the memory of the men who trekked through, including Private Quailey, lives on and has become part of the *genius loci*. As a reminder to mankind, in the words of Abraham Lincoln, 'There is nothing good in war. Except its ending.'

Analysis: Sabah Tea Resort

This chapter set out to conceptualise a relationship between SoP and niche tourism to explore how tea gardens can connect supply to demand in novel ways. This framework aids in diversification to attract alternative revenue streams from interrelated niche markets. This analysis is based on the Q-sort, supplementary task, and STR case. Results of Q-sort 1 are displayed in Table 19.4. Participants identified a total of eight niche markets: adventure, dark, educational, gastronomy, photography, rural, tea, and volunteer tourism.

Table 19.5 presents the results of Q-sort 2 identifying the perceptions of niche markets and REAL components (*rewarding, enriching, adventuresome, learning*). The results were used to conceptualise the relationships between SoP and REAL experiences at STR (Figure 19.2).

STR is a special place, well suited for niche tourism. Although tea tourism is the main target market, diversification into related niche markets has proven useful. The overall strategy depends on how well the supply (place) meets the expectations of demand (experience). It is evident that STR protects the unique qualities of place, its *genius loci*, by nurturing key elements of each aspect of the tea garden.

Table 19.4 Results of Q-sort I

No.	Product Categories	Niche Tourism
1	A Tea Adventure (4)	Tea
		Educational
2	Sporty and Motivational (8)	Adventure
3	Socially Responsible (2)	Volunteer
4	Educational (4)	Educational
		Dark
5	Fun time with Animals (3)	Educational
		Rural
6	Great Photo Shoots (Insta 'Famous' Sites) (3)	Photography
7	Treehouse Experience (2)	Adventure
		Photography
8	Sabah Tea Restaurant (1)	Gastronomy
9	Accommodation (1)	Rural

Table 19.5 Results of Q-sort II

No.	Product Categories	Niche Tourism	R	E	A	L
1	A Tea Adventure	Tea	√	√		√
		Educational	√	√	√	√
2	Sporty & Motivational	Adventure	√		√	√
3	Socially Responsible	Volunteer	√	√	√	√
4	Educational	Educational	√	√		√
		Dark	√	√		√
5	Fun time with Animals	Educational	√	√		√
		Rural	√	√		
6	Great Photo Shoots (Insta 'Famous' Sites)	Photography	√	√		√
7	Treehouse Experience	Adventure	√	√	√	√
		Photography	√	√		√
8	Sabah Tea Restaurant	Gastronomy	√	√	√	√
9	Accommodation	Rural		√		√

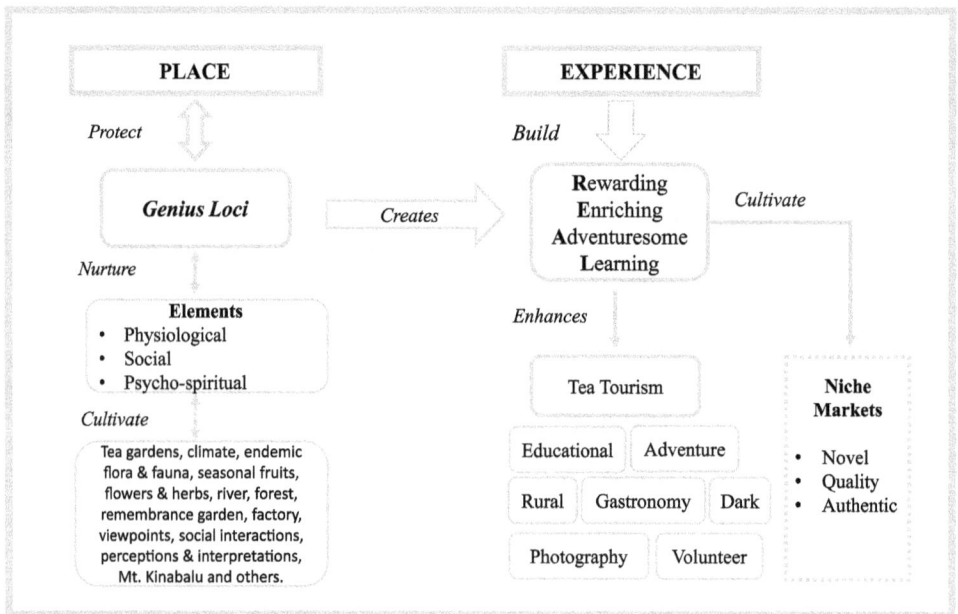

Figure 19.2 Conceptualising SoP and REAL experiences at Sabah Tea Resort
Source: Authors

All eight niche markets within STR (based on this analysis) are intricately linked to the tea landscape. The available activities allow visitors to appreciate the *genius loci* inherent in the physical and sensory environment (for example, endemic flora and fauna or the river). Social interactions with others and the psychospiritual elements of place also foster attachment. The *genius loci* are embedded within the activities. Cultivating these aspects, in turn, strengthens key elements and preserves place.

Nurturing SoP creates REAL experiences that, when applied to branding and marketing, enhance tea tourism and the other niche markets. REAL experiences also produce novel, quality and authentic experiences which every discerning niche tourist seeks. The framework in Figure 19.2 could be used to design or evaluate other tea gardens. Future work can test this framework as it is only contextual presently. In essence, the SoP at STR has been qualified, projected and utilised to build smaller niche markets linked to tea tourism. Not every visitor to the tea garden will be a tea tourist at the onset, but therein lies the potential to entice interest in tea tourism.

STR is located off the beaten international tea tourism track. Therefore, only attracting specialist tea tourists is not cost-effective. Alternative revenue is generated by niche markets utilising the landscape and activities within the tea gardens. In this way, STR lives up to its motto, 'Always Something New'. Most, but not all, other niches were developed purposefully. For example, the efforts of historians and people in the tourist trade brought recognition for Quailey's Hill. The Garden of Remembrance attracts tour groups on battlefield tours of Borneo's Second World War tragedy, the Sandakan-Ranau Death March.

The management's foresight to add interrelated niche markets has supported the operations and expansion of STR. The location of STR, in the shadow of Mount Kinabalu, with a temperate climate and picturesque rolling hills, further enhances its popularity. For tea aficionados, the knowledge that pesticide-free *Camellia sinensis* is cultivated and produced makes the site worthy of a visit. Desa Tea has also strategically added a floral and fruit tea series to attract more tea fans. The well-developed special events facilities at STR, supported by accommodation and food service, also offer further opportunities for expansion into other relevant niche markets in the future.

Conclusion

Tea landscapes are unique rural attractions. This chapter contributes a novel, conceptual interpretation of tea tourism by connecting SoP to the REAL model of experiences which is proposed to identify and cultivate suitable niche markets within one tea garden. A tea estate seeking a diversification strategy could utilise this interdisciplinary framework to ensure continuity and avoid fragmented markets. The framework builds on the strengths of tea tourism to develop interrelated markets. The *genius loci* needs to be considered and nurtured to generate authentic experiences that build on the estate's offerings. Expanding into related niche markets on-site could increase revenue.

Practitioners could consider developing a dynamic SoP toolkit (Jarratt *et al.* 2018) to cultivate SoP in tandem with niche markets designed to provide REAL experiences. The toolkit could be used for branding the tea estate and creating loyalty programmes for the domestic market. Consistent promotional messages merged with the tea tourism brand could also entice new followers. Ultimately, the framework asks that developers remain cognizant of both the supply (place) and demand (experience). The SoP-REAL experiences framework depicted in Figure 19.2 suggests that a successful visitor attraction brings both tangible and intangible attributes of a destination together. Above all, SoP, or *genius loci*, should be cultivated to develop, manage, and sustain tea estates as visitor attractions.

Discussion questions

1   How could other tea tourism destinations apply the REAL model?
2   How do REAL experiences translate into reality for tea tourists?
3   Is the *genius loci* model presented applicable for the tea experience in other tea destinations?

References

Barnes, R., 2019. *Expansion plans for Sabah Tea Gardens* [online]. Kuala Lumpur: Tourism Media Asset Base. Available from: https://tourmab.com/2019/01/29/expansion-plans-for-sabah-tea-gardens/ [Accessed 5 May 2021].

Chen, S.H., Huang, J. and Tham, A., 2021. A systematic literature review of coffee and tea tourism. *International Journal of Culture, Tourism and Hospitality Research*, 15(3), 290–311.

Christou, P.A., 2020. *Philosophies of hospitality and tourism: giving and receiving*. Bristol: Multilingual Matters.

Christou, P.A., Farmaki, A., Saveriades, A. and Spanou, E., 2019. The "genius loci" of places that experience intense tourism development. *Tourism Management Perspectives*, 30, 19–32.

Huang, Y., Qu, H. and Montgomery, D., 2017. The meanings of destination: a Q method approach. *Journal of Travel Research*, 56(6), 793–807.

Jarratt, D., Phelan, C., Wain, J. and Dale, S., 2018. Developing a sense of place toolkit: identifying destination uniqueness. *Tourism and Hospitality Research*, 19(4), 408–421.

Jolliffe, L., ed., 2007. *Tea and tourism: tourists, traditions and transformations*. Clevedon: Channel View Publications.

Mondal, S. and Samaddar, K., 2021. Exploring the current issues, challenges and opportunities in tea tourism: a morphological analysis. *International Journal of Culture, Tourism and Hospitality Research*, 15(3), 312–327.

Read, S.E., 1980. A prime force in the expansion of tourism in the next decade: special interest tourism. *In*: Hawkins, D.E., Shafer, E.L. and Rovelstad, J.M., eds. *Tourism marketing and management issues*. Washington, DC: George Washington University Press, 193–202.

Silver, L.R., 2007a. *Sandakan. A conspiracy of silence*. Kota Kinabalu: Opus Publications.

Silver, L.R., 2007b. *Annihilate them all. Sandakan, and the Death Marches to Ranau*. Ranau: Sabah Tea Garden.

Smith, M., Macleod, N. and Robertson, M., 2010. *Key concepts in tourist studies*. London: SAGE.

Stergiou, D. and Airey, D., 2011. Q-methodology and tourism research. *Current Issues in Tourism*, 14(4), 311–322.

Su, X. and Zhang, H., 2020. Tea drinking and the tastescapes of wellbeing in tourism. *Tourism Geographies*, 1–21.

20

PERCEPTIONS OF TEA TOURISM VALUE AND ITS IMPACT ON DESTINATION ATTRACTIVENESS

J.P.R.C. Ranasinghe, A.C.I.D. Karunarathne, U.G.O. Sammani, H.M.J.P. Herath and P.G.S.S. Pattiyagedara

Introduction

Tourism can rejuvenate destinations with innovative and viable offerings. Innovativeness and the quality of tourism experiences are hence energizing measures to ensure attractiveness of destinations. According to Mayo and Jarvis (1981), destination attractiveness is a set of concerns associated with the ability of a destination to generate a satisfactory level of tourists' benefits. This ability is strengthened by the unique characteristics of destinations such as attractions, facilities, services, and local residents. Thus, the more a destination can fulfil tourists' desires, the more likely the tourists will select the destination against competing destinations.

Apart from the development of infrastructure, agricultural resources also perform a significant role in tourism. Particularly, investment in the tea industry designates much potential for tourism growth in some developing economies. For example, tea is the fourth-largest foreign exchange earner for Sri Lanka. Considering novelty, uniqueness of resources, authentic history and culture, and potential to offer fascinating experiences, tea plantations in Sri Lanka play a prominent role in tourism testifying to a substantial social and economic payback, and their premium brand and expertise knowledge supports a win-win situation for both tea and tourism industries.

As the global tourism industry is moving towards nature-based and experience-based investments, many niche tourism segments have emerged in recent decades. Tea tourism is a good example which can offer innovative and authentic experiences for tourists. Therefore, in the context of tea tourism, destination attractiveness may not be merely relied on the destination attributes. The tourists' perceived experience appears as another substantial factor in deriving the value to visit a tea tourism destination. As the apparent tourists' perceived value of tea tourism in Sri Lanka is not yet academically settled, the study aims to model perceived tea tourism values and their impacts on destination attractiveness from the viewpoint of local tourists using a quantitative approach. Particularly, the impacts of tea tourism–driven activities, services, and product offerings on the attractiveness of destinations are discussed.

DOI: 10.4324/9781003197041-24

Literature review

Destination attractiveness

Attractions are vital elements within a destination that can enhance the visitors' travel intentions (Hu and Wall 2005). In other words, destination attributes are essential in attracting or motivating tourists to visit a specific zone. Diverse perspectives on destination characteristics are discussed by Mayo and Jarvis (1981) and Van Raaij (1986), including natural features (climate, landscapes, beaches, mountains, etc.) and human-made characteristics (hotel and transport facilities, package tours, sports and leisure facilities tailored to customer tastes, etc.). A tea tourism destination with unique attributes and sufficient facilities may inspire tourists to visit such a place. Thus, the attractiveness of a tea tourism destination can play a significant role in attracting more tourists and generating tourism revenues.

Perceived value of tea tourism

Perceived value refers to the consumer's own assessment on the utility of a product or a service (Zeithaml 1988). Furthermore, perceived value has a positive connection with consumer satisfaction of the tourism industry. The fascinating role of tea can create values for tourists and has made a significant contribution to induce visitors to tea estates. Studies have found that with a tremendous collaboration of tea and tourism, the perceived value can be created through the blend of local assets, resources, and attractions that set potential growth and business opportunities as they offer a variety of experiences such as processing, dining, tasting, entertainment, leisure, and education related to tourism. For example, Sotiriadis (2015) discusses value creation through experience, and the opportunities found in the aspects of tea tourism, which benefits local communities (business and employment opportunities, farmers/producers, retailers), natural and cultural resources (preservation and improvement), and visitors (getting better knowledge and understanding of places visited). In other words, the assortment of assets and resources attached to the tea industry is a wonderful heritage resource for tourism value creation (Jolliffe and Aslam 2009). In this study, we measure the perceived value of tea tourism with three components: (1) intrinsic destination resources and mix of activities, (2) perceived experience, and (3) supporting facilities. These components are discussed in the following sections.

Intrinsic destination resources and mix of activities

The demand for a destination is exclusively derived from an appealing resource base and potential experiences surrounding this, and the same factors will lead visitor engagement (Crouch and Ritchie 1999). In this perspective, diverse events and activities are considered as an important pull factor and the most critical aspects of destination appeal. Tourism destinations, however, should offer a unique experience for all guests using a bundle of facilities and services with a number of multidimensional attributes. Accordingly, all the destinations are made up of an amalgam of tourism products to provide an integrated tourism experience for the guests (Buhalis 2000). The attributes and elements facilitating the visitor experience at the destination in the form of tangible and intangible assets collectively endorse the attractiveness in a site (Inskeep 1991). Ritchie and Crouch (2010) also have highlighted the elements such as special events, physiography, environment, culture and history, activities, and entertainment as essential attributes in appealing to visitors within a destination. Jolliffe (2007) has intensively discussed the potentials of the tea industry blended with tourism that offers a diverse range of unique experiences for the visitors.

Perceived experience

The values that create the perceived experience of a visitor are explained in diverse studies with multiple approaches. Gentile *et al.* (2007) predicted the physiological values that are derived from the fulfilment of people's physiological needs (sleep, hunger, thirst, and sex) and the physical values derived from participating in physical activities. Alternatively, according to Fernández and Bonillo (2007), relaxation values are known as the recovering energies in a relaxing environment, while olfactory values are defined as values which a person gets by consuming some sweet or unique smell in a product. The visual beauty of natural and artificial scenes of objects, are preferred as aesthetic value, while functional values are those that derive from getting some work or feature done through certain offerings such as warm hospitality or affection provided in a strange place. The values derived from knowledge that appreciates novelty, surprise, and humour are demarcated as epistemic values (Fernández and Bonillo 2007). Hence, tea-based activities also incorporate values such as relaxation, enjoyment, novelty, and hospitality with the visitor experiences (Jolliffe 2007).

Supporting facilities

Attractiveness of a destination is heightened by its capabilities in providing facilities for tourists (Buhalis 2000). Without destination supporting facilities, tourist activities may be limited and the economic rent generated will be greatly compromised. Competitiveness of a destination is achieved when the provision of services and facilities are competitive against alternative destinations. Supporting facilities vary and depend on the objectives of destination management organizations. According to Buhalis (2000), core destination provisions have six unique entities: attractions (natural, human-made, artificial, purpose built, heritage, special events); accessibility (transportation comprising routes, terminals, and vehicles); activities (what consumers do during their visit at the destination); amenities (accommodation, catering facilities, retailing, other tourist services); ancillary services (services used by tourists such as banks, telecommunication, post, news agents, hospitals); and available packages (pre-arranged packages by intermediaries and principals).

Communication and promotion

Destination branding acts as an inducement for communication in streaming benefits to visitors and works as a device for collaborating functional and emotional benefits to those visitors (Krešić and Prebežac 2011). Tourists value the experiences associated with reputation, and positive reputation could mean increased visitation and longer length of stay by visitors (Roberts and Dowling 2002). This is completely led by the communication of value. The capability of a destination in attracting its target group of visitors and competing globally is also associated with its delivered reputation and image. Fernando (2000) suggested a destination image under the brand name "Ceylon Tea", comprehensive worldwide advertising campaigns using social media marketing tools, and the introduction of sustainable tourist experience packages should be implemented to boost tea tourism in Sri Lanka. Communication works as a mediator to share benefits with tourists. Mediators also manage the interaction between the parties and encourage constructive communication through sharing functional and emotional benefits aligned with perceived tea tourism value and destination attractiveness.

Hypothesis development and conceptual model

Based on previous studies, we proposed a theoretical model as Figure 20.1.

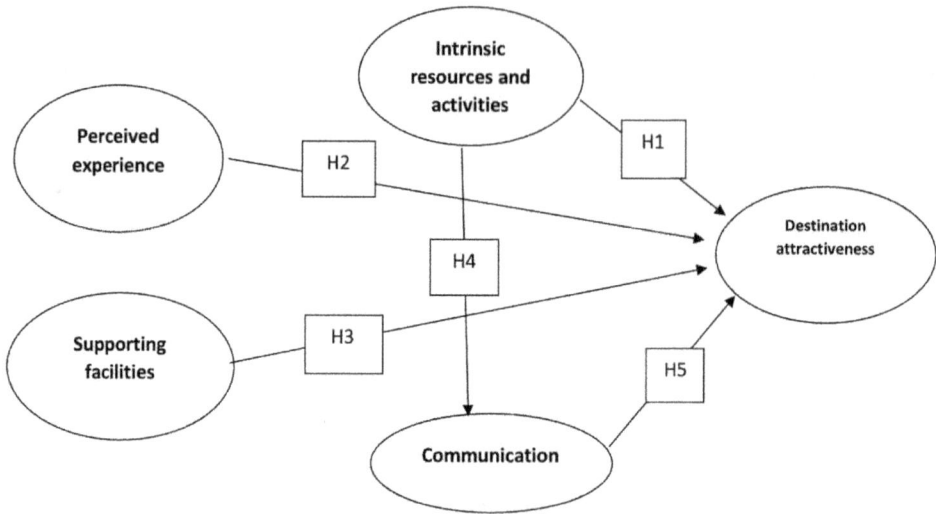

Figure 20.1 Proposed theoretical model
Source: Authors

The hypotheses are explained as follows.

Hypothesis 1 – The perceived value of intrinsic destination resources and mix of activities in tea tourism positively influences destination attractiveness.

Hypothesis 2 – The value of perceived experience in tea tourism positively influences destination attractiveness.

Hypothesis 3 – The perceived value of supporting facilities in tea tourism positively influences destination attractiveness.

Hypothesis 4 – The perceived value of intrinsic destination resources and mix of activities in tea tourism positively influences communication.

Hypothesis 5 – Communication generates a positive relationship between perceived tea tourism value and destination attractiveness.

Methods

Data was collected through an online questionnaire with 27 items. The questionnaire was developed to measure the variables on a 5-point Likert scale (1 = strongly disagree, 5 = strongly agree). The study was completed through a sample drawn for the purpose from tea tourists scattered in the most prominent tea tourism sites in Sri Lanka (Nuwara Eliya, Haputhale, and Ella). To test the proposed theoretical model, a structural equation modelling (SEM) technique was adopted. Partial least squares (PLS) is an approach to SEM that allows researchers to analyse relationships while emphasizing theory testing rather than theory building. All 391 valid responses were included for SEM with the application of PLS path modelling, an approach that is robustly component based and avoids estimation and identification issues. Most importantly, this study focuses on theory testing and analysis of perceived tea tourism value and its impact on destination attractiveness.

Data screening and preparation for PLS-SEM

According to the preliminary examination and analysis, to handle potential measurement problems, the data should be appropriate for PLS-SEM and confirm the reliability and unidirectionality of measurement scales. According to Hair *et al.* (2013), the rule of thumb for PLS path modelling is 1 to 10 times the arrows pointing to a variable in the model. Given the five arrows in the model this rule demands only a sample of 50 cases where we have a sample of 391, well above the minimum requirement. The Kaiser–Meyer–Olkin (KMO) value was 0.924, confirming sampling adequacy for the test. Bartlett's test of sphericity was significant (X2 = 21453.284, $p < 0.001$) confirming the item correlation standards required for the analysis. The results illustrate the outcomes of the preliminary examination and purification of data for PLS-SEM. Moreover, communalities of extraction were 0.689, 0.679, 0.656, 693, and 0.638, for intrinsic resources and activities (IRA), perceived experience (PE), supporting facilities (SF), communication (COM) and destination attractiveness (DA), respectively. Communality values for all the variables exceeded the Keiser critical value criterion (0.60) required for PLS-SEM analysis.

Case study context

Tea is currently a major source of GDP that covers 221,000 ha (324,900 acres), nearly 4% of the terrestrial area in Sri Lanka (Fernando 2000). The authentic history, legacy, and cultural identification of tea makes tea tourism a niche segment merged with education and leisure and generating value as a means of heritage tourism utilizing the existing investments of industry in Sri Lanka (Jolliffe and Aslam 2009). Technically, there are six major development regions in Sri Lanka, recognized as the "Isle of Tea" (Heiss and Heiss 2007): Dambulla, Galle, Kandy, Nuwara Eliya, Ratnapura, and Uva. The uniqueness of culture and scenic beauty blended in the tea industry in Sri Lanka creates novel and authentic experiences (nature walks, accommodation, tea tasting and retail, factory visits) in the travel and tourism routes of the island. Accordingly, a strong emphasis could be built on the potential of the tea industry to invest in value creation within the country for economic upgrades and tourism growth in Sri Lanka (Jolliffe and Aslam 2009).

Results

Demographic profile of the respondents

As per the analysed data, the majority of the respondents were women (55%), and younger (89%) in the age range of 20 to 30 years. Approximately 86% of respondents were single, and 55% were students. Thirty-one percent resided in the Western Province, followed by Central and Uva Provinces. Leisure, entertainment, and tea-related activities were the prominent reasons for visits, and 58% of respondents were travelling by personal vehicles, while 23% used trains during their travel. Social media, websites, and blogs were the most popular information sources among the respondents.

Evaluation of the measurement model

The reflective measurement models' robustness was established based on composite reliability, indicator reliability, convergent validity, and discriminant validity. High reliability of measures was indicated by all composite reliability values (0.808–0.918), which were well above the critical value of assessment (0.7). Three of the 20 indicators were slightly lower than the threshold value for their item outer loading, which was 0.708. These indicators were spared, given their contribution to retain the composite reliability of the construct. Greater loadings of indicators (0.70–0.96) specified the reliability of measures. Convergent validity of measures was established through average variance extracted (AVE), and this value should be greater than 0.50 (Hair *et al.* 2013). All AVE values (0.77–0.97) were above the standard requirement for robust PLS model loadings. Finally, to ensure discriminant validity, cross loadings of indicators were observed. Aligning with Hair *et al.* (2013), item loadings were compared and confirmed that the outer loadings on the related constructs are higher than all of their loadings on the other constructs.

Evaluation of structural model and hypothesis testing

The SEM illustrated in Figure 20.2 was assessed by the coefficient of determination (R2) of endogenous latent variables.

Aligning with Hair *et al.* (2013), the model explains 60.4% of destination attractiveness, indicating a moderate and substantial predictive power of endogenous latent variables (see Figure 20.2). To assess the model further, the authors implemented a PLS iterative bootstrapping procedure by generating 5000 sub-samples with 391 cases. As indicated in Figure 20.2, all proposed relationships were statistically significant. Table 20.1 illustrates the standardised path coefficients (predictive relevance) and relevant *t*-statistic of relationships obtained through the PLS bootstrapping performance.

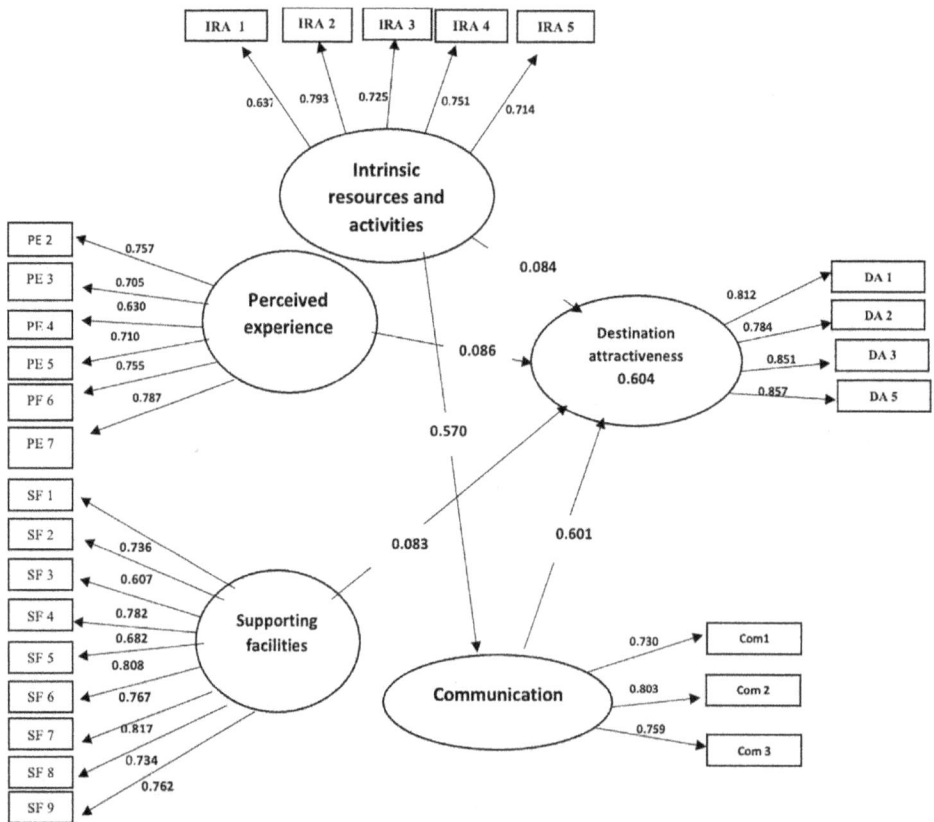

Figure 20.2 The results of PLS-SEM

Source: Authors

Table 20.1 Results of testing by the hypothetical model using PLS-SEM

Hypothesis	Variables	Path coefficient	t-statistic (significant at p < 0.01)	Status
H1	Intrinsic destination resources and mix of activities – Destination attractiveness	0.084	3.328	Supported
H2	Perceived experience – Destination attractiveness	0.086	6.174	Supported
H3	Supporting facilities – Destination attractiveness	0.083	11.973	Supported
H4	Intrinsic destination resources and mix of activities – Communication	0.570	8.687	Supported
H5	Communication – Destination attractiveness	0.601	12.583	Supported

Note: The significance levels are determined through bootstrapping analysis (Hair *et al.* 2013).

Source: Survey 2021

Table 20.2 Construct cross-validated redundancy

Endogenous variable	SSO	SSE	Q² (1-SSE/SSO)
Communication	297.000	244.132	0.178
Attractiveness	396.000	240.962	0.392
Intrinsic resources and activities	495.000	495.000	
Perceived experience	594.000	594.000	
Supporting facilities	891.000	891.000	

Source: Survey 2021

Hypothesis 1 proposes a positive relationship between intrinsic destination resources and mix of activities and destination attractiveness. H_1 was supported with a weak path coefficient ($\beta = 0.084$, $t = 3.328$, $p < 0.01$). The positive relationship proposed between perceived experience derived through experiencing the physical and social environment and destination attractiveness (H_2) was supported, indicating a weaker effect ($\beta = 0.086$, $t = 6.174$, $p < 0.01$). The proposed positive relationship between supporting facilities and destination attractiveness (H_3) is moderately supported by the results ($\beta = 0.083$, $t = 11.173$, $p < 0.01$). Equally, the proposed positive relationship between intrinsic destination resources and mix of activities and communication (H_4) is supported. This is indicated by a positive relationship with a greater effect ($\beta = 0.570$, $t = 8.687$, $p < 0.01$). The positive relationship hypothesised between communication and destination attractiveness (H_5) is accepted as indicated in the output ($\beta = 0.601$, $t = 12.683$, $p < 0.01$).

Blindfolding and predictive relevance of model (Q)

Aligning with Hair *et al.* (2013), the authors computed Q2 values for the three endogenous latent variables of the model to assess predictive relevance. As per the threshold points, all the Q2 values higher than zero ensured the path models' predictive relevance for respective endogenous constructs. To obtain Q2, the PLS blindfolding procedure was followed. The model's predictive relevance (Q2) for reflective endogenous variables, namely, destination attractiveness (Q2 = 0.392) and communication (Q = 0.178), is illustrated in Table 20.2.

Discussion

Contemporary global tourism industry directs visitors to focus on alternative tourism and unique experiences during visits. The diverse landscapes, colonial artefacts, tea-based livelihoods, and emerging tourism services in the central highlands of Sri Lanka provided a fertile ground for this study. The application of PLS-SEM to predict destination attractiveness in terms of tea-based tourism services and activities is a novel stance in tourism research.

H1 proposed a positive relationship between intrinsic destination resources and mix of activities and destination attractiveness. It consists of the functional value of the tea-based tourism product, culture, and history of the destination, entertainment and special events, and cuisine of the destination. Although it generates a weak path coefficient in H_1, the relationship supports denoting its importance of all items in the destination attractiveness. The overall perceived experience at the tea tourism destination consisted of aesthetic value, physiology of the surrounding, emotional relationships built by the visitors, social context, and relaxation. The positive relationship and the individual item loadings indicated a strong significance of these items for overall destination attractiveness, and this was further confirmed with the supported hypothesis (H_2).

Supporting facilities in the tourism industry such as shopping and retail, superstructures, accessibility, information availability, visitor management, emergency and crisis management, accommodation, and food and beverage indicate substantial loadings, proving their meaning in the model. Therefore, these indicators and their effect on the destination attractiveness (H_3) was strong in the case of the study setting. This may be due to the growing tourism-supported system available in the area where the tea-based colonial cultural aspects and properly established tourism management systems are effective. Indicators derived for the measurement of destination attractiveness (perceived satisfaction, revisit intention, overall appeal, and intention to recommend) showed robust item loadings and higher *t*-statistics, showing the visitor behaviour is closely related with the proposed variables in this study. Thus, communication and promotion, support facilities, destination attractiveness, and intrinsic resources and activities of the destination play a critical and invaluable role in shaping and building destination attractiveness. Tea and related destinations thus need to focus on the measured variables and items introduced in this study to reinforce destination attractiveness.

The overall SEM outcome elucidates the tourist destination attractiveness and perceived tea tourism value at a destination in the light of resources and activities, perceived experience, supporting facilities, and communication and promotion. Prominently, resources and activities of a terminus can predict the communication and promotion of a destination as per the final model derived. Moreover, destination attractiveness of a tea-based tourism value can be predicted to an extent of 60% as per the variables tested in this study. Further, the remaining 40% could be predicted from the other factors such as visitor-based reasons and individual factors (e.g., experience and education, other climatic conditions).

Conclusion

This study projected and empirically tested a model to investigate tea tourism value and its impact on shaping tourist destination attractiveness in a colonial tea heritage destination in Sri Lanka. All the directed hypotheses in the theoretical model were supported in the empirical study. Twenty-seven indicators were covered in the final model leading to the implications elaborated in this paper. The study concludes that tea-based tourism resources, products, services, supporting facilities, marketing and communication, and visitors' perceived experience can effectively be used to predict destination attractiveness. Thus, the perceived value of tea tourism could be an effective marketing and promotion tool, and the study further argues that tea-based tourism and services play a significant role in making tourist destinations attractive.

Theoretical contributions

Tea heritage has comparatively drawn lesser attention than other types of tourism, and the available literature is mostly based on a qualitative approach to understanding. This study proposes a SEM model in this field to predict tea tourism value and its impact on tourist destination attractiveness. The study affirmed that tea tourism value can be operationalized to predict tourist destination attractiveness. Moreover, variables operationalized and the final validated model consists of 27 indicators established a novel theoretical approach in the field of tea tourism research. Thus, the proposed hypotheses in this study, namely the relationships between tourist destination attractiveness and intrinsic resources, and the mix of tea tourism-based activities, experiencing physical and social environments, supporting facilities and communication and promotion are novel theoretical contributions in this study. Furthermore, future studies can be guided towards seeking other variables influencing destination attractiveness in terms of tea tourism, and studies can be conducted on each variable that is separately relevant to the destination attractiveness in tea tourism.

Managerial contributions

Empirically, this study should draw destination planners, policymakers, destination marketers, and tea tourism facility operators' attention. Specifically, marketing of tea-based adventurous tourism activities requires novel marketing channels such as social media and the web instead of traditional channels while focusing on the young crowd given their exclusive interest in such activities. Thus, marketers and operators need to take special interest in such market segments in their efforts. Moreover, the strong statistically significant relationship between resources and communication implies the relative influence they generate between each variable emphasizes that communication plays a critical role in intrinsic resources that are effective in destination attractiveness. Sustainable tourism practices with tea tourism development could be more marketable than the traditional marketing strategies. Thus, it is recommended that research focus on more global promotional campaigns, specifically on social media and web-based marketing. Visitors' perceived experience has a strong positive relationship with destination attractiveness, and it implies that operators need to make sure the visitor experience is up to the standard of expectation. Further, supporting facilities available for tea and related tourist activities demonstrate a strong positive relationship with destination attractiveness. Thus, policy planners and destination development agencies need to strongly focus on such supportive services.

Discussion questions

1   What are the prominent factors that motivate tourists to explore tea-based destinations other than the factors discussed in this chapter?
2   Elaborate on possible strategies to transform the product-oriented tea industry in Sri Lanka to a profitable service-driven tourism experience.
3   Compare and contrast the tea tourism products and experiences offered in different countries.

References

Buhalis, D., 2000. Marketing the competitive destination of the future. *Tourism Management*, 21(1), 97–116.
Crouch, G.I. and Ritchie, J.B., 1999. Tourism, competitiveness, and societal prosperity. *Journal of Business Research*, 44(3), 137–152.
Fernández, R.S. and Bonillo, M.Á.I., 2007. The concept of perceived value: a systematic review of the research. *Marketing Theory*, 7(4), 427–451.
Fernando, M., 2000. *Story of Ceylon tea*. Sri Lanka: Mlesna (Ceylon).

Gentile, C., Spiller, N. and Noci, G., 2007. How to sustain the customer experience: an overview of experience components that co-create value with the customer? *European Management Journal*, 25(5), 395–410.

Hair, J.F., Ringle, C.M. and Sarstedt, M., 2013. Partial least squares structural equation modeling: rigorous applications, better results and higher acceptance. *Long Range Planning*, 46(1–2), 1–12.

Heiss, M.L. and Heiss, R.J., 2007. *The story of tea: a cultural history and drinking guide*. Berkeley, CA: Ten Speed Press.

Hu, W. and Wall, G., 2005. Environmental management, environmental image and the competitive tourist attraction. *Journal of Sustainable Tourism*, 13(6), 617–635.

Inskeep, E., 1991. *Tourism planning: an integrated and sustainable development approach*. New York: Wiley.

Jolliffe, L., ed., 2007. *Tea and tourism: tourists, traditions and transformations*. Clevedon: Channel View Publications.

Jolliffe, L. and Aslam, M.S.M., 2009. Tea heritage tourism: evidence from Sri Lanka. *Journal of Heritage Tourism*, 4, 331–344.

Krešić, D. and Prebežac, D., 2011. Index of destination attractiveness as a tool for destination attractiveness assessment. *Tourism: An International Interdisciplinary Journal*, 59(4), 497–517.

Mayo, E.J., Jr. and Jarvis, L.P., 1981. *The psychology of leisure travel*. Boston, MA: CBI Publishing Inc.

Ritchie, J.R. and Crouch, G.I., 2010. A model of destination competitiveness/sustainability: Brazilian perspectives. *Revista de Administração Pública*, 44(5), 1049–1066.

Roberts, P.W. and Dowling, G.R., 2002. Corporate reputation and sustained superior financial performance. *Strategic Management Journal*, 23(12), 1077–1093.

Sotiriadis, M.D., 2015. Culinary tourism assets and events: suggesting a strategic planning tool. *International Journal of Contemporary Hospitality Management*, 27(6), 1214–1232.

Van Raaij, W.F., 1986. Consumer research on tourism mental and behavioral constructs. *Annals of Tourism Research*, 13(1), 1–9.

Zeithaml, V.A., 1988. Consumer perceptions of price, quality, and value: a means-end model and synthesis of evidence. *Journal of Marketing*, 52(3), 2–22.

21

REVITALISING A REGION USING TEA TOURISM

The case of Umegashima, Japan

Masako Saito

Introduction

Umegashima is located in the mountainous region at the northern end of Shizuoka City, 200 km west of Tokyo. With 1700 years of history, it is known for its hot springs, beautiful natural scenery formed by the Abe River source, the tea gardens on the mountains, shiitake mushroom, wasabi green mustard, and high-quality green tea.

Like most rural areas of Japan, Umegashima has been facing a rapid decline in its youth population, which has led to a decline in active personnel working in the local industry. Tea farmers in the region, especially, have suffered from constant decreases in the price of tea, forcing many to find other sources of income besides tea farming. The price of tea peaked in 2000 and has consistently fallen to almost half its peak value. Furthermore, the first harvest from the Umegashima region comes to market much later than those that are grown in lower altitude plains regions, due to the colder climate at higher altitudes. By the time the first harvest from Umegashima is in the market, the second harvest from the lower altitude regions has already hit the market and is sold for one-third the price of the first harvest from Umegashima (Tea Promotion Division 2021). Even though the quality of the tea from Umegashima's first harvest is excellent, its quality is not appreciated against the much cheaper second harvest tea in the market.

The tea gardens cultivated on the mountain ridges in Umegashima create awe-inspiring scenery. To pass this treasure on to future generations, the locals need to manage the tea business sustainably. Tea tourism is a good endeavour for sustaining the tea industry in Umegashima. Since 2009, the Association for Protecting Kakurecha Teas has promoted tea tour programmes to encourage interaction between urban tea consumers and supporters in the local community. Programs have been implemented that leverage the Japanese concept of urban-rural interaction in green tourism (i.e., sustainable agritourism) proposed by the Ministry of Agriculture, Forestry, and Fisheries of Japan. The urban-rural interactions concept encourages bringing youthful energy and revenue to rural communities. In conjunction with the concept of tea tourism, the Association organised learning programs such as tea picking, hiking tea trails, and participation in local workshops with tea farmers (see Figure 21.1).

Therefore, this chapter aims to present the tea community's revitalisation approach through the lens of tea tourism utilising participatory action research (PAR) conducted by the Association. Tea tourism programmes bring benefits to tea farmers and many other establishments in the community, including ryokan inns (Japanese-style hotels) where participants stay overnight. As most

DOI: 10.4324/9781003197041-25

Tea picking with the farmer at the Kakurehca tea garden

The hiking at the Abe River Source

Grilling river fish for a dinner

Lunch box: rice with tea leaves wrapped by Hoba leaves

Figure 21.1   Tea tourism–related activities

Source: Author

tea-producing regions in Japan face similar problems as Umegashima, the tea tourism approach presented in this chapter will be a good reference for those areas.

Literature review

Urban-rural interactions in tea tourism

Urban-rural interaction is one of the fundamental approaches to revitalise a rural area. The Ministry of Agriculture, Forestry and Fisheries of Japan, in the *White Paper on Food, Agriculture and Rural Areas 2011*, argues that urban-rural interactions are important in promoting the exchange of people, goods, and information. It also helps people living in cities and rural areas to share the attractions of each other's regions and deepen understanding. The participants of tea tourism, mainly from Tokyo, long to experience the rural environment which is abundant in Umegashima (e.g., greenery, streams, serene air, quietness, birds singing). On the other hand, examples of urban resources include young people's energy, money, new ideas, IT skills, communication skills, and flexibility. As the resources in Tokyo and Umegashima are completely different, it is beneficial for both sides when resources are exchanged. This is why urban-rural interactions are a focus in Umegashima. The importance of urban-rural interactions has been pointed out since the 1970s and was promoted in the 1980s (Saito 2014).

Applying the concept of urban-rural interactions in tourism can promote green tourism, which was introduced by the Ministry of Agriculture, Forestry and Fisheries in 1992. The concept of green tourism was inspired by urban dwellers in Europe who vacationed in rural areas, soaked up the rural idyll, and appreciated rural values. However, green tourism in Japan is different from in Europe (Khaokhrueamuang and Chueamchaitrakun 2019). It focuses on the agricultural experience of urban-rural interactions, which encourages urban residents to stay in an agricultural area in a rural environment and participate in farming activities. Both farmers and urban dwellers enjoy the interactions in the rural communities (Bixia and Zhenmian 2013). Thus, the term 'Japanese green tourism' came to be associated with the experience of urban-rural interactions, rather than simply staying and relaxing in a rural environment (Khaokhrueamuang 2021).

Furthermore, urban-rural interactions in green tourism involve building the relationship between urban people and farmers and strengthening the network of education and business through the transfer of people between urban and rural markets (Ohe 2014). For example, tea products are traditionally shipped from the factory to urban stores. When implementing a tea tour with the green tourism concept, urban consumers are invited to a tea-producing community to be a part of tea processing as farm volunteers and rural tourists (Khaokhrueamuang 2021). Tea tourism in Umegashima applies the concept of urban-rural interactions in tea tourism by addressing the participation of the local and urban personnel in running tea tour programmes.

All aspects of the programme underpin the process of participatory action research (PAR). PAR is a process in which individuals in the community being studied enthusiastically participate with the investigators throughout the research process – from the preliminary design to the concluding presentation of the outcomes and discussion of the action implications (Whyte 1991). PAR is conducted by local participants attempting to find new solutions (Karlsen 1991), such as guidelines for tea community revitalisation. The PAR cycle commonly revolves around four phases: plan, act, observe, and reflect. These four phases form a continuous quality improvement process (Child Family Community Australia 2015). Each round of PAR's cyclical process identifies ways to improve the programme and ideas for revitalising the community (Khaokhrueamuang and Chueamchaitrakun 2019).

Tea tourism has taken place every year at Umegashima since 2009. Implementing the concept of urban-rural interactions and following the PAR process improves the programme to better satisfy urban dwellers. According to Plog's (1973) tourist typology, tourists who are interested in learning about tea with tea farmers, tea processing, and tea culture are called 'allocentric travellers'. Allocentric tourists are self-confident, curious, adventurous, and outgoing, differing from a group of psychocentric tourists who tend to be anxious, self-inhibited, non-adventurous, and concerned with 'little problems' (Beeton 2006). The target tourists of the tea tour, therefore, are allocentric travellers.

Methodology

The participatory action research (PAR), with its four stages of plan, act, observe, and reflect, was used in this study. These cycles were observed every time a tea tour programme was organised. The programme used one round of PAR to identify the programme improvements and ideas for revitalising the community. The details of the four stages in tea tourism in Umegashima are as follows:

1   The *plan* stage consists of the following: plan the time frame of the tea tour; decide on what kinds of programmes to offer, considering the participants' learning objectives (described later), what kind of transportation is to be used from Tokyo to Umegashima, accommodation for lodging, and how to recruit the participants.
2   In the *act* stage, the planning is executed, keeping improvement points in mind.
3   For the *observe* stage, we observe the participants' enjoyment and note down any areas for improvement. At the end of every tea tourism programme, all of the participants and staff have a discussion to review the good and bad points of the tour, as well as improvements and ideas on what can be done to revitalise the area. The participants then answer questionnaires to record their responses.
4   In the *reflect* stage, before ending the programme, the staff discuss what improvements can be made for the next tea tour programme and subsequently integrate them.

Analysis

Since the pilot tea tour in 2009, the Association implemented 23 tea tour programmes between 2010 and 2019. These 23 tea tours can be categorised into three groups, according to the objectives and target participants. This chapter analyses three typical tea tours from each group to describe the tea tourism approach, through PAR, in revitalising the tea industry in Umegashima.

Group 1

This tea tour is considered an extension to public lectures on 'Learning about community-based revitalisation' by the Tokyo University of Agriculture. The objective is to visit Umegashima, where community revitalisation is practised using tea tourism. The targeted participants are people who previously attended the university extension lectures, as well as current students from the university.

It took place on 8–9 May 2010 with 21 participants (7 men and 14 women). Day 1 of the programme included listening to the tea and wasabi farmers, discussing and exchanging ideas with them, and exploring the forests in the area. Day 2 of the programme consisted of tea picking at the Kakurecha tea garden and hiking at the Abe River source. The details of the PAR of Group 1 are shown in Table 21.1.

Table 21.1 PAR in Group 1

Plan	Act
This tour was part of the public lectures organised by the Tokyo University of Agriculture. The plan is prepared by the Association, considering the university's intentions with advice from a professor from the university.	The professor specialised in landscape environmental studies at Tokyo University of Agriculture and attended the tea tours as an adviser. The students from the university attended the tour as support staff.
Observe	**Reflect**
We observed the programme's implementation for this stage, including a lecture, hearing the community members, and taking activities. Participants were asked to make comments and suggestions based on the activities in the action stage. Here are some examples of their comments.	After the observation of Group 1, the following improved strategies resulting from the reflection were proposed.
Theme 1 Desire to stay longer, do activities longer and more slowly	*Theme 1 Desire to stay longer, do activities longer and more slowly*
Too many programs in two days. I would like to take longer hours for tea picking (participant 1).	The number of events in one programme was reduced to allow participants some free time.
I would like to stay two nights instead of one and walk around slowly. I would like to attend a relaxed weekday course as well as a weekend course (participant 2).	*Theme 2 Communication with local people makes the tour attractive*
It would have been nice to spend longer time to summarize the tour and think about the direction of the revitalization of the village (participant 4).	The tea programme offered more chances to meet local people and communicate with them. We invited the tea farmers and community to head to the dinner and made time for them to talk with participants during and after dinner.
I would like to walk around the area more and see the place more (participant 10).	
Theme 2 Communication with local people makes the tour attractive	*Theme 3 Desire to have more local food*
The hearing from the local people made the tour more interesting (participant 3).	Lunch on the first day was prepared by the inns rather than asking participants to bring their own lunch. We asked the inn to serve tea dishes for dinner and lunch, which gave the participants more satisfaction and helped them remember Umegashima as a special tea place.

It is good to have hearings and exchanges with local people (participant 6).

Theme 3 Desire to have more local food

The lunch box on the first day prepared by participant individually now, it would be nice if it is prepared by the inn in Umegashima (participant 7, 8, 9).

Theme 4 Desire to have the professor speak more

I enjoyed the talk on flora and the lecture on community revitalization of the professor very much, so would like to have them longer (participant 11).

Theme 5 Desire to purchase Kakurecha tea during the tea tour in Umegashima

Since we come to Kakurecha tea garden, we would like to buy the Kakurecha tea here (participant 13).

Theme 6 Desire to have transportation provided

The transportation from Tokyo to Umegashima is not convenient (participant 12).

Theme 7 Desire for attire recommendations

It is not easy to find what kind of attire is adequate for hiking (participant 14).

Theme 8 Analysis of the participants

The participants in this tea tour are very enthusiastic on revitalization; that makes me impressed and interested.

Theme 4 Desire to have the professor speak more

Instead of using a car to return from the tea garden to the inn, we would walk with the professor, who would explain the flora on the way back.

Theme 5 Desire to purchase Kakurecha tea during the tea tour in Umegashima

Kakurecha tea is now available to purchase at the Kakurecha tea garden, but was previously only available online.

Theme 6 Desire to have transportation provided

A chartered bus is used between Tokyo and Umegashima. Some wanted to use public transportation from Tokyo to Shizuoka, so the transportation topic is still under consideration.

Theme 7 Desire for attire recommendations

The ideal attire is shown as a picture, not a written description.

Theme 8 Analysis of the participants

The Association encourages the participants to get more involved in the Association's activities.

Group 2

This group's objective was to activate urban-rural interactions, bring various benefits to the community, and bolster support for Umegashima. The target participants are people who love tea and who are interested in sustainable community development.

It took place on 9–10 June 2018 and included 15 participants (5 men and 10 women). Day 1 of the programme consisted of tea picking, making hand-rubbed green tea, and making vine baskets. Day 2 included hiking at the Abe River source and visiting the Tenku tea garden (the tea garden in the sky). The PAR details for Group 2 are shown in Table 21.2.

Table 21.2 PAR in Group 2

Plan	Act
The Group 2 plan was created by considering the opinions of former participants and instructors.	The main activity on the first day includes tea picking and making hand-rubbed tea with the tea farmers at Kakurecha tea garden. Activities on the second day include hiking at the Abe River source, Oya landslide, or Abe Otaki waterfalls. The professor and/or a professional tea master and/or a graduate of the Tokyo University of Agriculture and/or AG EcoTours takes the role of instructor.

(Continued)

Table 21.2 (Continued)

Observe	**Reflect**
Theme 1 Satisfaction of the tea tour in Umegashima	*Theme 1 Satisfaction of the tea tour in Umegashima*
I was surprised to know that there was such a place within half a day's drive from the central Tokyo (participant 3).	The Umegashima area has the charm and potential to attract people. In addition to the green tea tours in spring, black tea tours in summer and bancha tea tours in autumn are organised.
It was the first time for me to be able to pick tea and hike around in one tour (participant 1).	*Theme 2 The value of Kakurecha green tea*
The experience was more than I expected and imagined, and it was very stimulating and satisfying. Clean air, delicious tea, and homemade food; these make me like Umegashima now (participant 5).	At the orientation in Umegashima, the Association shows slides about Kakurecha tea and offers a short workshop to taste it before lunch. The tea farmers attend the orientation and communicate with the participants.
The hiking to the Abe River source was the most enjoyable. I would like to try picking wild vegetables next visit (participant 7).	*Theme 3 The professor makes the tea tour more valuable*
I am glad that vine basket making is not too difficult and beautiful basket is completed by DIY (participant 8).	After finishing the tea picking programme, we walk back to the inn with the professor while listening to a lecture on flora for one hour.
Theme 2 The value of Kakurecha green tea	*Theme 4 Transportation*
I could understand why the price of Kakurecha green tea is so high and why it is so valuable (participant 2).	The transportation from Tokyo to Umegashima is still under discussion.
The fact that you can snack on the tea while it is being picked and rubbed by hand is only possible because it is pesticide-free (participant 4).	*Theme 5 Duration of tea picking*
Theme 3 The professor makes the tea tour more valuable	There are very different desires on tea picking duration. The Association offers a standard programme, and participants change it based on their desires.
The professor's commentary on the vegetation was very valuable, and it was a special experience that I would never have experienced in an ordinal tour. I would like to come back here in the autumn (participant 6).	*Theme 6 Adequate shoes for hiking*
Theme 4 Transportation	Before hiking, the person in charge shows the participants the adequate attire and items to take.
It would be better to meet at Shizuoka Station instead of Shibuya in Tokyo in the future so that people can participate from Nagoya area as well (participant 9).	*Theme 7 Target/government*
It is difficult for people to get to Umegashima area and to tea gardens on their own, so chartered buses are needed (participant 11).	Under consideration.
Theme 5 Duration of tea picking	*Theme 8 Alternate programme for non-athletic people*
The tea picking time should be longer (participant 12). One hour is enough for tea picking (participant 13).	For the participants who are not good at hiking, alternate programmes are prepared, which involve walking in the community or dying scarves with tea colour.
Theme 6 Adequate shoes for hiking	*Theme 9 Desire a translator*
Hiking with rain boots was not comfortable, so should be announced to wear hiking shoes.	For the participants who do not understand Japanese, we assign a translator from the staff or supporters.
Theme 7 Target/government	
Should get the government more involved. Targeting children (participant 10).	
Theme 8 Alternate programme for non-athletic people	
It will be better for non-athletic people to have a program just walking around the community instead of hiking courses (participant 14).	
Theme 9 Desire a translator	
I need an English guide (participant 15).	

Table 21.3 PAR in Group 3

Plan	Act
This programme is tailor-made for tea masters in Tokyo. The tea masters would like their pupils to be exposed to the tea gardens and experience tea picking, making tea, observing tea gardens, and learning from tea farmers. The programme contains a tea workshop, communicating with tea farmers, eating tea dishes, and strolling the tea gardens.	This tourism event is a one-day programme, which leaves Shibuya in the morning and returns to Shibuya at night. The participants are mostly tea class pupils or tea-related professionals.
Observe	**Reflect**
Theme 1 Appreciation for the first experience of tea picking and tea workshop in Umegashima	*Theme 1 Appreciation for the first experience of tea picking and tea workshop in Umegashima*
I was glad to have my first tea picking experience and to learn how to make green tea for the first time at such a place where difficult to visit by myself (participants 1 and 2).	Although the participants are pupils of tea classes, we realised that most of them have never experienced tea picking and hand-rubbed tea. In the orientation at the beginning of the programme, we give a brief workshop on Umegashima tea.
I was able to compare 'fresh green tea and aged green tea' in the workshop (participant 6).	*Theme 2 The value of the tea tourism in Umegashima*
Theme 2 The value of the tea tourism in Umegashima	We start emphasising the specific charms of the area more in social networking services (SNS).
The rich nature of the area is very attractive and cannot visit as a normal tourist.	*Theme 3 The importance of meeting tea farmers*
Theme 3 The importance of meeting tea farmers	We offer more chances to meet the consumers and farmers, even in the Tokyo farmer's market.
I was able to listen to the real voice of farmers and learn a lot (participants 4 and 5).	*Theme 4 Desire to support Umegashima (revitalisation)*
Theme 4 Desire to support Umegashima (revitalisation)	We send newsletters about Kakurecha/Umegashima to the participants so they stay posted about events in Umegashima.
This visit made me want to support Umegashima even more (participant 3).	*Theme 5 Desire for dishes using tea leaves*
Theme 5 Desire for dishes using tea leaves	Eating dishes with tea leaves is very special for the people from Tokyo. We consulted with the local inn to serve green tea leaves with tofu and/or rice at lunch.
It was nice to eat tea leaf dishes (participant 7). I want you to sell Lapetso, fermented tea leaf salad!	

Group 3

This tour is designed for tea masters and their students in Tokyo. The objective is to learn about tea while observing real tea trees and the environment surrounding the tea garden. The tea masters recruited the participants and a tour bus attendant. It took place on 10 June 2019 with 19 participants (all women). The programme featured tea workshops, tea picking, and making hand-rubbed green tea. The PAR details for Group 3 are shown in Table 21.3.

Discussion

The observations of Group 1 highlighted how the academic nature of tea tourism offers greater satisfaction than the participants expected. The academic content included lectures on flora, fauna, geology, a tea workshop, forestry, vine basket making, and dying cloth with tea colour. Listening to the community members was another highlight for the participants. It gave tourists satisfaction and also helped build social capital in the community. The participants who attend this tour are considered *allocentric travellers* and are willing to learn something new during the tea tour. It is advantageous to collaborate with academic organisations or persons to make a unique tea tourism experience.

The observations of Group 2 showcased how these participants are enthusiastically willing to get involved in the revitalisation activities. Many become active participants and try to find ways to solve problems in the community, alongside the Association. Much repeat attendance was observed. Some of the participants have even taken on larger roles in the Association now. The observations of Group 3 foregrounded how participants evaluate Umegashima's tea. Their extensive knowledge of tea helps them disseminate the value of Umegashima tea, as well as Umegashima itself, to their friends and pupils. If the region has a special product, it may be beneficial to involve product specialists in the community. Li and Hunter (2015) argue that community involvement in heritage tourism is vital to its success. Similarly, as shown in each group above, the Association has strengthened relationships between the community through tea tourism.

In the three cases above, the participants enjoyed the local food in Umegashima. The dishes provided at the ryokan inn (even the lunch boxes) were made using local ingredients, including deer meat, wild boar meat, mountain fish, mountain vegetables, and fresh tea leaves (Figure 21.1). This increased participants' satisfaction beyond their expectations. Food plays a significant role in giving travellers a rich and authentic experience of the host community (Quaranta 2016). The Association cooperates with the cooks at the ryokan inns to create unique dishes (from a non-local point of view) such as tea leaves on tofu, tea leaves with rice, or lunch rice balls with tea leaves on top of traditional tea leaf tempura dishes.

The participants realised that the special experiences offered in Umegashima cannot be experienced in other places they may visit. Umegashima is unique in that you can enjoy various nature, agriculture, and athletic experiences in one place. The Association tries to sustain and extend Umegashima's resources so that the participants can appreciate Umegashima tea tourism and consequently realise sustainability for the community.

Conclusions

Ten years of tea tourism programmes in Umegashima have enabled the acceleration of various urban-rural interactions. The more urban-rural interactions occur, the more the community gains. The Umegashima community has received positive outcomes from its tea tourism–supported community revitalisation.

The first outcome is increasing tea farmers' confidence and motivation. The tea farmers usually do not meet their consumers. Therefore, they do not know how the consumers appreciate their tea. Tea tourism gives the farmers the opportunity to meet the consumers and receive compliments and rewards directly from them. These experiences encourage the farmers, making them confident and proud of their work.

Secondly, tea tourism formed a bond between tea farmers and consumers. The participants learn how to pick the tea and make hand-rubbed tea, gradually growing affectionate for Umegashima and becoming loyal consumers and supporters. As Joo *et al.* (2019) discussed, the residents who participated in the projects also create social capital in their community.

The third outcome was the additional human resources from outside of the community. One couple who attended a 'reclaiming abandoned tea field activity' became interested in tea and started producing chai products, which became a favourite in Umegashima. Another couple moved from Tokyo in 2020 after learning of Umegashima through tea tourism in 2013 and now take on important roles in the community. Tea tourism contributes to social capital in the community by giving the young people's group 'Community W' from Tokyo a chance to reclaim abandoned tea fields. Moreover, an idea for a new product called Lapetso (Myanmar-style fermented tea cuisine) was introduced by a Thai tea tourism participant in 2018. The experiment to produce the fermented tea has begun and includes tea farmers, the university, Shizuoka Prefecture, and others who became involved through tea tourism. The fermented tea will be the second product to gain the Fair Trade label (the first was Kakurecha green tea), which will further contribute to revitalising the community.

Discussion questions

1   How does the community benefit from active urban-rural interactions?
2   How can tea tourism contribute to revitalising the tea industry and the community through the PAR process?
3   What are the three resources of Umegashima's tea tourism?

Acknowledgement

I would like to express my deepest gratitude to Dr Amnaj Khaokhrueamuang, associate professor at the University of Shizuoka, for giving me the opportunity to introduce Umegashima's tea tourism and for his advice. I also appreciate Hajime Akiyama, the tea farmer in Umegashima who started the tea-related projects. Dr Megumi Aso, emeritus professor at Tokyo University of Agriculture, has given advice on revitalising Umegashima. Hiromi Hirano has supported us as a tea master. Yuko Isobe introduced Umegashima in her tea salon and encouraged tea farmers.

References

Beeton, S., ed., 2006. *Community development through tourism*. Collingwood: Landlinks Press.
Bixia, C. and Zhenmian, Q., 2013. Green tourism in Japan: opportunities for a GIAHS pilot site. *Journal of Resources and Ecology*, 4(3), 285–292.
Child Family Community Australia, 2015. *Participatory action research*. Melbourne: Australian Institute of Family Studies.
Joo, J., Choi, J.J. and Kim, N., 2019. Examining roles of tour dure producers for social capital and innovativeness in community-based tourism. *Sustainability*, 11(19), 5337.
Karlsen, J.I., 1991. Action research as method: reflections from a program for developing methods and competence. *In*: Whyte, W.F., ed. *Participatory action research*. Thousand Oaks, CA: SAGE, 143–158.
Khaokhrueamuang, A., 2021. International exchange in tea tourism: reconceptualizing Japanese green tourism for sustainable farming communities. *In*: Sharpley, R. and Kato, K., eds. *Tourism development in Japan: themes, issues and challenges*. London: Routledge, 140–159.
Khaokhrueamuang, A. and Chueamchaitrakun, P., 2019. Tea cultural commodification in sustainable tourism: perspectives from Thai and Japanese farmer exchange. *The 1st ICRU international conference on sustainable community development*, 18–19 February. Chiang Mai Rajabhat University, Rajabhat Chiang Mai Research Journal, 104–117.
Li, Y. and Hunter, C., 2015. Community involvement for sustainable heritage tourism: a conceptual model. *Journal of Cultural Heritage Management and Sustainable Development*, 5(3), 248–262.
Ohe, Y., 2014. Characterizing rural tourism in Japan: features and challenges. *In*: Diaz, P. and Schimtz, M.F., eds. *Cultural tourism*. Southampton: WIT Press, 63–75.
Plog, S., 1973. Why destination areas rise and fall in popularity. *Cornell Hotel and Restaurant Administration Quarterly*, 14(3), 13–16.
Quaranta, G., Citro, E. and Rosanna, S., 2016. Economic and social sustainable synergies to promote innovations in rural tourism and local development. *Journal of Sustainability*, 8(7), 668.
Saito, A., 2014. 都市農村交流に関する研究動向と今後の展開 (Research trends and development of the urban-rural exchange). *Journal of Rural Planning and Association*, 33(3), 343–348.
Tea Promotion Division, 2021. *Current status of Shizuoka Tea Industry*. Shizuoka: Agricultural Bureau, Shizuoka Prefecture Economy and Industry Department.
Whyte, W.F., 1991. *Participatory action research*. Thousand Oaks, CA: SAGE.

22
LINKING TEA, TOURISM, AND COMMUNITY USING PORTER'S DIAMOND MODEL

Imali N. Fernando

Introduction

Tourism is an assembly of activities, services, and industries that deliver travel experiences such as transportation, accommodation, dining, retail, entertainment, and hospitality services (UNWTO 2019). It links the natural environment, diverse cultures and heritages, and is a multiplicity of experiences, livelihoods, and entertainment of a destination (Fernando 2021). Furthermore, tourism can be the foremost provider of both direct and indirect employment to boost national income and to encourage economic growth, prosperity, and poverty eradication (WTTC 2020). However, contemporary demands for effective relaxation, combined with growing economic prosperity, have led to intense competition in the global tourism industry (Miller and Gibson 2005; Kvist and Klefsjo 2006).

Tea tourism, defined as "tourism . . . inspired by an interest in the history, traditions and tea consumption" (Jolliffe 2007, 9), is a popular niche market of food-based tourism (Cheng *et al.* 2010). Tea destinations are recognised for their tea-related history, cultivation and production, traditions, ceremonies, festivals, and events. Tea-related landscapes, cultural heritage, and scenic beauty also boost tourism demand (Su *et al.* 2019). Drinking tea is a touristic component of experiencing local tea traditions, cultures, services, and attractions (Jolliffe 2007) but also a noteworthy aspect of social interactions in host destinations (Zhang *et al.* 2017). Thus, destination marketing strategies towards tea tourism are significant (Cheng *et al.* 2012; Su and Zhang 2020).

This chapter uses Porter's Diamond Theory of National Advantage, often called Porter's Diamond model (Porter 1990), to comprehend how economic facets successfully interact. Further, the model recognises competitive positions within the tea tourism market (Esen and Uyar 2012; Ozer *et al.* 2012). Tourism scholars have previously used the model to analyse comparative and competitive advantage (Esen and Uyar 2012; Ozer *et al.* 2012; Arzesen-Otamisa and Yuzbasioglu 2013; Estevão *et al.* 2018; Luo 2019; Xu 2019; Fernando 2021). This chapter further adapts the model to explain how various economic facets interact in the context of tea tourism.

This research investigates tea tourism stakeholders' perceptions about development in the host community. The author uses the modified Porter Diamond model to examine the current state of tea tourism in the renowned destination of Sri Lanka (Jolliffe 2003; Cheng *et al.* 2010). The economic facets discussed include (1) conducive factor conditions, (2) domestic demand-driven forces, (3) strategy development with the host community and intensive rivalry among the players, and (4) industrial linkages/value chains. The chapter also examines governmental interventions, industry shockwaves during the COVID-19 pandemic, and potential future strategies.

DOI: 10.4324/9781003197041-26

Literature review

Tourism as a vital global industry

Tourism makes enormous contributions to the world's economy. The direct-indirect-and-induced impact of the travel and tourism industry accounted for 8.9 trillion USD in 2019, 10.3% of global GDP, and translates into $949 billion in capital investments and 330 million jobs (1 in 10 jobs globally; WTTC 2020). Tourism is widely considered a strategy for rural revitalisation and economic diversification, and can potentially address the challenges faced by rural communities in numerous developing economies. In addressing the socioeconomic challenges of peripheral rural areas, tourism can catalyse better public services like education and healthcare, local entrepreneurship, revitalised local culture, customs and heritage, flourishing natural and built environments, and improved rural infrastructure.

"Tea" as a symbol of hospitality

Tea is a beverage, plant, art, meal service, export, agricultural product, industry, religion, dedicated pastime, and integral part of food service (Jolliffe 2007). Though the tea plant originated in China, it has been introduced to many nations and adopted into diverse cultures. Offering tea is a universal sign of hospitality. In lodging settings, a "cup of tea" is an "experience" (Fernando *et al.* 2017), while within hospitality, tea is both a symbol and resource that demonstrates a noteworthy capability for commercial hospitality and uplifts the destination image.

Linking tea with tourism

Tea has affected many facets of global heritage as a traditional beverage, a culture and livelihood, and a plantation crop (Aslam and Jolliffe 2015). Furthermore, tea weaves together ancient and modern history, bridging the ancient Chinese and Japanese traditions with British colonial rule. As an agricultural crop, backed by social-cultural associations and unified with tourism, tea has the potential to sustainably enhance community livelihoods (Jolliffe 2007; Su *et al.* 2019). In short, tea and tourism are closely linked. According to Jolliffe (2007), tea tourism is developed with tourists' keen motivations for understanding tea history, tea traditions, and tea consumption. Recently, a niche market of tourism destinations specialising in unique tea traditions, customs, and cultural heritage has emerged. The potential benefits of tea tourism are not only reaped by merchandisers but also by agriculture, arts, handicraft producers, and local entrepreneurs. Linking tea and tourism can provide a variety of employment opportunities and boost the sustainable livelihoods of rural communities (Su *et al.* 2019).

Although its value to tea tourism destinations is considerable, tea tourism remains an under-researched area, with limited scholarly work linking tea tourism to positive community impacts (Cheng *et al.* 2010; Fernando *et al.* 2017; Su *et al.* 2019). More research is needed to recognise the diverse experiences and integrated partnerships of local tourism participants and key tea producers (Su *et al.* 2019). A paradigm shift is needed to boost community participation through the formation of "community associations", to gain the dual benefits of profitability and community, predominantly through a sustainable, participatory approach (Fernando 2021).

Porter's Diamond model

Porter's Diamond model (Porter 1990) incorporates economic variables to conceptualise how national competitive power is formed and sustained through highly localised processes. It explains why some countries or industries are efficacious at the global scale (Ozer *et al.* 2012), whereby countries which foster dynamic local markets are more successful and competitive (Esen and Uyar

2012). The Diamond model can be adapted to any industry or economic sector to better explain the factors affecting the competitive power among the local industries and predict future proceedings (Esen and Uyar 2012).

Tourism scholars have applied the Diamond model to many countries, regions, and industries, such as Turkey (Esen and Uyar 2012; Arzesen-Otamisa and Yuzbasioglu 2013); Turkey and Spain (Ozer *et al.* 2012); China (Luo 2019); Portugal (Estevão *et al.* 2018); Central Asia (Xu 2019); Sri Lanka (Fernando 2021); and the broader tea industry (Yu *et al.* 2019; Jiang 2019). Based on previous studies, a place's natural and cultural assets are the most important factor in becoming a preferred tourism destination (Esen and Uyar 2012).

This chapter applies the Diamond model's four main economic determinants: factor condition; demand conditions; firms' strategies, rivalries, and structures to tea tourism; and related and supporting industries in Sri Lanka. Two sub-determinants – chance/situational forces and government interventions – are also analysed (Ozer *et al.* 2012). Each of these determinants is explained in the following sections.

Factor conditions

These conditions include the inherited and created endowments of nature, culture and heritage, skilled HR, infrastructure, and investment-conducive capital market structures (Fernando 2021). The determinant is divided into *source-based factors* (historical, cultural, mineral, agricultural, and forest resources) and *usage-based factors* (human resources, work ethics, infrastructure, accommodation, and transportation network; Esen and Uyar 2012). Resource endowments, natural and created, also generate "specialisations with uniqueness" (Nordin 2003; Fernando and Long 2012).

Demand conditions

The majority of developing countries lack home-market demand conditions (Miller and Gibson 2005), meaning emerging segments must stimulate conducive demand conditions (Lin and Juan 2010; Fernando 2021). Competitive home-market demand and niche-marketing segmentations should be instigated (Nordin 2003).

Business strategy, structure, and rivalry

This factor includes proper resource management, vigorous competition and healthy rivalry among the local players, and sustainable administrative strategies (Fernando 2021). It acknowledges community links with "core tourism" such as accommodation establishments, leisure and entertainment, local entrepreneurship, and micro-to-small and medium-scale enterprises.

Related and supporting industries

Countries must develop healthy industrial linkages through diverse related and supporting industries, local bases, and minimised economic leakages. In the developing economy context, tourism leakages are significant (Fernando 2021). Local industry linkages and community involvement are necessary conditions for sustainability; therefore, the destination should support core industries and minimise economic leakages (Miller and Gibson 2005).

Situational forces/chance

Tourism is vulnerable to external shocks and chance occurrences such as ethnic and civil wars, economic downturns, natural disasters, epidemics, and global terrorism (Lin and Juan 2010; Fernando *et al.* 2016). Situational forces include the Asian tsunami (2004), the 2019 Easter bombing in Sri Lanka, and the COVID-19 pandemic. These events required the tourism industry to adapt with outside-the-box strategies.

Government actions

This determinant conceptualises how the government enhances or slows industry. This includes industrial infrastructure and facilitation, monetary policies to attract foreign investments, identifying priority sectors, transportation, communication, health and sanitation, education, public infrastructure, and national economic policies. To enact development, governments should grow equity, extend resource endowments, harmonise economic policies, and promote equality among the public (Miller and Gibson 2005; Lin and Juan 2010).

Case study context: Sri Lanka

Sri Lanka is a small, teardrop-shaped tropical island (approximately 65,610 square miles) located in the Indian Ocean off the southeastern coast of India and 880 km north of the equator. The island's many historical names – *Taprobane* (ancient Greek), *Serendib* (Arabic), and *Ceylon* (British) – relay its strategic location along global sea routes. The country has a variety of climates and weather conditions, ranging from 22°C–33°C in the lowlands to 7°C–21.6°C in the central highlands, with an average annual rainfall of 1270 mm. Tea plantations thrive in the wet zone, where the high elevation contributes to one of the best quality black teas on the global market (Encyclopaedia Britannica 2021).

Tea has been part of Sri Lanka's history and economy since the colonial period in the late 19th century. As the fourth-largest tea producer in the world, tea has been a primary economic contributor for several decades. The "Ceylon Tea" brand has also helped promote the country as a tourism destination. The climatic and geographic areas that are best for cultivation are also conveniently located in the attractive central highlands, surrounded by misty mountains and diverse tourist attractions (Aslam and Jolliffe 2015; Fernando *et al.* 2016).

Known as the "Isle of Tea", Sri Lanka mainly produces "black tea" in seven major tea production regions: Dimbula, Galle, Kandy, Nuwara Eliya, Uda Pussellawa, Ratnapura, and Uva (Heiss and Heiss 2007; Aslam and Jolliffe 2015; Tea Exporters Association 2021). Tea factories are located on estates, many of which contain historic planters' bungalows built during the British colonial period which have now been converted into hotels (Aslam and Jolliffe 2015). Though these estates employ many people, directly and indirectly, the estate communities live in poverty (Dishankaa and Ikemotob 2014; Sanjeewani 2015). Tea tourism has the potential to uplift the community while sustainably harmonising relations between plantations and tourism. Despite the promise of "linking tea and tourism" to generate double economic benefits (Aslam and Jolliffe 2015), there has been only minimal attention dedicated to developing the niche market (Fernando *et al.* 2017; Koththagoda and Dissanayake 2017). This has hindered Sri Lanka's ability to become a strategic global market performer (Cheng *et al.* 2010).

Tea-growing states are clustered amid the central mountains and southern foothills, which are renowned for an inimitable climate and terrain (Sri Lanka Tea Board 2014; Tea Exporters Association 2021). The scenic tea-producing hill country offers many historic accommodation venues that combine colonial-era heritage (Aslam and Jolliffe 2015), varied topography, mountain ranges with breath-taking views, misty hilltops, and a salubrious cold climate (Fernando *et al.* 2016). Nature walks in tea gardens, accommodation in colonial bungalows, and tea processing and plucking experiences could all become popular tourist activities (Fernando *et al.* 2016). However, in practice, most tourists only visit tea factories and shops located near the renovated plantation houses to purchase tea and souvenirs (Koththagoda and Dissanayake 2017). "Tea trails" are also included in most tour itineraries.

Tourism is a sensitive industry that is susceptible to external environmental shocks. Tourism development in the tea region was hindered by 30 years of civil war and ethnic conflicts which ended in 2009, natural disasters like the Asian tsunami, terrorism, and the COVID-19 pandemic (Cochrane 2008; Fernando 2021). The cross-cutting nature of tourism could entail potential socio-economic community costs within the industry, whereas the downturn could be long-term.

Methods

To understand current tea tourism practices in the Central and Uva Provinces of Sri Lanka, this chapter draws on survey research and exploratory site visits to tea-related establishments. Both provinces are located in hill country and host a comprehensive assemblage of tea plantations, accommodations, attractions, and scenic points (Aslam and Jolliffe 2015). Snowball and purposive sampling techniques were adopted. Thirty respondents from tea tourism–related establishments, hotels, tea eco-lodges, retail outlets, scenic locations, tea factories, and estate bungalows were interviewed. The sample includes owner-managers, front office managers, assistant superintendents, and sales and marketing executives (see Table 22.1).

Table 22.1 Sample Profile

Tea tourism–related establishment	Province	Designation
Dambethenna Tea Estate (Lipton's Seat)	Uva	manager
Blue Field Tea Garden	Uva	senior sales and operations manager
98 Acres Resort: Ella	Uva	front office manager
The Secret: Ella the Secret	Uva	manager
Heritance Tea Factory Hotel	Uva	front office manager
Stafford Estate Bungalow	Uva	front office manager
Sherwood Bungalow	Uva	manager
Thotalagala Tea Plantation Bungalow	Uva	sales and marketing manager
Chill Ville Hotel	Uva	manager
Living Heritage	Uva	front office manager
Will Guest Homestay	Uva	owner manager
Serenite Ella	Uva	owner manager
Marabedda Resort	Central	front office manager
Mapakanda Village	Uva	manager
Dutch House Bandarawela	Uva	manager
The Romance Valley	Uva	manager
Demodara Estate	Uva	assistant superintendent
Mandira Dickoya Bungalow	Uva	manager
Hewaheta Estate	Central	assistant superintendent
Mooloya Estate	Central	assistant superintendent
Loolkandura Estate	Central	assistant superintendent
Lucky Star – Ella	Uva	front office manager
The Symbol of Ella	Uva	manager
Tea Hills Bungalow	Uva	manager
Walimada (Malwaththe) Estate	Uva	assistant superintendent
Samanala Farm	Central	executive
Highforest Estate	Central	assistant superintendent
Udapussellawa Plantations	Central	assistant superintendent
Siang-Yung Restaurant	Central	executive

Source: Sample profile from Central and Uva Provinces, Sri Lanka

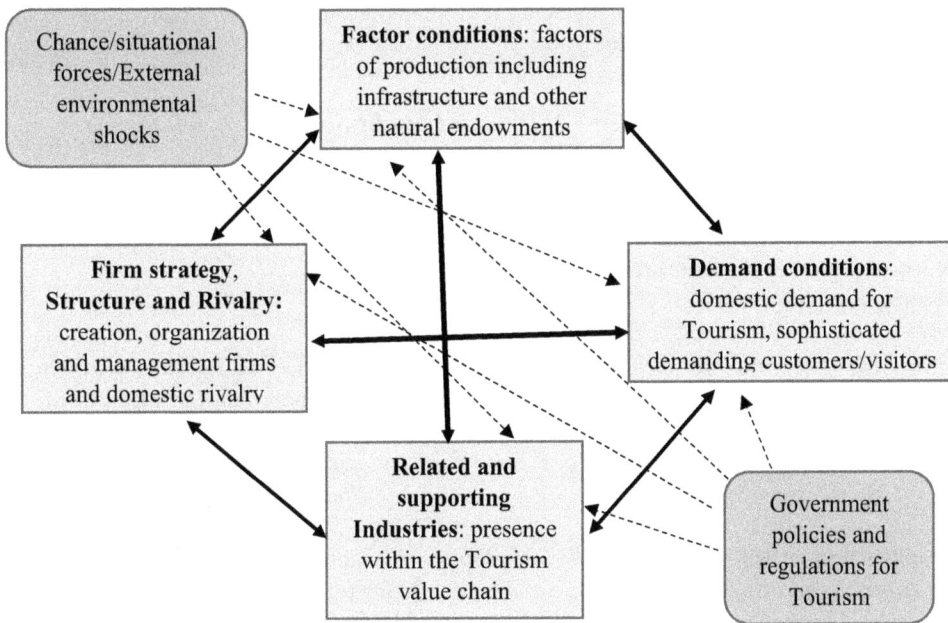

Figure 22.1 Modified Diamond model
Source: Adapted from Porter, 1990

A structured interview guide was used to investigate each determinant of the Diamond model. Interviews were conducted over the phone and online from March to June 2021 (during the second and third waves of COVID-19). The majority of interviewees participated in telephone interviews, email was used as a supplementary method, and three respondents were interviewed via Zoom video calls. The qualitative interview sessions were recorded, transcribed, and thematically analysed using a modified version of the Diamond model.

Analysis

The following sections discuss the current status of tea tourism in Sri Lanka, framed using the modified Diamond model themes (see Figure 22.1).

Theme 1: Factor condition

Sri Lanka is blessed with rich natural and cultural resources comprising both comparative and competitive advantages: (1) natural: tea gardens, topography, and scenic beauty; (2) tea traditions and local hospitality; (3) heritage: colonial tea bungalows, museums, factories, and heritage hotels; (4) skilled talents: experienced staff; (5) infrastructure: well-built transportation and railroad networks; and (6) Investment-conducive market structure: attractive to foreign investments with few financial constraints. Moreover, tea tourism would diversify the food and lodging industry, preserve and conserve heritage properties (colonial-era estate bungalows), and increase participatory community development. Interviewees emphasised the potential community impacts of tea tourism, specifically the benefits to estate workers, truck drivers, tour guides, and local entrepreneurs. Ideas for sustainable development included renovating rural schools and worship places, funding housing schemes, improving sanitary facilities, and encouraging local entrepreneurial ventures such as selling local food, vegetables, fruits, and handicrafts.

Theme 2: Demand conditions

Domestic demand is affected by the structure and size of the market, desires, and tastes. Respondents focused on the need to recognise the "guest" first and prioritise customised services that are tailored to "individual needs". COVID-19 has pushed domestic demand toward nature travels, and therefore, could positively impact tea tourism.

As one participant narrated:

> You have to identify the client who is genuinely interested in Ceylon tea, then the client wants to know how this tea is being produced. It has to be more individualised but the problem is our people do not study the guest individually and are not customised enough. I believe we can add more value to the economy and community if this is more customised.
>
> *(Respondent 14, Field survey 2021)*

The following strategies were identified to develop tea tourism: (1) popularise nature-based travel with domestic tourists, (2) emphasise the benefits of drinking tea, (3) increase community involvement in local entrepreneurship, and (4) develop sustainable community-related projects on micro-to-medium and mass scales.

Theme 3: Business strategy, structure, and rivalry

Respondents agreed that regions with tourism establishments and more industry cooperation with competitive focus produce continuous community benefits. For example, as a respondent stated:

> Yeah, competition is always healthy. Because rather than being in a monopolistic market we have to add more competitors, otherwise you won't grow within. If we have enough players in the industry, everybody will try to outsmart the other person, which will be beneficial for everyone
>
> *(Respondent 21, Field survey 2021)*

A healthy rivalry among tourism actors, partnering within the value chain network and forming local community associations, can enhance the bargaining power of the community and could be sustainable win–win strategies.

Theme 4: Related and supporting industries

Respondents perceived current community involvement to be inadequate. Sustainable community approaches are required to create value chain linkages. As one respondent stated,

> Our hotel was originally a superintendent's bungalow. Our community has organised a walk called "tea trail" which is good community participation and very rarely interacts with the management. Some foreigners, especially [from the] UK, inquire about the historical background of the estate. Some of their forefathers were in Ceylon during that colonial era, and they collect memories and help community entrepreneurship.
>
> *(Respondent 6, Field survey 2021)*

Local residents should form industrial linkages through community associations. Additionally, other supportive industries like food and beverage, transportation, tour agencies, accommodation, home-stays, and tourist guides could be cluster partners within the value chain.

Theme 5: Situational forces/chance

The industry's economic vulnerability to external shocks was highlighted by the COVID-19 pandemic. However, the increasing global acceptance of tea as an immune booster was considered an opportunity. One interviewee speculated,

> [The] global pandemic is something which has . . . affected . . . the entire world, and tourism [is] highly affected. We should intervene and support to enhance the community involve[ment] because without community, "Ceylon Tea" will not be a global brand. Our responsibility [is] to look after the entire network connected to the tea tourism.
>
> *(Respondent 13, Field survey 2021)*

Future tourism efforts should focus on nature-based travel. One silver lining potentially allowing the "Ceylon Tea" brand to recapture tourism market share is the increased recognition that drinking tea leads to a healthy life.

Theme 6: Government actions

Most of the respondents did not support government interventions and recent economic policies. For example, most of the tea planters (in both the plantation and tea-tourism sectors) expressed uncertainty about a recent government policy on "organic fertiliser". As one tea planter put it:

> the other thing is fertiliser. The quality of tea will drop, and the brand image of Ceylon Tea maintained for centuries won't be there anymore and meanwhile, Kenya and China will overtake us. Ultimately when everything is done for tea tourism there won't be a tea industry to base it on.
>
> *(Respondent 6, Field survey 2021)*

Consequently, the foreseeable downturn for the tea plantations resulting from certain government interventions could negatively impact the sustainable community benefits.

Conclusion

In Sri Lanka, tea is a major component of the economy, and the Ceylon Tea brand attracts many tourists. While a noteworthy economic actor within the plantation economy, the sector rarely links "tea and tourism" together into a sustainable niche market. This chapter conceptualises tea tourism and community sustainable development using the Diamond model. The findings also review stakeholders' vulnerability to external shocks and the effects of government interventions. Interviewees emphasised the importance of local entrepreneurship, "Ceylon Tea" branding, forming community associations and industry linkages, diversifying tea-related experiences with community participation, and encouraging domestic demand. Tea tourism will help the tea industry rejuvenate and flourish, while also enhancing community livelihoods and encouraging sustainable development.

This chapter hopes to inspire further studies on tea tourism across different destinations, regions, and comparative contexts. It offers lessons for practitioners and marketers by identifying factors that will increase tourists' lengths of stay in the region, incorporate diversified activities, and sustain the local community. Specifically, community and collaborative government interventions should link the cultural heritage and colonial era, market the global brand (Ceylon Tea), and strengthen industrial linkages and local entrepreneurship.

Finally, this chapter modifies the Diamond model and links it to tea tourism and host community visions for sustaining the economic facets of tourism (during a pandemic). This framework will be beneficial to tourism marketing students and tea tourism planners in different regions. Future researchers can link the model to marketing philosophies, brand management, and destination marketing. Tourism planners might incorporate the following lessons from Sri Lanka: (1) unambiguously linking factor endowments while collaborating with the community; (2) forming associations to enhance bargaining power, strategically connecting cultural heritage resource endowments; (3) encouraging local entrepreneurship; and 4) rejuvenating domestic demand. While this study focused on the tea industry, the approach can also be applied to other niche markets – actual and potential – that are interwoven with tourism.

Discussion questions

1   Can you use Porter's Diamond model to analyse other tea tourism destinations?
2   What impacts might community-tourist linkages have on tea tourism destinations?
3   Are the tea planters' ideas about linking "tea with tourism" a sustainable strategy?

References

Arzesen-Otamisa, P. and Yuzbasioglu, N., 2013. Analysis of Antalya tourism cluster perceived performance with structural equation model. *Social and Behavioral Sciences*, 99, 682–690.
Aslam, M.S.M. and Jolliffe, L., 2015. Repurposing colonial tea heritage through historic lodging. *Journal of Heritage Tourism*, 10(2), 111–128.
Cheng, S., Hu, J., Fox, D. and Zhang, Y., 2012. Tea tourism development in Xinyang, China: stakeholders' view. *Tourism Management Perspectives*, 2(3), 28–34.
Cheng, S., Xu, F., Jie Zhang and Yu-ting, Z., 2010. Tourists' attitudes toward tea tourism: a case study in Xinyang, China. *Journal of Travel & Tourism Marketing*, 27(2), 211–220.
Cochrane, J., 2008. *Asian tourism: growth and change*. Oxford: Elsevier.
Dishankaa, S. and Ikemotob, Y., 2014. Social development and labour productivity: the problem and a solution for the tea plantation sector of Sri Lanka. *Colombo Business Journal*, 5(1), 67–80.
Encyclopaedia Britannica, 2021. Chicago, IL: Encyclopedia Britannica.
Esen, S. and Uyar, H., 2012. Examining the competitive structure of Turkish tourism industry in comparison with diamond model. *Procedia-Social and Behavioral Sciences*, 62, 620–627.
Estevão, C., Nunes, S., Ferreira, J. and Fernandes, C., 2018. Tourism sector competitiveness in Portugal: applying Porter's Diamond. *Tourism & Management Studies*, 14(1), 30–44.
Fernando, I.N., 2021. Assessing the competitiveness of Sri Lanka's tourism in the Covid period by Porter's Diamond Model. *In*: Baporikar, N., ed. *Handbook of research on strategies and interventions to mitigate COVID-19 impact on SMEs*. Hershey, PA: IGI Global, 1–22.
Fernando, I.N. and Long, W., 2012. New conceptual model on cluster competitiveness: a new paradigm for tourism? *International Journal of Business and Management*, 7(9), 75–84.
Fernando, P.I.N., Kumari, K. and Rajapaksha, P., 2017. Destination marketing to promote tea tourism socio-economic approach on community development. *International Review of Management and Business Research*, 6(1), 68–75.
Fernando, P.I.N., Rajapaksha, P. and Kumari, K., 2016. Tea tourism as a marketing tool: a strategy to develop the image of Sri Lanka as an attractive tourism destination. *Kelaniya Journal of Management*, 5(2), 64–79.
Heiss, M.L. and Heiss, R.J., 2007. *The story of tea: a cultural history and drinking guide*. Berkeley, CA: Ten Speed Press.
Jiang, N., 2019. Research on the improvement of the core competitiveness of tea industry in famous tea town: a case study of Huangshan City, Anhui Province. *IOP Conference Series: Earth and Environmental Science*, 346, 012082.

Jolliffe, L., 2003. The lure of tea: history, traditions and attractions. *In:* Hall, C.M., Sharples, L., Camborne, B., Macionis, N. and Mitchell, R., eds. *Food tourism around the world: development, management and markets.* London: Butterworth-Heineman, 121–136.

Jolliffe, L., ed., 2007. *Tea and tourism: tourists, traditions and transformations.* Clevedon: Channel View Publications.

Koththagoda, K.C. and Dissanayake, D.M.R., 2017. Potential of tea tourism in Sri Lanka: a review on managerial implications and research directions. *In:* Misiak-Kwit, S. and Wiścicka-Fernando, M. eds. *Equality and management.* Szczecin: University of Szczecin, 51–68.

Kvist, A.K.J. and Klefsjo, B., 2006. Which service quality dimensions are important in inbound tourism? *Managing Service Quality*, 1(5), 520–537.

Lin, C.T. and Juan, P.J., 2010. Measuring location selection factors for international resort parks. *Qualitative and Quantitative*, 44, 1257–1270.

Luo, X., 2019. Coastal tourism commodity industry cluster based on diamond model and ecological niche. *Journal of Coastal Research*, 94(1), 828–832.

Miller, M.M. and Gibson, L.J., 2005. Cluster based development in the tourism industry; putting practice into theory. *Applied Research in Economic Development*, 47–63.

Nordin, S., 2003. *Tourism clustering and innovation path to economic growth and development.* Östersund: European Tourism Research Institute Publication.

Ozer, K., Latif, H., Sarusik, M. and Ergun, O., 2012. International competitive advantage of Turkish tourism industry: a comparative analyse of Turkey and Spain by using the diamond model of M. Porter. *Procedia – Social and Behavioral Sciences*, 58, 1064–1076.

Porter, M.E., 1990. *The competitive advantage of nations.* New York: The Free Press.

Sanjeewani, S.K., 2015. The impact of women empowerment programs on upcountry tea estate women in Sri Lanka. *International Journal of Scientific Research and Innovative Technology*, 2(7), 36–47.

Sri Lanka Tea Board (SLTB), 2014. *Sri Lanka Tea Board.* Colombo: Sri Lanka Tea Board. Available from: www.srilankateaboard.lk/ [Accessed 15 September 2021].

Su, M.M., Wall, G. and Wang, Y., 2019. Integrating tea and tourism: a sustainable livelihoods approach. *Journal of Sustainable Tourism*, 27(10), 1591–1608.

Su, X. and Zhang, H., 2020. Tea drinking and the tastescapes of wellbeing in tourism. *Tourism Geographies*, 1–21.

Tea Exporters Association, 2021. [online]. Colombo: Tea Exporters Association. Available from: teasrilanka.org [Accessed 1 July 2021].

United Nations World Tourism Organization (UNWTO) Annual Report, 2019. *Progress report on the 10-year framework of programmes on sustainable consumption and production patterns.* Geneva: United Nations.

WTTC, 2020. *Economic impact reports.* London: World Travel and Tourism Council.

Xu, J., 2019. Analysis on the tourism competitiveness of five central Asian countries based on diamond model. *Advances in Social Science, Education and Humanities Research*, 286, 566–569.

Yu, C., Li, X. and Tian, X., 2019. Thoughts on the competitiveness of Hainan tea industry under the background of free trade port construction. *In: 3rd international conference on education, management science and economics*, 14–15 September, Dalian, China. Singapore: Atlantis Publishers.

Zhang, J., Xu, H.-G. and Xing, W., 2017. The host-guest interactions in ethnic tourism, Lijiang, China. *Current Issues in Tourism*, 20(7), 724–739.

23

MARKETING GREEN TEA TOURISM DESTINATIONS

Kunihiko Iwasaki and Amnaj Khaokhrueamuang

Introduction

Research background

In Japan, consumer demand for green tea leaves is currently low. Many tea producers, wholesalers, and retailers are experiencing stagnant or declining business situations. The tea industry is also suffering in Shizuoka, the largest tea-producing region in Japan (Ministry of Agriculture, Forestry and Fisheries 2020). In 2020, the total amount of land devoted to tea plantations nationwide decreased by 1500 ha, 700 ha of which were in Shizuoka. According to a survey of tea business conditions in Shizuoka conducted by Iwasaki in 2016, 45.8% of tea dealers said their businesses were 'slumping' (becoming 'sluggish' or 'slightly sluggish'), and 42.7% were already 'stagnating'. In contrast, only 11.5% of dealers found themselves in 'good conditions' ('good' or 'slightly good'). These results depict a worryingly negative trend in the tea business.

In addition to the decreased demand for tea leaves, another factor driving the decline of Shizuoka's tea industry is its poor cost competitiveness compared to other production areas in Japan. Many tea plantations in Shizuoka are small and located on the sides of mountains, which makes it difficult to use large machines, such as riding-type plucking machines, and reduces production efficiency. In contrast, most tea plantations in Kagoshima Prefecture, the second-largest tea-producing region in Japan (21.4% of the total tea plantation area in 2020), are located on flat land and readily use plucking machines (Ministry of Agriculture, Forestry and Fisheries 2020). In Kagoshima, the area devoted to tea cultivation increased steadily until 2010 due to the increased demand for tea products. Although it has decreased slightly since 2015, the rate of decline has been more gradual than in Shizuoka (Ministry of Agriculture, Forestry and Fisheries 2020). Moreover, Kagoshima can produce first-flush tea more rapidly than Shizuoka due to its warm weather conditions.

Considering the changing external factors, such as a decline in the demand for green tea leaves and intensifying competition amongst tea-producing regions, Shizuoka will continue to lose tea-growing areas if it fails to implement an effective survival strategy. According to Iwasaki's (2017) survey, 58.4% of tea dealers felt future trends for the tea industry in Shizuoka were 'dark' or 'slightly dark'.

Research purpose: creating demand via green tea tourism

It is difficult to sell the same volume of Shizuoka green tea as in the past. Innovation is, therefore, a useful approach for addressing these decreases, due in part to high competition with the South

DOI: 10.4324/9781003197041-27

Kyushu region (Kagoshima and Miyazaki Prefectures). The cost of fresh tea leaves per 10 ha is 386,391 yen in Shizuoka and 276,455 yen in Kagoshima (70% of Shizuoka's tea production costs). In terms of yield per 10 ha, Shizuoka earns 1413 yen, whereas Kagoshima earns 1925 yen – nearly 1.4 times more (Ministry of Agriculture, Forestry and Fisheries 2003).

In contrast, Shizuoka has several inimitable strengths that support using green tea to sell tourism in the prefecture (e.g., its unique tea production areas, the beautiful scenery of mountain tea planta-tions, and its proximity to high consumption areas like Tokyo and Nagoya). Visiting tea-producing regions has a positive effect on consumer behaviour related to tea. For example, Iwasaki's (2017) survey found that the experience of visiting a green tea-producing region enhanced attachment to tea and the image of the tea-producing region. Therefore, it is possible to brand Shizuoka with green tea tourism. This chapter proposes the innovative approach of 'selling Shizuoka with green tea', replacing the conventional notion of 'selling Shizuoka's green tea'.

Understanding the characteristics of tourists and the factors that influence them is crucial for investigating green tea tourism. Green tea tourism, in this study, is based on the Japanese green tour-ism concept proposed by the Ministry of Agriculture, Forestry and Fisheries. Japanese green tourism focuses on the agricultural experience of urban-rural interactions between tourists and farmers, not just relaxing stays in a rural environment (Hasegawa 2005; Khaokhrueamuang 2021; Ministry of Agriculture, Forestry and Fisheries 2013). Thus, green tea tourism is defined as tourism that draws on green tea, nature, culture, relaxation, and interactions between locals and visitors to create leisure activities in a tea-producing region.

This study aims to explore the demand for green tea tourism in Japan's largest tea-producing region, Shizuoka, via a quantitative analysis of consumer data to understand the characteristics of tourists and the factors influencing their decisions when selecting tea tourism destinations. Further-more, it proposes suggestions for green tea tourism marketing.

Literature review

Tea tourism destinations and tea tourists

According to Robinson and Novelli (2007), niche tourism destinations, such as wine regions, can be established using different approaches: a geographic and demographic approach, a product-related approach, or a customer-related approach. This notion can apply to tea regions with different loca-tions and types of tea production based on geography, culture, and consumption. Jolliffe (2007) indi-cates tea tourism can be promoted in both tea-producing and tea-consuming countries, such as the UK, Sri Lanka, India, China, Taiwan, Kenya, and Canada. Furthermore, tea tourism is a 'niche' – a specific product tailored to meet the needs of a particular market segment: tea tourists. Tea tourists are often tea lovers, driven by their interests in tea and tea culture. Cheng *et al.* (2010) identified potential tea tourists in China and explored their attitudes toward tea tourism. The results suggest that tea tourists and non-–tea tourists have significantly different attitudes about tea drinking and willingness to purchase tea as a souvenir. Therefore, a tea-growing region can be marketed as a niche destination, offering a tea tourism experience with its specific teas.

Tourist motivations

According to the Travel Career Pattern (TCP) developed by Pearce and Lee (2005), the novelty of the destination is the most important factor for attracting visitors. The TCP considers 14 motiva-tional factors that influence changes in motivation patterns during travel careers. In order of impor-tance, these factors are (1) novelty, (2) escape/relax, (3) relationship (strengthen), (4) autonomy, (5) nature, (6) self-development (host-site involvement), (7) stimulation, (8) self-development (personal development), (9) relationship (security), (10) self-actualisation, (11) isolation, (12) nostalgia, (13) romance, and (14) recognition. This study considers four crucial factors (escape/relax, relationship [strengthen], nature, and self-development [host-site involvement]) to form two categories of green tea tourism: relaxation-based green tea tourism and interaction-based green tea tourism. Escape/relaxation and nature are important for relaxation-based green tea tourism, while relationship (strengthening) and self-development (host-site involvement) are the most important factors for interaction-based green tea tourism.

Lack of marketing research in tea tourism

In Japan, green tea is the main tea product. However, no previous studies have been conducted on Japanese green tea tourism marketing. Additionally, no quantitative studies on consumers' perspectives on green tea tourism have been conducted. This finding aligns with international tea research, which generally focuses on the supply side, not demand (Cheng *et al.* 2010). According to the International Conference on Tea Culture and Science organised in Shizuoka, most research centres on tea production and the health benefits of tea, rather than tea marketing strategies and tea tourism (The Executive Committee World O-Cha (TEA) Festival 2010). Chen *et al.* (2021) conducted a meta-analysis of 33 research articles on tea tourism, finding that few studies have used interdisciplinary frameworks to study tea tourism. Furthermore, compared to wine tourism (Dodd 1995; Hall 2003; Hall *et al.* 2000), tea tourism lacks marketing research, particularly on the branding of tea tourism destinations using a region's tea products (Khaokhrueamuang *et al.* 2021). This study aims to understand the strategy to marketing a green tea–producing region for tourists and thus can fill the research gap in the mainstream tea studies.

Methodology

Data collection

This study draws on a quantitative analysis of primary data from an online survey of Japanese consumers conducted by the Iwasaki Laboratory, School of Management and Information, University of Shizuoka. Questionnaires were disseminated to 1000 Japanese people living in Tokyo who were randomly selected from more than one million consumers registered in an internet research agency's (Neo Marketing) database. The company emailed the questionnaire to selected consumers in May 2021. The sample was evenly gender-distributed, male (50%) and female (50%), with 20% representation from each age group: 20s, 30s, 40s, 50s, and 60s. The survey subjects were asked to access the online questionnaire and select one option for each question. The main question topics were as follows:

1   Perception of Japan's tea-producing regions (Table 23.1)
2   Intention to visit the green tea–producing area (Table 23.2)
3   Tourism activities in the green tea–producing area (Table 23.3)
4   Factors emphasised when choosing where to go on a general sightseeing trip (Table 23.4).

Data analysis

Data collected from the questionnaire survey were analysed and classified into three sections: (1) perceptions and potential of green tea tourism, (2) categorisation of green tea tourism, and (3) factors influencing green tea tourism destination selection. The perception of Japan's tea-producing regions used a percentage to rank tea-producing areas' images with consumers' perception of Japan's finest tea (Table 23.1). The potential of green tea tourism used a 5-point Likert scale to measure the demand to visit the tea-producing region; the results were presented as frequencies and percentages (Table 23.2). Categorisation of green tea tourism used a 5-point Likert scale to measure green tea tourism activity; this was followed by a two-factor analysis of relaxation and interaction (Table 23.3). This was an exploratory factor analysis to find factors supporting the selection of tourist destinations based on two types of green tea tourism: interaction-based activities and relaxation-based activities.

To explore the relationship between consumer awareness of tourism in general and consumer awareness of tea tourism, we conducted two-factor analyses: on question items related to tourism in general (Table 23.4) and on tea tourism (Table 23.3). The results of factor analysis were presented as factor loading scores. Finally, factors influencing respondents' destination selection, as well as green tea tourism characteristics, were analysed. A 5-point Likert scale was used to clarify factors extracted from the question items about tourist destination selection. These factors were then analysed using multiple regression (stepwise method) to determine each factor's impact on green tea tourism intentions.

Case study 23.1 Green tea tourism destinations in Shizuoka Prefecture

Shizuoka Prefecture is a leader in the production of Japanese green tea; tourists can partake in tea tourism in the eastern and western parts of the region. Tourism Shizuoka Japan (TSJ), the Destination Marketing Organization (DMO), responsible for marketing and promoting tourism in Shizuoka Prefecture, introduced tea tourism destinations on tea farms near the following four main bullet train railway stations.

1. Mishima Station

Twenty minutes south of Mishima station, on the Izu Hakone line, is the small town of Izu Nakaoka in Izunokuni City, Izu Peninsula. This is the eastern gateway from Tokyo to Shizuoka's green tea fields. Although Izu is not a large green tea–producing region, some smaller producers offer green tea tourism experiences. For example, the Kuraya Narusawa tea farm runs a tea-picking programme featuring traditional tea farmer dress. It is a small plantation, but on a clear day, it offers a spectacular view of the majestic Mt. Fuji.

2. Shin Fuji Station

Pictures of Mt. Fuji with rows of green tea in the foreground usually come from Fuji City, particularly the Obuchi Sasaba and Imamiya tea plantations. Both are a 20-minute taxi ride from Shin Fuji Station and are popular destinations for budding photographers. The best season to take a memorable photo is mid-April through May, due to the vibrant first tea leaves of the year and the snow at the peak of Mt. Fuji.

3. Shizuoka Station

One of the most striking green tea plantations in Shizuoka City, Moriuchi tea plantation, is located about 20 minutes away from Shizuoka Station. Situated on steep land where green tea has been cultivated since the Edo period, the Moriuchi plantation allows visitors to experience all aspects of green tea production (picking, steaming, and hand-rolling). Visitors can relax in the 150-year-old traditional house and learn about tea culture, an integral part of Japanese life.

4. Kakegawa Station

Featuring Makinohara, Shimoda, Kakegawa, and Kawane, this area is considered the green tea capital of Shizuoka and the world. This region was designated as a Globally Important Agricultural Heritage System (GIAHS) and is the site of the traditional tea grass-integrated system. Kakegawa City is recommended for tea lovers who do not have time to visit many places in Shizuoka or are nervous about travelling off the beaten track. There are many green tea experiences within walking distance of Kakegawa Station, including remote villages and towns. Most of the area is flat, creating the perfect opportunity for cycling through the tea fields. Other unforgettable experiences include taking a steam locomotive along the Oi River up into the mountains, lodging with tea farmers, and visiting the region's hot springs.

Results

Perception and potential of green tea tourism

Shizuoka ranks first (81.4%) in consumers' perceptions of fine green tea in Japanese tea-producing regions (see Table 23.1).

Connecting the image of Shizuoka green tea with tourism and confirming the existence and level of green tea tourism potential is essential to implementing green tea tourism in Shizuoka. The percentages of consumers attracted to green tea tourism (demand) are shown in Table 23.2. Nearly one-quarter of consumers (24.3%) wanted to spend a holiday in Shizuoka's tea-producing region, indicating that demand for green tea tourism does exist ('agree', 5.9%; 'somewhat agree', 18.4%).

Table 23.1 Tea-producing regions ranked by consumers' perceptions of tea quality

Rank	Region	Percentage
1	Shizuoka	81.4
2	Kyoto	5.9
3	Kagoshima	3.5
4	Saitama	2.4
5	Tokyo	1.3

Source: Authors

Table 23.2 Intention to visit Shizuoka's tea-producing region

Opinion	Number of participants	Percentage
Agree	59	5.9
Somewhat agree	184	18.4
Neither agree nor disagree	348	34.8
Somewhat disagree	234	23.4
Disagree	175	17.5
Total	1000	100.0

Source: Authors

Table 23.3 Results of the two-factor analysis on green tea tourism

Green tea tourism--related activities	Relaxation-based green tea tourism	Interaction-based green tea tourism
Spending a relaxing time in a tea-producing region	0.867	
Relaxing in a tea-producing region	0.866	
Enjoying the greenery of tea leaves and nature on the tea plantation	0.833	
Being healed whilst at the tea plantation	0.816	
Enjoying the scenery of the tea plantation	0.812	
Interacting with tea producers		0.905
Interacting with people from the tea-producing region		0.864
Visiting the tea factory		0.672
Experiencing tea picking		0.663
Learning about the history of tea and of tea-producing regions		0.653
Cumulative contribution rate (%)	43.6	79.0

Notes: Factor extraction method: principal component analysis (varimax rotation); factor loading set at ≥ 0.5.

Source: Authors

Categorisation of green tea tourism

Participants preferred various tea tourism experiences: tea picking, tea factory tours, relaxing, enjoying the scenery, and interacting with tea producers. A 5-point scale was used to measure intention, and the resulting two-factor analysis is displayed in Table 23.3.

The first significant demand factor for green tea tourism is relaxation. This factor emerges from the desire for different green tea–related leisure activities, such as resting, relaxing, and enjoying the greenery of tea leaves and nature in the plantation. The second factor is interaction, which emerges from different green tea-related learning activities like interacting with tea producers and tea farmers, visiting a tea factory, or watching/participating in tea picking.

Factors influencing green tea tourism destination selection

The factor analysis yielded nine factors from the question items about tourist destination selection (Table 23.4). The first factor was 'learning,' identified from education-related questions measuring

Table 23.4 Factors influencing respondents' selection of tourist destinations

General tourism factors	LN	IT	LP	EP	NT	RL	SP	AT	TS
Learning	0.82								
Knowledge expansion	0.81								
Historical consciousness	0.76								
Experiencing local culture	0.74								
Exchanging knowledge and experience		0.90							
Interacting with local people		0.89							
Experiencing rural life		0.81							
Cheap travel expenses			0.90						
Reasonable price accommodation			0.90						
Cheap transportation cost			0.88						
Special experience				0.88					
New experience				0.82					
Unique experience				0.80					
Rich natural surroundings					0.90				
Interaction with nature					0.82				
Beautiful landscape					0.71				
Relaxation						0.87			
Peacefulness						0.85			
Leisure						0.78			
Unique souvenirs							0.83		
Shopping enjoyment							0.76		
Special attractive products							0.74		
Convenient to go								0.87	
Good accessibility								0.85	
Various tourist spots									0.78
Famous tourist attractions									0.61
Attractive places or historical sites									0.57
Cumulative contribution rate (%)	11.9	21.8	31.4	40.6	49.4	58.3	66.4	72.6	78.5

Abbreviations: LN = Learning, IT = Interaction, LP = Low price, EP = Experience, NT = Nature, RL = Relaxation, SP = Shopping, AT = Accessibility, TS = Tourism spot.

Note: Factor extraction method: principal component analysis (varimax rotation); factor loading set at ≥ 0.5.

Source: Authors

interest in deepening education and expanding knowledge. The remaining factors were interaction, price, operational experience, nature, relaxation, shopping, accessibility, and other tourist attractions. All factors contained in Table 23.4 have high factor loading scores (greater than 0.5), indicating influence over tourists' destination selection.

However, it remains unclear whether these factors influence intentions to undertake green tea tourism and, if so, how much influence each factor exerts. To address this question, a multiple

Table 23.5 Factors influencing green tea tourism intentions

Green tea tourism destination selection factors	Level of influence
Interaction	0.323
Learning	0.244
Nature	0.172
Shopping	0.162
Experience	0.162
Relaxation	0.119
Low price	0.057
Tourism spots	0.054

Note: The numbers are standardised regression coefficients; they indicate the degree to which each factor influences green tea tourism intentions.

Source: Authors

regression analysis (stepwise method) was performed: the factors related to tourism destination selection served as independent variables, while the factors related to green tea tourism intentions were dependent variables. The results of the multiple regression analysis show that factors such as interaction, learning, exposure to nature, shopping, and experience had a statistically significant impact on green tea tourism (Table 23.5).

The standardised regression coefficient revealed that the interaction factor had the most significant impact on green tea tourism intentions. The next-highest factors (in descending order) were learning and nature. The interaction and learning factors were related to interaction-based green tea tourism, while nature was important for relaxation-based green tea tourism. In particular, the impact of the interaction-based green tea tourism factors was relatively high with standardised regression coefficient 0.531, compared to relaxation-based tourism 0.367.

Discussion: considerations for green tea tourism marketing

Based on the analysis, the following three premises were identified:

1   Consumer demands for green tea tourism in Shizuoka's tea-producing areas rely on the perception that they produce Japan's finest tea.
2   Interactions, learning, and nature affect consumers' selection of tea tourism destinations (for those attracted to green tea tourism).
3   There can be two types of green tea tourism: relaxation-based green tea tourism and interaction-based green tea tourism. However, interaction-based green tea tourism is in relatively higher demand.

While interest in green tea tourism exists, it remains a smaller, niche market. This finding suggests that marketing for green tea tourism should focus on green tea consumers and use Robinson and Novelli's (2007) product-related and customer-related approaches, which are suitable for niche markets.

Following the Japanese concept of green tourism, interaction-based green tea tourism should provide learning activities to increase its attractiveness. This could include exchange programs between tourists and green tea producers, seminars on green tea history, and learning about tea itself. This suggestion supports Cheng *et al.*'s (2010) finding that promoting tea knowledge is as important as promoting tourism products in marketing tea tourism. Marketing for interaction-based green tea tourism should focus on green tea lovers (Cheng *et al.* 2010) who enjoy learning about tea

231

and visitors who run businesses selling tea products. Furthermore, the interaction activities between tourists and green tea producers could also target educators (e.g., researchers and students). This group of visitors may promote tea-producing communities through the exchange of knowledge with tea academics to increase volunteer activities and urban-rural interactions. For example, volunteer tourists can harvest tea leaves, helping farmers reduce labour costs, while researchers can provide advice on how to improve tea products.

Tourists who purchase tea products at the tea farms may choose to buy them again at urban stores returning home. As Ohe (2014) and Khaokhrueamuang (2021) note, interaction activities based on the Japanese concept of green tourism can stimulate the growth of rural and urban markets due to the transfer of goods, people, and knowledge. In contrast, relaxation-based green tea tourism should target both tea lovers and non–tea lovers (Cheng *et al.* 2010). The factors of nature and relaxation positively impact green tea tourism intentions, so relaxation tours should take advantage of nature and the beautiful landscape of the tea plantation. 'Relaxation' is often a primary travel motivation factor (Pearce 2005). This group is composed of both green tea lovers and non–tea lovers (non–green tea consumers) who seek activities related to the beautiful nature of tea, such as photography and drinking tea.

As Kraftchick *et al.* (2014) explain, tourism niches can assist destinations and attractions in market segmentation (i.e., motivation, demographics, or interests). Implementing green tea tourism in this tea-producing region may increase demand for green tea consumption and support the revitalisation of declining tea-growing communities. The tea plantations surrounding Mishima Station and Shizuoka Station are the most suitable for interaction-based tourism, which targets green tea lovers and educators since the sites are comparatively difficult to access and offer engaging learning activities. Meanwhile, the plantations around Shin Fuji Station and Kakegawa Station are more accessible and should be used to target non–tea lovers seeking relaxation-related activities, such as photography and travelling by steam locomotive.

Conclusion

As the demand for green tea decreases and competition among tea-producing regions intensifies, new ideas for the tea industry's revitalisation in Shizuoka will become indispensable. This chapter shifts the focus away from the conventional idea of 'selling Shizuoka's green tea' toward 'marketing Shizuoka with green tea' using green tea tourism. The quantitative analysis of the online survey of 1000 consumers in Tokyo facilitated several valuable suggestions regarding green tea tourism marketing and revitalisation strategies. Specifically, green tea tourism must go beyond tea drinking – it should combine interaction or exchange, learning, and relaxation.

Shizuoka should capitalise on the demand for green tea tourism. The tea, tourism, and local industries can collaborate on branding to take advantage of this opportunity. Future research should focus on the practical approaches to branding Shizuoka with green tea and on tourists' revisits. This study of Japanese consumers is limited to domestic tourism, so further research should also address inbound tourism segmentation across different factors, such as demographic factors, geographic factors, and psychographic factors. For example, geographic factors might reveal differences between the tea-producing areas preferred by Asian visitors or Westerners, while psychographic factors can address differences between attracting tea lovers and non–tea lovers. According to the Japan Tourism Agency's survey on visitor repeat ratio, tourists are not interested in revisiting Shizuoka (Ishii 2018). Therefore, branding the green tea tourism destination to target repeat tourists is crucial. Green tea branding for Shizuoka could use images of Mt. Fuji and tea fields; however, small-scale branding for particular villages should use distinctive features of the community area.

Green tea tourism relies on interaction-based and relaxation-based activities. Targeting tourists in these two categories requires understanding their different characteristics and market segmentation.

This concept can be applied to national and global tea regions with different types of tea production. For example, at the national level, tea tourism destinations in Kyoto can focus on branding the region as the capital of Japanese matcha (green tea powder). Branding global tea-producing areas as tea tourism destinations in different countries should consider their unique aspects of tea production and link the two types of tea tourism activities: interaction and relaxation. This concept contributes to the tea tourism literature and the development and revitalisation of tea-producing regions through tourism to counter the tea industry's decline and take advantage of the increasing consumption of regional tea products.

Discussion questions

1   How could tea plantation landscapes impact the tea industry and tea tourism in Shizuoka?
2   In what way could green tea tourism in this tea-producing region revitalise the declining tea industry?
3   What are the differences between relaxation-based green tea tourism and interaction-based green tea tourism? How does this relate to tea tourism potential in this tea-producing region in terms of marketing strategy?

References

Chen, H.-S., Huang, J. and Tham, A., 2021. A systematic literature review of coffee and tea tourism. *International Journal of Culture, Tourism and Hospitality Research*, 15(3), 290–311.
Cheng, S.W., Xu, F.F., Zhang, J. and Zhang, Y.T., 2010. Tourists' attitudes toward tea tourism: a case study in Xinyang, China. *Journal of Travel & Tourism Marketing*, 27(2), 211–220.
Dodd, T., 1995. Opportunities and pitfalls of tourism in a developing wine industry. *International Journal of Wine Marketing*, 7(1), 5–16.
Hall, C.M., 2003. *Wine, food, and tourism marketing.* New York: The Haworth Hospitality Press.
Hall, C.M., Sharples, L. and Macionis, N., 2000. *Wine tourism around the world: development, management and markets.* Oxford: Butterworth-Heinemann.
Hasegawa, H., 2005. *Rediscovering Japan: green tourism.* Tokyo: Japan for Sustainability. Available from: www. japanfs.org/en/news/archives/news_id027801.html [Accessed 1 September 2021].
Ishii, I., 2018. *Sekai ga kyōgaku gaikokujinkankōkyaku o yobu Nihon no kachi patān (Inbound tourism strategy of Japan: astonishing the world).* Tokyo: Nikkei BP.
Iwasaki, K., 2017. *Nōgyō no māketingu kyōkasho: Shoku to nō no oishī tsunagi kata (Agricultural marketing textbook: a delicious connection between food and agriculture).* Tokyo: Nikkei Publishing.
Jolliffe, L., ed., 2007. *Tea and tourism: tourists, traditions and transformations.* Clevedon: Channel View Publications.
Khaokhrueamuang, A., 2021. International exchange in tea tourism: reconceptualizing Japanese green tourism for sustainable farming communities. *In*: Sharpley, R. and Kato, K., eds. *Tourism development in Japan: themes, issues and challenges.* London: Routledge, 140–159.
Khaokhrueamuang, A., Chueamchaitrakun, P., Kachendecha, W., Tamari, Y. and Nakakoji, K., 2021. Functioning tourism interpretation on consumer products at the tourist generating region through tea tourism. *International Journal of Culture, Tourism and Hospitality Research*, 15(3), 340–354.
Kraftchick, J.F., Byrd, E.T., Canziani, B. and Gladwell, N.J., 2014. Understanding beer tourist motivation. *Tourism Management Perspectives*, 12, 41–47.
Ministry of Agriculture, Forestry and Fisheries, 2003. *Nogyo Keieitokei Chosa (Agricultural management statistics survey)* [online]. Tokyo: Ministry of Agriculture, Forestry and Fisheries. Available from: www.maff.go.jp/j/tokei/kouhyou/noukei/index.html [Accessed 1 September 2021].
Ministry of Agriculture, Forestry and Fisheries, 2013. *Green tourism: enjoy Japanese rural areas.* Tokyo: JTB Tourism Research and Consulting Company.
Ministry of Agriculture, Forestry and Fisheries, 2020. *Current outlook of Japanese Tea, December 2020* [online]. Tokyo: Ministry of Agriculture, Forestry and Fisheries. Available from: www.maff.go.jp/e/policies/agri/attach/pdf/tea_202012.pdf [Accessed 1 September 2021].
Ohe, Y., 2014. Characterizing rural tourism in Japan: features and challenges. *In*: Diaz, P. and Schmitz, M., eds. *Cultural tourism.* Southampton: WIT Press, 63–75.
Pearce, P., 2005. *Tourist behaviour: themes and cultural schemes.* Bristol: Channel View Publications.

Pearce, P. and Lee, U., 2005. Developing the travel career approach to tourist motivation. *Journal of Travel Research*, 43(3), 226–237.

Robinson, M. and Novelli, M., 2007. Niche tourism: an introduction. *In*: Novelli, M., ed. *Niche tourism*. London: Routledge, 1–11.

The Executive Committee World O-Cha (TEA) Festival, 2010. 4th international conference on O-Cha (tea) culture and science [online]. Available from: www.ochafestival.jp/archive/2010/english/10_tea_conference/index.html [Accessed 1 September 2021].

PART IV

Innovation and practice in tea tourism

24

GASTRONOMY AND TEA TOURISM

Tea-oriented gastronomy tours in Rize, Turkey

Gulsun Yildirim

Introduction

In Turkey, tea is produced in the provinces of Rize, Trabzon, Giresun and Artvin in the Eastern Black Sea region and is widely consumed throughout the country. About 201,000 farmers in the area of 787,000 ha in the Eastern Black Sea region are engaged in tea farming. The tea crop harvest has increased and reached 1.25 to 1.3 million tons on average in the last four years, and in 2020 the harvest production had increased to 1.4 million tons. There are 207 tea factories in Turkey. Most of these factories are in the province of Rize due to the production of tea in Rize and its surroundings (Rize Mercantile Exchange 2021). Being the place of main tea production, Rize Province is attractive for tea-oriented gastronomy tours. Gastronomy tourists will have the opportunity to taste and experience the brewing and service of tea there. They can also witness tea-processing stages and the harvest of tea in this province.

This chapter considers the development process of tea-oriented gastronomy tours in Turkey and identifies the problems of developing the tour program. This study is original in terms of providing in-depth information about the entire process of organising tea-oriented gastronomy tours in Turkey for the first time. Tea-oriented tourism activities are new in Turkey, and there is limited literature on this topic. In this respect, this chapter will extend the existing literature on the emergence of tea-oriented gastronomy tours by providing information about the development process of such tours in Turkey, a developing country as to tea tourism activities.

Literature review

Tea tourism

Camellia sinensis is grown in different countries of the world (e.g., China, India, Iran, Sri Lanka, Malaysia, Turkey, Indonesia) and widely consumed as a beverage (Chen *et al.* 2017). In addition to being a drink for pleasure, tea is used in some countries for therapeutic purposes or in religious rituals (Kurt 2020).

Tea tourism is defined as "tourism that is motivated by an interest in the history, traditions, and consumption of tea" (Jolliffe 2003, 136). India, Nepal, China, Sri Lanka and Japan are tea tourism hotspots where both personalised and commercial forms of tea tourism are practiced (Jolliffe

DOI: 10.4324/9781003197041-29

and Zhuang 2007). Jolliffe (2007, 10) describes the tea tourist as "a tourist who experiences tea consumption in relation to history, culture and traditions". Tea tourists may stay in bungalows inside tea gardens and participate in tea-related activities such as picking tea leaves, tasting tea, walking in tea gardens, visiting tea factories, and purchasing tea and tea-related products (Chen *et al.* 2017).

Tea-oriented gastronomy tourism

Gastronomy tourism was described by Wolf (2002) as travel in order to search for, and enjoy prepared food and drink and unique and memorable gastronomic experiences. The term "tea-oriented gastronomy tourism" used in this chapter refers to travelling in order to search for and enjoy different tea cultures through picking tea leaves, processing and tasting tea. Tourists participate in tea-oriented tours for different reasons. These include viewing tea gardens, visiting tea museums, buying tea and tea-related products, experiencing tea processing and tasting, according to Jolliffe (2007). It is particularly important for some tourists to taste the tea produced in different cultures, geographies and soils. The main purpose of these dedicated tourists is to taste and experience different teas (Cheng *et al.* 2012; Sohn *et al.* 2014). Examples are the teas of the West Bengal, Assam and Darjeeling regions of India (Goonwalla and Neog 2011), the Ceylon Tea of Sri Lanka, and the tea of the Suzhou, Hangzhou and Jiangnan regions of China, all of which are highly noted around the world.

In China, tourists can taste tea by getting information about the history of tea, types of tea, traditions about tea and tea stories during their visit to the National Tea Museum of China (Weber 2019). In Sri Lanka, tourists can also participate in activities such as visiting tea gardens, watching tea pickers in tea gardens, tasting tea and buying tea-related souvenirs (Zhou 2011). Utilising tea within gastronomy tourism does not only mean tea tasting, but also it is important for tourists participating in this tourism activity to see and experience tea pairings with food.

Method

For this chapter, a case study method was used with the emphasis on a single case with multiple embedded units. Case study models are used to provide detailed descriptions of a particular phenomenon. While only one case can be analysed within the single case pattern, a detailed evaluation of all layers and layers related to the situation can be provided with the single case with embedded units. If more than one case is examined in the research, the multi-case pattern is more suitable (Şimşek 2012). When the phenomenon is considered important, it is possible to reveal the variables that make up the situation, to determine the interactions between variables or to compare different situations with data collection techniques such as document review, observation and interview. A single case with embedded units has been identified as being suitable for this research (Beeton 2005).

The first tea-oriented gastronomy tour in Turkey at Rize is the case studied. This tour was examined from the perspective of both the supplier and demand side. Focused interviews with the manager of the Tamitur Tourism and Travel Agency and five customers of this tour were conducted between 11 June and 6 July 2021. The author also participated in several field trips with the manager of the travel agency on 20 March and 12 April 2021 at the stage of organising the tour, and after that the author participated in the prepared tour offered to customers. When participating in the field trips, the author took field notes and photographs. Thus, in addition to the focused interview, the participant observation method was also used as a data collection tool.

This particular travel agency was chosen because the manager has been trying to develop the first tea-oriented gastronomy tour in Rize for more than a year. Interviews with the manager of the travel agency and five tourists were conducted. The interviews were recorded with prior permission granted by interviewees. The interviews with the tourists who participated in the tour were conducted to examine their perspectives towards the tea-oriented gastronomy tour. In addition, written feedback after the tour was requested from customers. In the study, the descriptive analysis method was used to analyse the data obtained. During the data analysis process, participant data recorded in the audio files were transferred to a Word document. During this transfer, a code number was given to the participants. Direct quotations were used in the text. The research design focuses on both supply and demand side and presents a holistic view about newly developing tea-oriented gastronomy tours in Turkey.

Case study context

Tea started to be adopted in Anatolian culture through Central Asian Turks. Information about the trade of tea in Anatolia in 1631 is included in Evliya Çelebi's travel book (Kuzucu 2012). During the reign of Abdülhamid II, there were attempts to cultivate tea in Anatolia. The first attempts for tea cultivation were in the province of Bursa in Anatolia, but this province was found to be not very suitable for tea cultivation (Caykur Statistical Bulletin 2019). In the last period of the Ottoman Empire, Hulusi Bey realised that the climatic conditions of Rize Province and Georgia/Batumi, where tea was grown, were similar. Thus he planted the tea seed he bought from Batumi in his own garden in 1912. Again, a report was prepared on tea plant and tea cultivation in the Ottoman period (Türkyılmaz 2015). After the establishment of the Republic of Turkey, legal regulations were put in place regarding tea cultivation in 1924, and the first black tea harvest was made in 1938. With the increase in tea production, the General Directorate of Tea Enterprises (CAYKUR) was established (Alikılıç 2016).

In Turkey, tea is mainly produced in the Eastern Black Sea region and in the province of Rize located in this region. Most of the tea fields in Turkey are in Rize Province with fields of 527,715 decares, followed by the provinces of Trabzon (150,237 decares), Artvin (91,343 decares), Giresun (16,332 decares) and Ordu (66 decares; Tea Industry Report 2018). There are eight clones registered as green leaves in this region. In production, Turkey uses the Orthodox System, and around 97% of its production is black tea. Black tea is also classified into the categories of first, second, and third harvests. In addition, in the classification, tea is in powder form both in the first and second harvest. Although different tea production has been started in recent years, this is not very high in volume. There are also boutique teas processed, such as green tea, white tea, oolong, and both green and black tea production by hand. Tea is a part of life in Turkey and is the most consumed beverage after water, reaching the status of a national beverage. CAYKUR, a public organisation, is responsible for half of the production, and private sector factories produce the other half. The tea plant reaches maturity in 45–50 days on average. There are three harvest times in Turkey. The first harvest begins in May, the second in July and the third in September (Tea Industry Report 2018).

Turkish tea that is brewed in a teapot and drunk in a tulip-shaped glass is a roasted powder black tea. In Turkish culture, tea is an indispensable part of breakfast. Turkish tea, which has an important place in Turkish culture, can be consumed not only at breakfast but at any time of the day. The food accompanying tea and the traditions of drinking tea show differing characteristics in different regions of Turkey. Turkish tea is not utilised enough in the tourism sector. While there are various tea-oriented tours around the world, such tours are not organised in Turkey. Recently, a tea garden in Rize has been recognised on social media, and as visits to that tea garden have increased, the activity of such visits has been added to the cultural tour or eco-tour programs.

Case study 24.1   Development of tea-oriented gastronomy tours supplier perspective

Burak Avcı (41 years old), who organised the first tea-focused gastronomy tour in Turkey's Rize Province, is the manager and owner of the travel agency named Tamitur. He has a master's degree in tourism and has been working in the tourism sector for 16 years. Avcı's agency mainly serves the domestic tourism market and generally organises boutique and eco-tours. He thinks that every detail of tea has a cultural value and tea is an important touristic product in terms of a niche market, at every stage from the field to being put into a glass to drink. He realised tea should be included in the tourism offer in Rize, like wine and coffee, and decided to organise tea tours with the aim of taking a priority place in the market. His primary goal is to make money and to contribute to the development of the local people by increasing the business volume with the money he earns. Burak Avcı encountered many challenges in organising the tea-oriented tour and received support from university academics and CAYMER (Tea Research Center) officials during this process. The difficulties faced by Burak Avcı in organising the tea-oriented tour were identified as follows:

- sectoral competition, due to the fact that the perception and culture of tourism in the province and the region is not yet at a conscious level;
- the difficulties of adjusting the balance between service quality and commercial gain;
- difficulties in finding a consistent and conscious service or operation provider enterprise;
- the fact that the tea product is seasonal and that it is produced and processed in limited times during the season creates an obstacle to continuity; and
- the possibility of developing and diversifying the organised tour with activities for tea is very limited.

There are three different tours organised by the agency with a focus on tea covered in the research. The first tour is a one-day tour and it includes "tea picking, tea processing, tea training and tea tasting". Support is received from CAYMER (Tea Research Center) for these four activities. CAYMER is an institution that started its activities in 2018 under the management of Rize Commodity Exchange and Rize Chamber of Commerce and Industry, with the joint financing of the European Union and the Republic of Turkey in order to improve the competitiveness capacity of the tea industry concentrated in and around Rize. The travel agency gets a lot of services (tea picking, tea processing, tea training, tea tasting) from CAYMER in the delivery of the tea-oriented tour, since there are not enough qualified personnel and facilities to host the tea culture and tea-related tourists in Rize or the provinces where tea is produced. In the first tour, there are four different tea tastings. In the second tour, there is no tea-picking activity in the tea garden, and the participants in the tour experience the processing and tasting of the picked tea. In addition, training on tea is also included in the tour. In this tour six types of tea are tasted and this tea tasting lasts for one hour. In the third tour, the tea is not collected from the garden and the tea is not processed by hand, but education of tea culture and a tea-tasting workshop are conducted. This tour offers 12 different tea tastings (Figure 24.1).

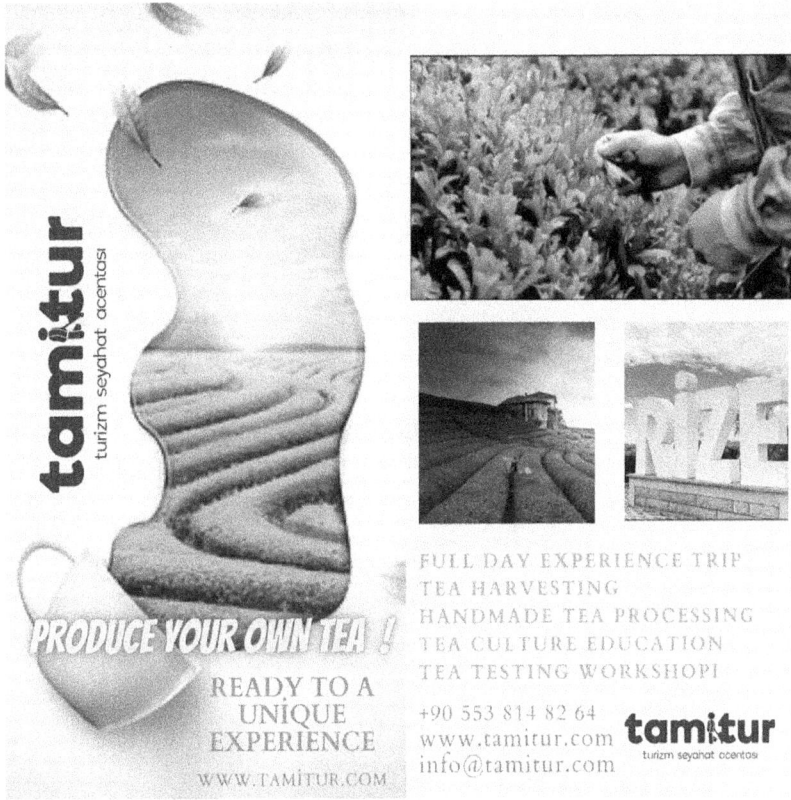

Figure 24.1   Tea tour poster

Source: Burak Avcı

Case study 24.2   Tea-oriented gastronomy tour experience customers' perspective

The tour set off at 8:30 a.m. with a group of five people. First, they went to CAYMER and were provided information on various subjects such as what tea is and how it is consumed in the world and in Turkey, and the importance of Rize in terms of tea. After half an hour of introductory information, they set out for the tea garden. When they arrived in the tea garden, the local tea-picking clothes, gloves, basket and such materials to be used were given to the tour participants (Figure 24.2). Then, practical training was provided on how to pick tea. Information on subjects such as two leaves and one bud, five leaves, white tea, stem and garbage, tea plant, wild grass, and garden care is also given. During the tea plucking, conversations were also held about local culture. Tea was collected in the tea garden for an hour and a half, and then they had a lunch break. After a one-hour lunch break, customers went to CAYMER to visit the tea-processing and production area and to process and taste the collected tea.

First, the tour participants were shown the method of hand-processing the tea, and then they were processed the tea they collected themselves. After about two hours of hand-processing, the ready-made tea was brewed and served to the tour participants (Figure 24.3). Then, the stages of processing in the factory of the tea collected in the region were shown and the tea production area in the factory was visited. Finally, one of the most important stages of the tour, information about all types of tea in this region (brewing method, water, water temperature, teapot feature, brewing time, infusion) was provided and the tea-tasting stage was started. Tasted teas were handmade Turkish green tea (Turkish premium green), handmade Turkish oolong, handmade Turkish black tea, sifter Turkish black tea, ice Turkish green tea and ice Turkish oolong. Before the tasting, tour participants were given a form to evaluate the tea offered for tasting. The participants were asked to write their comments about the appearance of the tea, the smell of the tea and the taste of the tea and to note down the points they liked and anything they wanted to add, if any. After the tasting, the participants of the tour gathered in the meeting room and the participants were presented with a packaged form of the teas that they had themselves collected. Each participant was given a certificate showing that they had experienced the whole journey of tea "From Tea Plant to Cup".

Figure 24.2 Tea picking in the tea field
Source: Author

Two academics from the field of gastronomy, two doctors and the researcher who conducted this study, participated in the tour, where the number of participants was limited to 4–12 people. Two of the participants in the tour were undergraduates, one of them held a doctorate and one of them was a graduate. All four of the participants participated in the tour from the provinces of Turkey other than Rize. Three of the tour customers had not participated in any tea-themed tour before. Only one of the customers had participated in a tea garden landscape tour. Tour participants joined the tour to experience the tea during the process from the garden to the glass. The expectations of three of the participants were met. In fact, one of these three people stated that the tour was far beyond their expectations. One of the tour participants stated that 90% of his expectations from the tour were met, and there were some minor deficiencies in the stages of tea picking and tea tasting. He stated that there are some difficulties in directing which leaves (two leaves and one bud) of the tea plant to collect during tea collection. In addition, it was noted that the necessary guidance was not given to the participants about wearing gloves and bonnets during the hand-processing phase of tea when they came to CAYMER. During the tasting, it was emphasised that the time allocated for tasting the teas was insufficient, and that more time should be given in order to examine the colour of the teas, to consider the smell, to feel the taste completely in the mouth and to take notes. In addition, in order to prevent mixing of the flavours of the tea varieties, crackers and so forth, as well as water, should be given to the tasters to neutralise the taste in the mouth. It was also stated that it is not correct to taste more than six products in one session. Even if more than six products are to be tasted, it is stated that a break of three to five minutes should be taken after each tasting.

Some suggestions were thus received from the participants to improve the tour. Since it is an intense and active tour, this includes increasing the rest breaks, brewing tea after the collection process in the tea field and drinking tea with local food of Rize Province. Participants liked the tasting part of the tour the most. Picking tea leaves from the tea garden with their own hands and tasting the tea after processing it themselves was the most valuable part of the tour for the participants. All of the participants stated that they liked the tour very much and would recommend it to others.

Figure 24.3   Tea tasting
Source: Author

Discussion

Tea-oriented gastronomy tours are offered especially in countries where tea varieties are abundant (Weber 2019). While Turkey is not at a level that can be compared with countries such as China and India, where tea history is very old and tea varieties are abundant, Turkish tea has an important potential as a touristic product and tea needs to be integrated into the tourism sector. In this context, the process of organising the first tea-oriented tour has been examined as a case study, and it has been seen that some problems have arisen in the context of integrating tea into tourism. Among these problems, the scarcity of service providers in tea-oriented tour areas and the inadequacy of trained personnel about tea came to the fore. Another prominent problem in the organisation of tea-oriented tours was the pricing of the tea tour. Tea tours are generally aimed at tea lovers or the niche market for tea experts (Jolliffe 2003).

In addition, limited capacity opportunities in Turkey such as with tea museums, accommodation establishments and restaurants prevent tea tours from attracting large groups. Therefore, in the attempt to organise a high-quality tour for a limited market, the high operational costs increase the tour price. This caused the agency manager to worry about whether he would be able to find enough tour customers. In addition, while tea cultivation is carried out for 12 months in tropical and equatorial regions, tea production is carried out only six months of the year in countries with high latitudes such as Turkey. This causes the organised tour to be held within a certain time frame. The problems of how this tea-oriented tour organised in Rize will be promoted in Turkey and abroad and whether sufficient demand can be created are also of importance. For a travel agency development and design of a new product as with this tea-oriented gastronomy tour is very risky and requires big costs. Hence marketing is very crucial for attracting new customers to the new product (Dudić, Dudić and Mirković 2016). A marketing department of a travel agency should be included in the new product development process (Kotler and Armstrong 2001). However, the travel agency studied is not a big company that has a marketing department.

In the analysed case, the tea gardens used by tour customers are the tea gardens of the local people. The owner of the garden is paid some money for the incoming tour customers to use the tea garden. In addition, tour customers participate in weeding the gardens and picking tea in the province of Rize, which has a labour shortage of tea workers. In a sense, tour customers support the local people living in rural areas both economically and in terms of labour. Therefore, this type of tourism provides important contributions to rural development. In this tea-oriented tour, local cultural elements such as welcoming the tour customers with local music, collecting tea by wearing local clothes in the tea garden and explaining the rituals of the local people while collecting tea were highlighted. In addition, the fact that the organised tea-oriented tour appeals to small groups rather than mass tourism will have a positive effect in preventing natural areas from being damaged. For these reasons, it would not be wrong to say that this tour carries the elements of sustainable tourism.

Conclusion

The main aim of the study was examining the emergence and development and problems of the first organised tea-oriented gastronomy tour in Rize, Turkey, a developing country in terms of tea tourism. A case study model was used as a method, and a single case with embedded units was used as a design. The first tea-oriented gastronomy tour here is the case in point. This tour was examined from both the supplier and demand side perspectives. Focused interviews with the manager of the travel agency and the customers of the tour were conducted. In addition to the focused interview, the participant observation method was also used to collect data.

The first participants of this tea-oriented tour, which was held for the first time in Rize, consisted of people who are not experts in tea tasting but want to experience tea "from the garden to the glass". The tour was highly appreciated, and only a few participants emphasised that the tour should be further developed. It has been revealed in this study that this tea-oriented tour can be attractive to people who want to experience tea, other than those who have a special interest in tea or are tea experts. However, it is necessary to conduct research on how many people will be willing to pay for this tour, which provides a high-level tea experience, and what the demand for it will be. The manager of the travel agency is not well informed about new product development processes and does not have a marketing department in the agency. Because of this, he has some concerns about marketing and tour pricing.

This chapter contributes to the gap in the literature in the context of tea-oriented gastronomy tourism in Turkey. In addition, this study highlights the problems encountered in the organisation of such tours in practice. In this way, the study provides important contributions in practice in terms of being a guide for sector entrepreneurs. A holistic perspective has been taken in this study, in which the experiences of an entrepreneur and travel agency owner trying to gain a place in the field of tea-oriented touristic activities and tour experiences have been explored.

This study has been limited in that it has been carried out with only one travel agency in one country and one province with data collected from the customers of a tour organised by that agency. In future studies, comparisons can be made by examining the contents of tea-oriented gastronomy tours in different countries and different regions. Solutions can also be developed in terms of overcoming problems by revealing practical difficulties and the differences that exist in such obstacles both between countries and regions. Finally, it is important for further research to examine tea-oriented gastronomy tours in the context of sustainable tourism, as well as the perceptions and attitudes of the local people towards these tours.

Discussion questions

1   What is the relationship between tea-oriented gastronomy tours and tea tourism?
2   Are there any tea-oriented gastronomy tours in different countries of the world?
3   What are the roles of tea-oriented gastronomy tours in terms of sustainable rural development?

References

Alikılıç, D., 2016. Çay'ın Karadeniz Bölgesi için önemi ve tarihi seyri (The importance and historical course of tea for the Black Sea Region). *Karadeniz İncelemeleri Dergisi*, 11(21), 269–280.
Beeton, S., 2005. The case study in tourism research: a multi-method case study approach. *In:* Ritchie, B.W., Burns, P.M. and Palmer, C.A., eds. *Tourism research methods: integrating theory with practice*. Wallingford: Cabi, 37–48.
Caykur Statistical Bulletin, 2019. Çay İşletmeleri Genel Müdürlüğü, Yıllık İstatistik Bülteni, Rize. Available from: www.caykur.gov.tr/Pages/Yayinlar/YayinDetay.aspx?ItemType=1&ItemId=701 [Accessed 5 October 2020].
Chen, Y., Jafar, R.M.S., Morley-Bunker, M., Lin, C., Chen, L., Wu, R. and Zhuang, P., 2017. On the Marketing Mix of Fujian Tea Tourism. *In: International conference on social science, public health and education (SSPHE 2017)*. Amsterdam: Atlantis Press, 127–137.
Cheng, S., Hu, J., Fox, D. and Zhang, Y., 2012. Tea tourism development in Xinyang, China: Stakeholders' view. *Tourism Management Perspectives*, 2, 28–34.
Dudić, Z., Dudić, B. and Mirković, V., 2016. An important determinant of the success of innovations – the commercialization of innovations. *Contemporary Issues in Economics, Business and Management*, 21.
Goonwalla, H. and Neog, D., 2011. Problem and prospect of tea tourism in Assam-a SWOT analysis. *In: 2011 International conference on advancements in information technology with workshop of ICBMG*, 20, 243–248.
Jolliffe, L., 2003. The lure of tea: history, traditions and attractions. *In:* Hall, M., Sharples, L., Mitchell, R., Macionis, N. and Cambourne, B., eds. *Food tourism around the world: development, management and markets*. London: Butterworth-Heinemann, 121–136.
Jolliffe, L., ed., 2007. *Tea and tourism: Tourists, traditions and transformations*. Clevedon: Channel View Publications.

Jolliffe, L. and Zhuang, P., 2007. Tourism development and the tea gardens of Fuding, China. *In:* Jolliffe, L., ed. *Tea and tourism: tourists, traditions and transformations.* Clevedon: Channel View Publications, 133–144.

Kotler, P. and Armstrong, G., 2001. *Principles of marketing.* 9th ed. Upper Saddle River, NJ: Prentice Hall.

Kurt, T., 2020. Dünyada Çay Kültürü ve Seramik Çay Kapları (Yüksek Lisans Tezi) (Tea Culture in the World and Ceramic Tea Pots, Master Thesis [online]. Available from: https://tez.yok.gov.tr/UlusalTez-Merkezi [Accessed 13 June 2021].

Kuzucu, K., 2012. *Bin yılın çayı: Osmanlı'da çay ve çayhane kültürü* (Tea of the millennium: Tea and teahouse culture in the Ottoman Empire). Kapı Yayınları.

Rize Mercantile Exchange, 2021. Türk Çay Sektörü Güncel Durum Raporu (Turkish Tea Industry Current Status Report) [online]. Available from: www.rtb.org.tr/tr/cay-sektoru-raporlari]. Available from: https://tez.yok.gov.tr/UlusalTez-Merkezi [Accessed 13 June 2021].

Şimşek, A., 2012. *Sosyal Bilimlerde Araştırma Yöntemleri (Research Methods in Social Sciences).* Eskişehir: Anadolu Üniversitesi Açık Öğretim Fakültesi Yayınları.

Sohn, E., Yuan, J. and Jai, T.M., 2014. From a tea event to a host destination: linking motivation, image, satisfaction and loyalty. *International Journal of Tourism Sciences*, 14(3), 1–23.

Tea Industry Report, 2018. Çay Sektörü Raporu (Tea Industry Report) [online]. Available from: www.caykur.gov.tr/Pages/Yayinlar/YayinDetay [Accessed 5 October 2020].

Türkyılmaz, K., 2015. *Türkiye'de Çaylık Alanların ve Üretici Sayılarının İstatistiksel Analizi.* (Statistical Analysis of Tea Areas and Number of Producers in Turkey). Gıda Sanayii Özel İhtisas Komisyonu Raporu, 1–113.

Weber, I., 2019. Tea for tourists: cultural capital, representation, and borrowing in the tea culture of mainland China and Taiwan. *Academica Turistica-Tourism and Innovation Journal*, 11(2), 15–22.

Wolf, E., 2002. Culinary tourism: A tasty economic proposition [online]. Available from: www.culinarytourism.org [Accessed 13 September 2021].

Zhou, M.I., 2011. *Exploration of factors associated with tea culture and tea tourism in United States, China, and Taiwan,* Master's Thesis. Available from: https://libres.uncg.edu/ir/uncg/listing.aspx?id=8331 [Accessed 10 October 2021].

25

INTERNATIONAL EXCHANGES AND GASTRONOMICAL TEA TOURISM

Amnaj Khaokhrueamuang, Piyaporn Chueamchaitrakun and Kazuyoshi Nakakoji

Introduction

In recent years, interest in gastronomic tourism has grown through the promotion of regional identity, economic revitalisation and heritage conservation. Destinations worldwide are now developing strategies to develop a culinary identity that will significantly influence tourists' decision to travel. Tea is among the agricultural products used to create distinctive gastronomic tourism experiences and can be a tool to revitalise both the tourism industry and farming communities. Therefore, gastronomical tea tourism has raised a critical application for tea-producing villages with declining tourism to change their tea culture into food and tourism products.

This chapter presents how tea culture and tea products can be created in gastronomic tourism to revitalise small enterprise tea communities in Thailand and Japan, specifically, Phaya Phrai village in Chiang Rai Province, Thailand, and Umegashima in Shizuoka Prefecture, Japan. The study seeks to find innovative ideas for creating gastronomical tea tourism through an international exchange programme between the two cases in different positions in Butler's (2006) tourism area life cycle (TALC). Its purpose is to create a conceptual model of innovative gastronomical tea tourism as a grounded theory derived from the international exchange of the cases.

Umegashima, a highland tea-growing community in Shizuoka, has nurtured hot spring tourism for over a century but is encountering a decline in domestic tourist numbers, which are assumed to have reached the stagnation stage. Marketing inbound tourism may be required to make up the gap. With the international exchange project, the village's strategy to rejuvenate both tourism and tea industries is to use gastronomical tea tourism to target Thai visitors. Meanwhile, Phaya Phrai, a highland tea-growing village in Chiang Rai, requires domestic tourists and can benefit from the lessons Umegashima learned at the exploration stage of the tourism business. Thus, international exchange in gastronomical tea tourism can offer tourism development guidelines for Phaya Phrai.

Literature review

Relationship of gastronomy, tea tourism and international exchange

Gastronomy is the art of cooking and eating fine food and the study of the relationship between culture and food (Gets *et al.* 2014; Carpio *et al.* 2021). Gastronomic tourism is a type of tourism activity characterised by the visitor's experience linked with food and related products and activities

DOI: 10.4324/9781003197041-30

while travelling. Along with authentic, traditional and/or innovative culinary experiences, gastronomic tourism may also involve related activities, such as visiting local producers, participating in food festivals and attending cooking classes (UNWTO 2019).

According to the UNWTO (2019), gastronomic tourism can provide the following benefits:

1   Differentiation and unique positioning of a region;
2   New values and experiences for visitors;
3   Implementation in less developed regions and those lacking in tourism resources (possible even in small villages);
4   High revenue for the region and a desire to return or loyalty among visitors.

Since local gastronomy can be easily accessed by tourists, as it is an inseparable part of any destination's products, local food has become an integral element of the tourist experience. Local food also represents intangible customs and heritages that result in the unique characteristics of the destination (Carpio *et al.* 2021). Based on Butler's TALC model (2006), gastronomy has an important leading power in the destination and life curve, from the discovery of the destination to the period of regression when demand decreases (Sahin 2015, 98). It also enhances tourists' intention to revisit the destination and the branding of the destination (Guruge 2020). Thus, several countries consider gastronomic tourism as an opportunity to develop rural areas, achieve sustainable production, reduce rural-urban migration and poverty, promote social entrepreneurship, effectively use resources and protect social heritage (Basaran 2020).

Similar to tea-growing communities, gastronomical tea tourism is a concept that aims to not only promote tourism but also sustain the tea industry. However, limited attention has been paid to the potential of supporting tea-growing communities in rural areas through gastronomy and tea tourism, especially in rural Japan. In particular, participation in tea tourism by international visitors is minimal (Khaokhrueamuang 2021). In the case of Japan, inbound repeat tourists seek unique experiences in rural areas and, consequently, their increased interest in Japanese tea may offer a potential resolution to the decline of the tea industry.

One strategy for targeting inbound tourists to farming communities, such as tea villages, is international exchange. However, international exchange projects related to tea and tourism have garnered limited attention. Online research data reveal that the main research focus is agriculture and entrepreneurial exchange. For example, the US Department of State has sponsored a two-way global exchange programme for young entrepreneurs from African countries and the United States to promote mutual understanding, enhance leadership skills and build lasting, sustainable partnerships between mid-level emerging leaders (Jayaratne *et al.* 2017). Higher education institutions and work-study abroad programmes have been instrumental in internationalising sustainable agriculture education, which addresses organic farming experience internships in the United States of participants from Peru and Ecuador (Cody 2017).

Regarding gastronomical tea tourism, the tea farmer exchange project between Japan and Thailand granted by the Center for Tourism Research, Wakayama University (Khaokhrueamuang and Chueamchaitrakun 2019), shows that the tea cuisine exchange activity is a highlight of tea tourism in tea-growing communities.

Description of the International Exchange Programme and Case Studies

The international exchange on gastronomical tea tourism is funded by the Toyota Foundation International Grant Program 2020, 'Cultivating Empathy through Learning from Our Neighbors: Practitioners' Exchange on Common Issues in Asia'. The programme focuses on depending mutual understanding and knowledge sharing among people in East and Southeast Asia. The international exchange programme aims to obtain new perspectives and expand the potential of future generations. The Toyota Foundation International Grant Program 2020 received 140 applications from targeted countries and approved nine grants for fiscal year 2020. Titled 'Revitalizing Tea Industry Community through Gastronomical Tea Tourism', the project will last from November 2020 to October 2022. It focuses on tea and tourism business exchange between Thailand and Japan by reviewing two cases: Phaya Phrai and Umegashima.

Phaya Phrai is a tea-growing village in the highland areas of Chiang Rai Province, Thailand. The village population is Akha, an ethnic group that migrated from southern China. Tea products are processed from three types of tea variety grown and harvested by the Akha: Assam (*Camellia sinensis* var. *assamica*), Chinese (*C. sinensis* var. *sinensis*) and oil tea (*C. sinensis* var. *oleifera*). Most of the tea products are for a tea-processing company to produce ready-to-drink tea. Ten years ago, villagers earned high income from the tea leaves sent to the factory. Recently, however, revenues from tea production have gradually decreased due to the low-price rate of tea leaves in Thailand. Therefore, the village leaders are interested in developing the village as a tea tourism destination to generate more income and provide jobs for young generations. According to Butler's (2006) TALC model, Phaya Phrai's situation is at the exploration stage, where tourism resources are first examined to create tourism products. The village needs to learn how to start and manage tourism from a successful case study. It can learn from Umegashima, a tea-growing community in Shizuoka's hot spring tourism destination.

With its aging population and low birth rate, Japan has encountered a lack of young farmers in rural communities. This problem causes the decline of a farming community like a tea-growing village and the regression of the local economy in rural tourism destinations. The problem is seen in Umegashima, a famous hot spring tourism village in the old days. Based on Butler's TALC model, Umegashima is currently at the stagnation stage, describing the halt in growth of the tourism business due to the decreased number of domestic tourists, who are the main target. A strategy proposed to solve this problem and rejuvenate the tea industry and tourism is to target inbound tourists, such as Thai visitors, with gastronomical tea tourism. The Thai tourist market ranks sixth among international tourists visiting Japan, with approximately one million people annually. Most Thai people prefer hot springs and Japanese cuisine.

Furthermore, there is a new trend of drinking Japanese green tea in new generations influenced by Japanese tea products in Thailand, such as the Shizuoka Green Tea brand (Khaokhrueamuang *et al.* 2021). This phenomenon is an excellent opportunity for Umegashima to take gastronomical tea tourism and attract Thai people to visit the village. By exchanging ideas with Phaya Phrai, Umegashima can build a tourism network with Thailand to help target Thai tourists.

Conceptual framework for international exchange in gastronomical tea tourism

The conceptual framework used for the exchange programme is based on four elements of tea culture commodification in tourism identified by Khaokhrueamuang and Chueamchaitrakun (2019), namely, tea spaces, tea communities, tea products and services, and tea-related activities.

Tea spaces involve tea plantation areas connecting with other farming systems that can be commoditised as places for tea-related tourism activities, such as tea plucking, a tea festival and tea folk music performances. Tea communities mean groups of tea producers or farmer households in the same environment of tea spaces and living with tea ways of life. These communities represent both tangible and intangible cultures regarding commodification. The tangible culture concerns tea farmers or producers' real estate, such as farmhouses or tea factories, commoditised as tea lodges, homestays, farm-stays or restaurants. The intangible culture represents customs, beliefs, jobs and lifestyles, including tea making and processing methods, tea cuisine and tea-telling stories. On the basis of the tea spaces and tea communities, products, services and tourism activities related to tea culture are created when commodification progresses. Examples are tea souvenirs, tea farmer restaurants and package tours of tea trails.

Regarding the gastronomical tea tourism project, these elements will be examined and merged as the content of exchange activities. Tea spaces concern the management of tea gardens and the related farming system as the attraction and sources of ingredients for tea cuisines, present in the harvesting calendar. Tea communities focus on managing accommodation and food services. Tea products involve creating a varied menu of tea cuisines and tea-related products. Tea activities relate to the creation of tea tourism routes linking with gastronomy.

Methods

The international exchange programme on gastronomical tea tourism in Umegashima and Phaya Phrai highlights the tea industry's importance in strengthening tourism communities. It is conducted by a project representative based in Japan and members of the two countries. The project representative takes responsibility for managing the programme. The participants of each country include a team director and coordinators in the agricultural network, tourism and media, tea-producing communities and accommodation and food business. The project also involves young farmers when the programme begins. It is implemented through a staged process of participatory action research (PAR) with three phases: pre-exchange, in-exchange and post-exchange.

The pre-exchange programme concerns an orientation the programme, selecting two young farmers to participate in each team, surveying tourism resources; undertaking a strengths, weaknesses, opportunities and threats (SWOT) analysis of the potential of tourism resources; and brainstorming ideas on using the areas' tourism resource potential to create gastronomical tea tourism routes. The study employs Godfrey and Clarke's (2000) tourism resource audit and the harvesting calendar to collect data. The first phase of the project took six months, from November 2020 to May 2021.

The second phase involves exchanging experiences and ideas on creating gastronomic tourism products: tea spaces, accommodations, food services, tea cuisines, tea pairing menus (which combine a specific tea with a particular food) and gastronomical tea tourism routes. This in-exchange programme took six months, from June to October 2021.

The post-exchange programme focuses on organising a gastronomical tea tourism familiarisation trip with pilot tourists and inviting tourism-related organisations and media to observe and evaluate tourist products, such as gastronomic tourism routes and tea cuisine. The final phase takes a year, from November 2021 to October 2022.

This chapter reports the achievement of the first phase. The study employs steps in PAR, which is relevant to both research and action that aim to find new knowledge by the project representative and in-country directors as researchers and with participants' activities leading towards the new solution (Karlsen 1991), such as tea cuisine menus. The research aims to conceptualise ideas on innovative gastronomical tea tourism as a grounded theory derived from the action process of international exchange. The aim of participants' activities is to find the outcomes of the projects, such as tea cuisine and gastronomical tea tourism routes, as tourism products provided for the tourism business. The study employs steps in an active process where action and research are integrated with shared roles and objectives (Karlsen 1991). It takes one round of a cyclical process divided into four stages: plan, action, observation and reflection. The results are discussed with the concept of tea culture commodification in creating tea tourism products.

Results

During the first period, the pre-exchange programme activities were implemented following the steps of PAR.

Thailand's Phaya Phrai village, Chiang Rai

Step 1: Plan

In Japan, the project representative explained the project details, including the implementation plan and budget, via an online orientation to Thai team members and villagers who voluntarily joined the project. In this meeting, the Thai team selected two young tea producers to participate in the programme.

Step 2: Action and observation

The tourism resource survey was conducted between December 2020 and January 2021.

The distinctive natural and agricultural resources related to tea spaces are Assamica tea gardens, Sinensis (Chinese) tea fields, tea oil plantations and rice terraces. These resources can be used to create tea tourism activities throughout the year (Table 25.1).

According to Table 25.1, tea tourism activities involve culture, service and event resources in the tea spaces. Akha culture is the most attractive resource, comprising food, traditional costumes, handicrafts, language, rituals and festivals emerging in the tea community. Service resources are the small number of accommodations, restaurants, tea and coffee shops. Ancestor-respecting rituals and the Akha New Year Festival are also resources.

The SWOT analysis revealed that the strengths are the three categories of tea gardens and Akha culture; weaknesses are the lack of attractive tourist attractions, such as waterfalls, and hospitality skills in the tourism business; and the threat is the high intensity of other attractions' competitiveness. Meanwhile, the opportunity to promote the community and gastronomical tea tourism is the award-winning gold prize from the World Tea Competition 2020 organised by the World Green Tea Association in Japan.

Along with the SWOT analysis, Thai team members conducted a focus group with villagers to brainstorm ideas on creating gastronomical tea tourism routes. However, these routes have not been completed at this stage of the project owing to the lack of attractions and tourism product development. Thus, improvement is required in the project's second period, including tourist spots and information development, tourism promotion and tea products.

Table 25.1 Agricultural resources in Phaya Phrai's tea spaces

Resources	Resource features	Harvesting season	Tourism activities
Assam tea gardens	Natural wild tea plantations growing in the forest or alongside other crops	January–December	Tea picking, tea trail trekking to the forest, visiting Assam green tea factories, tasting green tea and having tea cuisines
Chinese tea gardens	Chinese tea plantations on mountain slopes	January–December	Tea picking, tea trail trekking to see the panoramic view of the tea landscape, visiting tea factories and tasting oolong tea
Tea oil gardens	Tea oil plantations under royal patronage located near the border of Thailand and Myanmar's high mountains	January–December	Tea trail trekking to tea tree oil plantations with splendid mountain views, learning tea oil cultivation and production
Rice terraces	Rice terraces of the tea valley, growing and harvesting by hands	October–November	Rice harvesting, trekking and sightseeing the rice terrace

Source: Authors

During the first period, the programme implemented an additional two activities:

1   *Fieldwork on tourism management at Tha Khan Thong Village, Chiang Rai Province*: This fieldwork was conducted to take the lessons learned to implement from Tha Khan Thong village, the successful case study. Thai team members and the Japanese team director learned and observed how the villagers manage their community tourism business, such as foodservice and accommodation. The experience and ideas of managing community-based tourism contribute towards starting the tourism business for Phaya Phrai village.

2   *Workshop on tea cuisine by inviting a local chef to create the tea cuisine recipe*: This workshop was moved from the second period to the implementation phase because the local chef (instructor) who could create the tea cuisine recipe was only available during this period. The Thai team members and Phaya Phrai villagers participated in the workshop organised by Locus Native Food Lab. Kongwuth Chaiwongkachon, a famous local chef, taught them to create Akha tea cuisines using local ingredients by comparing their differences and similarity with Japanese culture.

Step 3: Reflection

After implementing the plan to develop gastronomical tea tourism, the Thai team took their knowledge and experiences from the fieldwork and workshop to create a unique tea cuisine reflecting Akha food identity. The cuisine comprises four recipes: (1) pork wrapped with tea leaves (Figure 25.1), (2) tea leaves salad, (3) tea leaves in peanut chilli paste and (4) steamed tea leaves in spicy minced pork. These recipes are served with traditional Akha dishes, such as soup, egg and locally

Figure 25.1   Pork wrapped with tea leaves: an innovative tea cuisine of Phaya Phrai village presenting the Akha identity

Source: Authors

grown organic vegetables. The Thai team was likewise encouraged to run the tea tourism business and develop the accommodation, tea café and restaurants through the lessons learned from Tha Khan Thong tourism village.

Japan's Umegashima community, Shizuoka

Step 1: Plan

The orientation was organised at the Yunoshimakan Ryokan, a Japanese style inn with hot springs. It is operated by Mr. Tsuneji Akiyama, tourism and media coordinator of the Japanese team. The project representative explained the project details, such as implementation plan, budget and selection of young farmers to join the programme. However, the young generation's lack of interest in the tea industry made it challenging to select qualified people for the programme.

Step 2: Action and observation

The tourism resource survey was investigated along with interviews, field observation, documents and websites. Interviews were conducted with local people, including accommodation business owners, restaurant owners, tea producers and tourist information officers. The majority of agricultural resources related to tea spaces are traditional tea gardens, modern tea gardens, organic tea gardens, wasabi plantations and shiitake mushroom farms. These resources created a tea tourism season between April and October. However, wasabi-related tourism activities are available throughout the year (Table 25.2).

Table 25.2 Agricultural resources in Umegashima's tea spaces

Resources	Resource features	Harvesting season	Tourism activities
Traditional tea gardens	Zairai, the original botanical subdivision of the Sinensis variety before breeding, has been conserved in bush shape due to a handpicking method	April–May, June–July, September–October	Learning the history of Japanese tea cultivation, tea picking, tea trail trekking across the suspension bridge to the tea landscape
Modern tea gardens in high mountains	Tea plantations on mountain slopes, shaped in rolls due to the portable harvesting machine used	April–May, June–July, September–October	Tea picking, tea trail trekking to see the panoramic view tea landscape
Organic tea gardens	Natural tea plantations in the modern tea landscape producing Wakocha (Japanese black tea)	April–May, June–July, September–October	Tea picking, Japanese black tea factory visit, tea trail trekking to the waterfall
Wasabi farms	Wasabi cultivation areas designated as a Globally Important Agricultural Heritage System connecting with tea farming and the irrigation system	January–December	Wasabi cultivation learning, wasabi picking, trekking along the river
Shiitake farms	Shiitake mushroom plantations in tea-producing areas	April–October	Learning shiitake cultivation, shiitake picking, camping, cooking (shiitake pizza)

Source: Authors

Apart from agriculture, the area is home to many attractions related to nature, culture, services and event resources. Natural resources are hot springs, waterfalls, rivers, forests and mountains. Cultural resources include Shinto shrines and tea factories. Service resources are various accommodations, from western-style hotels and Ryokan, Minshuku (family-operated, Japanese-style bed and breakfasts) to pension houses, including public hot spring baths, camping grounds, tennis courts, fishing ponds, restaurants and shops. Events include the Ume (Japanese plum) and Sakura blossom and firework festival.

According to the SWOT analysis, abundant natural resources with good facilities are the strength of the area, attracting nature lovers and adventure tourists such as cyclists; the decrease in tourist numbers due to the decline of domestic tourism is the weakness; and the high competition of hot spring destinations is the threat. Meanwhile, wasabi cultivation in the area is the opportunity due to its designation as a Globally Important Agricultural Heritage System by the Food and Agriculture Organization of the United Nations.

The focus group on creating gastronomical tea tourism routes to accompany the SWOT analysis has not been conducted due to the COVID-19 pandemic. This implementation will continue in the project's second period.

Step 3: Reflection

An interesting notion found during the field observation was that attractive tea innovation can add value to the general food menu. This is reflected in the available cuisine served in hot springs, a fried pork menu (green tea tonkatsu) related to the tea story provided at the Kogane hot spring restaurant, part of Umegashima. The pork used is different in that the pigs always drink green tea. Thus, they are promoted as 'Shizuoka Brand Tea Pork'.

Regarding the tea cuisines proposed by the project, Miwa Shimura, the Japanese team's accommodation and food business coordinator, cooked some dishes served in her Japanese style-inn, Onokisou, and presented these in a Thai television programme. The menu created was a Wasabi pasta set served with tea cuisine side dishes, green tea beverage and black tea syrup on sweets such as ice cream. The side dishes comprise fried green tea Shitake mushroom, fried green tea tofu and steamed tea leaves served on seasonal vegetables such as eggplant. The sweet was served with Umegashima Chai black tea syrup, the innovative tea product of the community. This lunch set menu represents the local identity and innovative ideas emerging from the fusion of Japanese and Western-style cuisines (Figure 25.2). Additionally, the local interview reflects that it is possible for accommodations participating in the project to create more tea cuisine recipes.

Exchange activities at the end of the pre-exchange stage

Before starting the second phase of the project, an online exchange was conducted with each team's participants and observed by the staff of the Toyota Foundation International Grant Program. Participants shared potential tourism resources, creative tea cuisine menus and exchanged ideas on developing tourism activities linked with tea tourism and gastronomy.

Discussion

Achievements and findings from the pre-exchange programme reflect Khaokhrueamuang and Chueamchaitrakun's (2019) four elements of tea culture commodification in tourism concerning the value creation of tea-related tourism resources (Table 25.3).

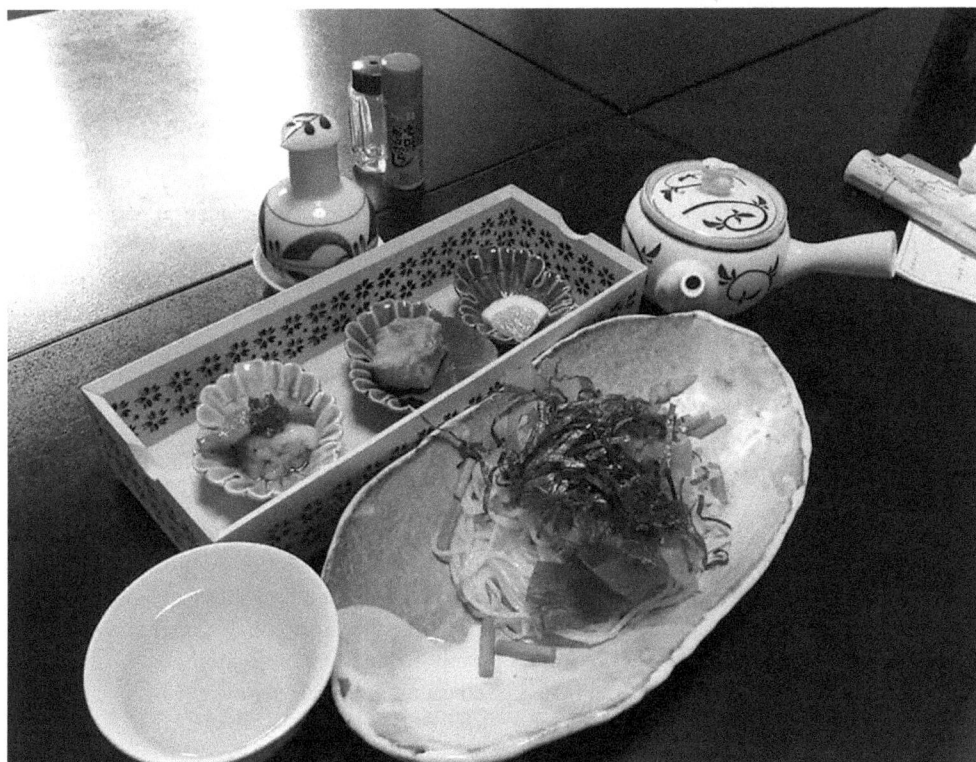

Figure 25.2 Innovative tea cuisines of Umegashima presenting the harmonious tastes of Japanese diet and Western food

Source: Authors

Table 25.3 Outcomes of the pre-exchange programme and premises on innovative ideas

Gastronomical tea tourism elements	Action programme outcomes	Innovative ideas from the international exchange programme
Tea spaces	Using tourism resources	Seasonal harvesting calendar
Tea communities	Exploring community identity and developing culinary skills	Community branding through innovative food from local identity
Tea products	Creative tea cuisine recipes	Innovative storytelling
Tea-related activities	Creating tourism activities related to tea spaces	Gastronomical tea tourism routes

Source: Authors

Tea spaces

Tea spaces play a significant role in attracting tea tourism activities and as the source of ingredients for tea cuisines. Thus, offering a tea- and gastronomy-based tourism experience relies on the harvesting calendar. A harvesting calendar is a crop calendar that indicates the harvest period of different farming products (Agricultural Musa 2020) and guides tourists to participate in farming activities, such as tea picking. It can be innovative when it is provided in tourist information sources such as websites or brochures.

Tea communities

Developing gastronomical tea tourism requires skills in culinary and service management that the young generation of tea producers will practice. This development concerns local engagement in supporting all movements of tourism and the tea industry, for example, providing a unique tea cuisine menu at accommodations or restaurants or operating tea tourism programmes in tea gardens and tea factories. Creating innovative food from local identity is crucial for branding tea communities towards the gastronomical tea tourism movement because people are attracted to differences rather than to similarities (Martins 2016). This notion is derived from efforts to create unique Akha tea cuisines and menus using Umegashima seasonal ingredients.

Tea products

Storytelling is an innovation when it is created with resources or products. This notion is seen in the Shizuoka Brand Tea Pork story of pigs drinking green tea. The creation of tea products such as tea cuisine requires a local identity that not only focuses on tea culture but includes an innovative story that interprets the place's meaning and values the products rather than merely presenting facts (Bassano *et al.* 2019). Storytelling can thus become a powerful branding tool to motivate tourists to visit the destination (Bassano *et al.* 2019) and influence consumer purchase decisions.

Tea-related activities

As gastronomic tourism can be implemented in less developed regions and for those lacking in tourism resources (UNWTO 2019), Phaya Phrai is ideal. Gastronomical tea tourism–related activities create the attractiveness of the community. Even in a declining tourism destination like Umegashima, gastronomical tea tourism plays a significant role as a new strategy to attract visitors with new values and experience. Accordingly, creating tourism routes for different types of tourists is essential to market gastronomical tea tourism. Tourism routes enable tea tourists to visit the attractions and perform activities following the introduced programme or adjust the plan to meet their interests.

Conclusion and implications

Creating gastronomical tea tourism for developing tea communities requires innovative ideas on the harvesting calendar, tea community branding through local food identity, storytelling and tea-related tourism routes. These assumptions are based on four elements of tea cultural commodification: tea spaces, tea communities, tea products and tea-related activities (Figure 25.3). This concept can be applied with tea tourism communities at the exploration stage to the rejuvenation stage of the tourism life cycle. The proposed innovative approaches in this chapter are examples describing the model. Different case studies may seek more diverse ideas.

Discussion questions

1   How can gastronomical tea tourism provide benefits to tea industry communities?
2   Why is an innovative food product crucial for developing gastronomical tea tourism?
3   What do you think about the innovative approaches proposed from the case studies?

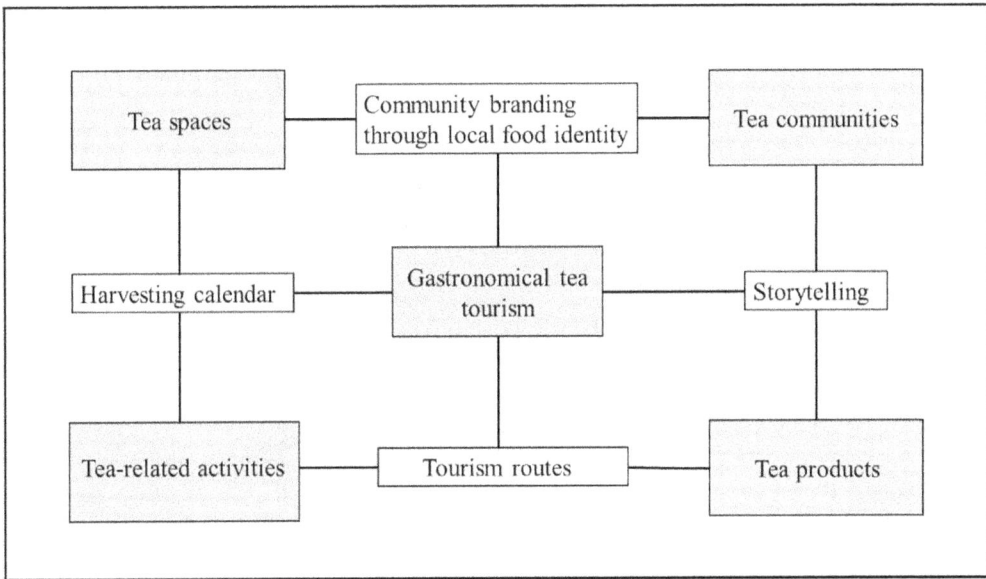

Figure 25.3   Conceptual model of innovative gastronomical tea tourism
Source: Authors

References

Agricultural Musa, 2020. Cropping calendar definition, objectives, and importance [online]. Available from: https://agriculturistmusa.com/everything-about-cropping-calendar/ [Accessed 15 August 2021].

Basaran, B., 2020. Perceptions, attitudes and behaviours of consumers towards traditional foods and gastronomy tourism: the case of Rize. *Journal of Tourism and Gastronomy Studies*, 8(3), 1752–1769.

Bassano, C., Barile, S., Piciocchi, P., Spohrer, J.C., Iandolo, F. and Fisk, R., 2019. Storytelling about places: Tourism marketing in the digital age. *Cities*, 87, 10–20.

Butler, R.W., 2006. *The tourism area life cycle, vol. 1: applications and modifications.* Cleveland: Channel View Publications.

Carpio, M.N., Napod, W. and Do, W.H., 2021. Gastronomy as factor of tourists' overall experience: a study of Jeonju, South Korea. *International Hospitality Review*, 35(1), 70–89.

Cody, K., 2017. Organic farming and international exchange: participant perceptions of North-South transfer-ability. *International Journal of Agricultural Sustainability*, 15(1), 29–41.

Gets, D., Robinson, R., Andersson, T. and Vujicic, S., 2014. *Foodies & food tourism.* Oxford: Goodfellow Publishers Limited.

Godfrey, K. and Clarke, J., 2000. *The tourism development handbook: a practical approach to planning and marketing.* London: Cassell.

Guruge, B.C.M., 2020. Conceptual review on gastronomy tourism. *International Journal of Scientific and Research Publication*, 10(2), 319–325.

Jayaratne, U., Taylor, K., Edwards, C., Sitton, S., Cartmell II, D., Watters, E. and Henneberry, R., 2017. Evaluation of an international entrepreneur exchange program: impact, lessons learned, an implication for agricultural development. *Journal of International Agricultural and Extension Education*, 24(2), 50–64.

Karlsen, J.I., 1991. Action research as method: reflections from a program for developing methods and competence. *In*: Whyte, W.F., ed. *Participatory action research.* Thousand Oaks, CA: SAGE, 143–158.

Khaokhrueamuang, A., 2021. International exchange in tea tourism: reconceptualizing Japanese green tourism for sustainable farming communities. *In*: Sharpley, R. and Kato, K., eds. *Tourism development in Japan theme, issues and challenges.* Oxon: Routledge, 140–159.

Khaokhrueamuang, A. and Chueamchaitrakun, P., 2019. Tea cultural commodification in sustainable tourism: perspectives from Thai and Japanese farmer exchange. *Special Issue of Rajabhat Chiang Mai Research Journal*, 104–117.

Khaokhrueamuang, A., Chueamchaitrakun, P., Kachendecha, W., Tamari Y. and Nakakoji, K., 2021. Functioning tourism interpretation on consumer products at the tourist generating region through tea tourism. *International Journal of Culture, Tourism, and Hospitality*, 15(3), 340–354.

Martins, M., 2016. Gastronomic tourism and the creative economy. *Journal of Tourism, Heritage & Services Marketing*, 2(2), 33–37.

Sahin, G.G., 2015. Gastronomy tourism as an alternative tourism: an assessment on the gastronomy tourism potential of Turkey. *International Journal of Academic Research in Business and Social Science*, 5(9), 79–105.

UNWTO, 2019. *Gastronomy tourism: the case of Japan*. Madrid: World Tourism Organization.

26

TEA CAFÉS AND COMMUNITY DIVERSIFICATION

Amnaj Khaokhrueamuang, Haruna Yagi, Mutsumi Yokota and Sousuke Goto

Introduction

Tea café experiences with tea beverages are among the various tourism activities representing meaningful drinks for tourists, cultural heritage, local spirit and identity (Cheung *et al.* 2021). Tea cafés are generally located in urban and suburban communities and feature creative concepts that make them exclusive and priced at a premium. However, the role of these cafés in community development through tourism promotion has not received much attention. Therefore, this chapter examines the characteristics of tea cafés in Japan's tea-processing regions to clarify the functions of community diversification through the creation of tea tourism routes. As a case study, tea cafés in the city of Shizuoka, the capital of Japanese green tea, are investigated.

Literature review

Travel styles of tea tourists and tea café routes

Tea tourism is defined as tourism that is motivated by an interest in many aspects of tea, including its history, growth, production, processing, blending, traditions and consumption (Jolliffe, 2007, 9–10), and offers activities that embrace both urban and rural forms of tourism. For example, buying and drinking tea in a café may be considered as urban tourist activities related to tea culture, whereas learning about tea growing and processing at a tea farm falls within the scope of rural tourism (Khaokhrueamuang 2021). Even though tea tourism is a small, specialised niche product for dedicated tourists, it can draw the attention of tea tourists to visit the urban communities where tea products are provided and the rural tea communities where teas are grown and produced. With this notion, the tea café is a factor in motivating tea tourists to visit both areas. Thus, in creating tea tourism routes that focus on linking tea cafés with other tourist attractions, understanding the travel styles of tea tourists is the preliminary step.

According to Plog's (1974) psychographic segmentation, tourists are roughly classified into two types: psychocentric and allocentric. Psychocentric tourists do not want to seek different travel experiences, tend to be anxious, are non-adventuresome and prefer mass tourism such as group tours. Meanwhile, allocentric travellers prefer different forms of travels and types of destinations and tend to be self-confident, adventurous and outgoing in special interests, individual trips and small group tours. These types are comparable with Grey's sunlust and wanderlust (1970) typology of

DOI: 10.4324/9781003197041-31

tourists. Sunlust tourists prefer travelling for relaxation, similar to psychocentrics, whereas wanderlust is in line with allocentric travellers because of its learning motivation. This premise supports the notion that learning and relaxation motives are important in determining the attitudes of consumers towards tea tourism (Yeap *et al.* 2021).

When considering such psychographic factors, tea tourists can then be categorised into relaxing tea tourists and learning tea tourists. Relaxing tea tourists are primarily motivated by the peaceful nature and the beautiful scenery of the tea landscape, tea shops and urban tea cafés, including staying in comfortable places. Meanwhile, learning tea tourists prefer experiencing tea-related activities, such as tea-making, hiking tea trails and learning tea culture. Therefore, tea tourism routes should be created that utilise tea cafés and are based on matching the styles of tea tourists with the characteristics of the cafés, their activities and other attractions.

Targeting tea tourists to Shizuoka's tea cafés

While many people drink tea before or during travel, only some are motivated as dedicated tea tourists to visit tea tourism destinations where tea cafés are located. Shizuoka Prefecture, the largest tea-producing region in Japan, provides many tea-related tourism attractions. Tea cafés are one such place for tea lovers, particularly in Shizuoka City. However, tea cafés have paid scant attention to tourist perceptions owing to the lack of tea café linkage with community resources.

The Tea Town Promotion Section of Shizuoka City has published the Shizuoka 'Ocha-Café' guidebook, which includes 23 recommended tea cafés (as of February 2020, fifth edition) in Shizuoka's urban area (Figure 26.1). This guidebook only provides information in Japanese, because international tourists are not a target at this development stage. Nevertheless, this chapter reviews the potential for tea tourists for both inbound and domestic tourism markets.

The inbound tourism market is divided into the tea-producing region and the tea-consuming region. The tea-producing region is the majority market share of Chinese and Taiwanese consumers

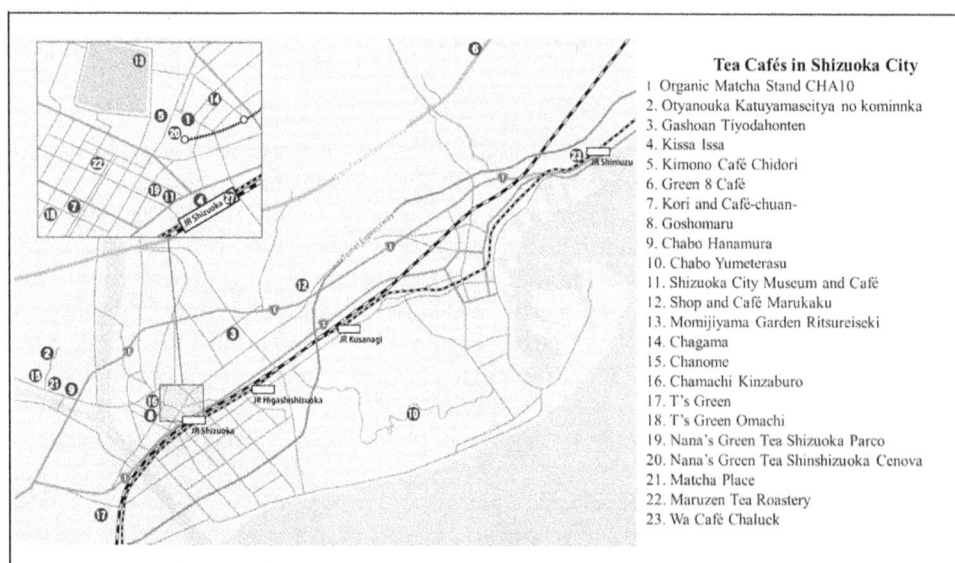

Figure 26.1 Shizuoka Tea Café map with 23 tea cafés
Source: HYZ Studio, 2021

who are highly interested in tea tourism and enjoy tea tourism activities (Yeap *et al.* 2021). Tourists from China and Taiwan also play a significant role as the foreigners who are the priority target to visit Shizuoka. On the other hand, the tea-consuming region, which is prominent among westerners, is the exploration market of Japanese inbound tourism. With regard to tea tourism, American tourists may target tea cafés in Shizuoka City because of the Japanese green tea boom in the United States (Tanaka and Uchida 2016). According to the Ministry of Finance, the United States accounted for approximately 46% of the export value of green tea in 2012, increasing 3.4 billion yen from 2014 to 2018. As a result, the demand for green tea in the United States is expected to increase in the future. The reason is related to the growing popularity of green tea, black tea and matcha (green tea powder) due to the American health boom (Yoshida 2015). Accordingly, linking tea trails and learning activities is essential in targeting American tourists to visit tea cafés. This perspective is supported by Liu's (2019) online survey on the intentions of American tourists to participate in Chinese tea tourism, where their primary motivation was revealed to be tea garden services, followed by extra activities and cultural activities.

For domestic tourism, targeting women as tea tourists to visit tea cafés is an interesting marketing perspective. In a survey of university students and visitors to the Sasama International Ceramics Festival in Shimada City regarding the tea tourism experience programme, Young (2018) found that both groups are highly interested in tea-related food and drinks, such as buying sweets and tea confectionery and drinking tea at a nice café. The survey also shows that compared to men, women are relatively more interested in tea-themed tourism. This result supports some research findings that most tea consumers are women with a college education (Fernando *et al.* 2016; Vieux *et al.* 2019; Khaokhrueamuang *et al.* 2021).

Based on the psychographic segmentation of tea tourists – namely, learning and relaxation motives – creating a tea café route requires a consideration of the potential target tourists. For Shizuoka's tea cafés, tourism stakeholders such as destination marketing organisations and travel agents can take advantage of this market trend.

Methodology

The study applies a qualitative research method by reviewing the Shizuoka 'Ocha-Café' guidebook provided by the Tea Town Promotion Section of Shizuoka City and conducts a field survey of selected tea cafés and their surrounding community tourism resources.

The content of tea cafés in the guidebook was categorised to understand their attributes and use these for selecting tea cafés to examine through an on-site survey. The field survey was conducted in urban communities of Shizuoka City, comprising the Shizuoka railway station area, Sumpu Castle Park, Sengen Shrine community area, Cha Machi (tea town) area and Gofuku-cho commercial street area. Data were collected by photographing, mapping and interviewing the owners of three businesses in Sengen Shrine community area: an organic tea shop, a travel agent and a tea café. The analysed data were used to propose tea tourism model courses linking tea cafés and other community attractions. The research was undertaken from October 2020 to March 2021.

Results

Content analysis of the Shizuoka 'Ocha-Café' guidebook

Tea cafés in the city of Shizuoka are divided into three groups: family-run, chain and cooperative. Family-run tea cafés mainly use tea grown locally in Shizuoka. Chain tea cafés are few in number

Table 26.1 Analysis of tea cafés introduced in the Shizuoka 'Ocha-Café' guidebook

Café no.	Café type	Tea beverages			Tea-related gastronomy and activity			Tea-related products and services		
		Green tea	Black tea	Matcha	Food	Sweets	Tea experience	Tea leaves	Tea ware	Take out
1	F	1	0	1	0	1	1	1	0	1
2	F	1	1	0	1	1	1	1	1	0
3	F	1	1	1	1	1	1	1	1	1
4	CO	1	0	0	0	1	0	1	1	1
5	F	1	0	0	0	1	0	0	0	0
6	F	1	1	0	0	1	1	1	1	1
7	F	1	0	1	0	1	0	0	0	1
8	F	0	0	1	1	1	1	1	1	0
9	F	1	0	1	1	1	0	1	1	1
10	F	1	1	1	1	1	0	1	0	0
11	CO	0	1	0	0	1	0	1	0	1
12	F	1	0	0	0	1	1	1	1	1
13	F	1	1	1	0	1	0	1	0	0
14	F	1	1	0	0	1	1	1	1	1
15	F	1	0	0	0	1	1	1	1	1
16	F	1	1	1	0	1	1	1	1	1
17	F	1	1	1	1	1	0	1	1	1
18	F	1	1	1	1	1	0	1	1	1
19	CH	1	0	1	1	1	0	1	0	1
20	F	1	0	1	1	1	0	1	0	1
21	F	0	0	1	0	1	0	1	1	1
22	F	1	0	0	1	1	1	1	1	1
23	F	1	1	1	1	1	1	1	1	1
Total (%)		86.95	47.82	65.21	47.82	100.00	47.82	91.30	65.21	78.26

Abbreviations: F = Family-run tea café, CH = Chain tea café, CO = Cooperative tea café.

Note: Café number with names is run by the 'Ocha-Café' guidebook shown in Figure 26.1

Source: Authors

and use tea from Uji in Kyoto Prefecture. Cooperative tea cafés are organised by a group of citizens or a local government and are primarily established as tourist attractions.

According to the three groups of icons shown in the guidebook, tea products and services are divided into tea beverages, tea-related gastronomy and activity and tea-related souvenirs and services. The tea beverages group comprises green tea, Japanese black tea and matcha. The tea-related gastronomy and activities group consists of food, sweets and tea experience (e.g., tea tasting and tea brewing). The tea-related souvenirs and services group involves tea leaves, tea wares and take-out service. Among these groups, tea sweets (mostly made from matcha) are the most provided tea product (100%), followed by tea leaves as souvenir (91.30%) and green tea beverage (86.95%), as shown in Table 26.1.

The three categories of cafés have different spatial characteristics based on their products and services and appear in both modern and traditional styles. Modern-style tea cafés are largely based on four functions (though traditional ones share some of these characteristics): (1) a space to communicate, (2) a place to relax, (3) a place to taste tea and (4) a space to learn about and exchange tea knowledge and culture. However, tea cafés that are not in areas attractive to tourists have not yet been included in tea tourism routes because most of their consumers are residents.

Survey of community resources

Tea cafés

Five tea cafés were investigated out of the 23 tea cafés introduced in the guidebook and categorised into family-run, chain and cooperative tea cafés.

FAMILY-RUN TEA CAFÉS

Cha-machi Kinzaburo   This is a third-generation tea business. It is located in Cha-machi, an old tea town since the Edo period (17th century), which has a concentration of tea industry, business and culture. Cha-machi Kinzaburo is a self-service café that offers 11 kinds of tea for free in the eat-in space with the atmosphere of a Japanese-style room on the second floor. The first-floor tea shop has an assortment of teas, tea sweets, tea wares and other tea-related souvenirs. The café's unique product is tea sweets, such as waffles and ice cream, using tea powder from different tea-growing regions in Shizuoka City. A tea event and tea learning course are also provided.

T's Green Omachi   This is a new tea café opened in February 2020, located on the first floor of the Omachi building in the Shizuoka Station. T's Green Omachi is the second branch of T's Green, which is situated in the Abegawa Station, outside the city. The café offers green tea and Japanese black tea (wakoucha), food, tea sweets and tea wares. Its distinctive tea-related product is the traditional Japanese tea sweet.

CHAIN TEA CAFÉS

Nana's Green Tea   This is a chain tea café located on the first basement floor of Cenova Department Store, Shin-Shizuoka Station. It provides a wide variety of meals and sweets with tea beverages. Tea lovers can enjoy high-quality matcha, green tea and roasted green tea from Uji, Kyoto, the famous matcha-producing region in Japan.

COOPERATIVE TEA CAFÉS

Kissa Issa   This café is situated on the first basement floor of JR Shizuoka Station. It is managed by the Tea Commerce and Industry Association of Shizuoka Prefecture. Fifty tea products from the association's members are available in the tea shop, all at the same price of 500 yen. In the tea café, three products for tasting are introduced weekly to help promote Shizuoka's tea products. Tea is served with the local wagashi (Japanese traditional sweet). The staff brews tea for the first set and the second and subsequent sets. Customers can learn how to make delicious tea from the staff.

Momijiyama Garden Ritsureiseki   This traditional Japanese-style tea room is located in the Momijiyama of Sumpu Castle park, a garden created with the scenic beauty of four seasons in the historic atmosphere. Tourists can drink tea at the Unkai (a tea room whose name means "sea of clouds") with cherry blossoms in spring, hydrangeas in summer, colourful leaves in autumn and camellias in winter. The tea is served with local wagashi cakes on a short table, while tourists sit on chairs called 'Ritsureiseki' (means a standing bow) versus a typical tatami mat setting. A large house is used for various occasions, including tea ceremonies, flower arrangements and exhibitions. Thus, tourists can also experience other tea-related activities, such as matcha making.

Urban community resources related to tea café community

Based on the four categories of tourism resources (Godfrey and Clarke 2000), the tea café community in Shizuoka's urban area comprises nature, culture, services and events. The natural tourism resource is Mt. Shizuhata in the north of the community, where the Shizuoka Sengen Shrine and an ancient burial site are located. It was developed as a nature trail for studying archaeology and seeing the city view and Mt. Fuji. Cultural resources consist of the Shizuoka Sengen Shrine, the ruins of Sumpu Castle as a historical park, the Shizuoka City Museum of Art and the Shizuoka O-cha (Tea) Plaza (Table 26.2). The distinctive tea-related service resources include tea shops, restaurants and a travel company (Table 26.3). Four festivals, namely, World O-Cha (Tea) Festival, Shizuoka Festival, Shizuoka Sengen Shrine Hatsukae Festival and Nagamasa Festival, are event resources held in the community area (Table 26.3).

Creating tea café routes

According to the analysed data of tea café and community resources, three routes for a day trip in Shizuoka City were created as examples for both inbound and domestic tourism. The inbound tourism market focuses on targeting Western visitors and is divided into two types: relaxing tea tourists (psychocentric tourists) and learning tea tourists (allocentric travellers).

Route 1: Linking urban and rural tea cafés

This route targets inbound relaxing tea tourists, particularly Chinese, Taiwanese and other Asian tourists, by combining various urban and suburban attractions with tea cafés. The route starts at Kissa Issa,

Table 26.2 Nature and culture tourism resources in the tea café community of Shizuoka City

Resource categories	Resource names	Resource features	Activities
Nature	Mt. Shizuhata	The 171-metre-high mountain in the north of urban areas; the place of a sixth-century ancient tomb and Sengen Shrine in the forest park	Trekking nature trails and seeing the city view with Mt. Fuji Learning Japanese history and archaeology at the burial site
Culture	Shizuoka Sengen Shrine	A sacred Shinto shrine in prehistoric time with 26 buildings designated as nationally important cultural assets	Paying respect to the gods, learning and religious-related architecture and Japanese history at the museum
	Ruins of Sumpu Castle	A lush green park in the ruins of Tokugawa Ieyasu's castle constructed in 1585 and rebuilt in 1607	Learning the history of Suruga (former name of Shizuoka City), walking and drinking tea at the tea room of Momijiyama Garden
	Shizuoka City Museum of Art	A place for art exhibition from a wide variety of genres; offers workshops, a souvenir shop and café corner	Watching art exhibitions and drinking Japanese black tea (wakoucha) at the museum's café
	Shizuoka O-cha (Tea) Plaza	A tea learning and promotion centre managed by the World Green Tea Association; close to the south of Shizuoka Station	Brewing and tasting the different taste of Shizuoka's green tea from different tea regions and participating in tea study programmes

Source: Authors

Table 26.3 Distinctive services and events in the tea café community of Shizuoka City

Resource categories	Resource names	Resource features	Activities
Services (distinctive shops and a local travel agent)	Chikume-do Tea Shop	An old tea shop founded in 1781 selling high-quality tea leaves such as Gyokuro and Sencha; located in the Gofukucho commercial street	Buying various types of tea leaves and learning Shizuoka's tea culture through the long history of the company
	Naoki Maeda Ceramics Studio	A private art studio offering the experience of making ceramics; a 13 minute-walk from Shizuoka Station	Learning and making ceramic works related to tea products such as teacups, teapots and plates
	Real Food Market Akustu	An organic food shop providing local farming products in the Sengen Shrine commercial street	Buying organic tea and farming products
	Sofuto Kenkyu Shizu Ryoko Center	A local travel company; organises a tailor-made tea tourism programme	Having tea tourism information and tea tourism programmes
Events	World O-Cha (Tea) Festival	The international tea festival organised in spring and autumn every three years	Participating in tea exhibition and culture, tea brewing contest, tea products competition, tea conference, tea tours and tea experience programmes
	Shizuoka Festival	The flower-viewing festival of the Sumpu Castle held during the Sakura blooming season	Watching a procession entering the castle town of Sumpu with the cherry blossom dance
	Shizuoka Sengen Shrine Hatsukae Festival	The festival of Sengen Shrine with Chigo-mai (children's dance); designed as intangible folk asset; held every April 1–5	Watching parades and entertainment contests

Source: Authors

a cooperative tea café in Shizuoka Station, before moving to Nihondaira Mountain outside the city for a rest with tea in the Terrace Café and panoramic views of Suruga Bay and Mt. Fuji. It is followed by a visit to the tea garden on the mountain slope for taking photos. Afterwards, there is a shopping trip at tea shops in the city, such as Chikume-do Tea Shop in the Kofukucho commercial street.

Route 2: Linking urban tea cafés with nature

This route targets inbound learning tea tourists, particularly Americans and other westerners, with the theme 'tea + nature'. It begins at Momijiyama Garden, a traditional tea café in Sumpu Castle Park. Then, a visit to the Sengen Shrine involves hiking a mountain nature trail, resting with sweet tea at Kinzaburo, a tea café in the old tea town of Cha-machi, and learning the history of the Japanese tea trade (Figure 26.2).

Route 3: Linking urban tea cafés with art

This route is designed for the domestic tourism market and concentrates on female Japanese tea lovers as learning tea tourists with the theme 'tea + art'. It starts at the Shizuoka O-Cha (Tea) Plaza to

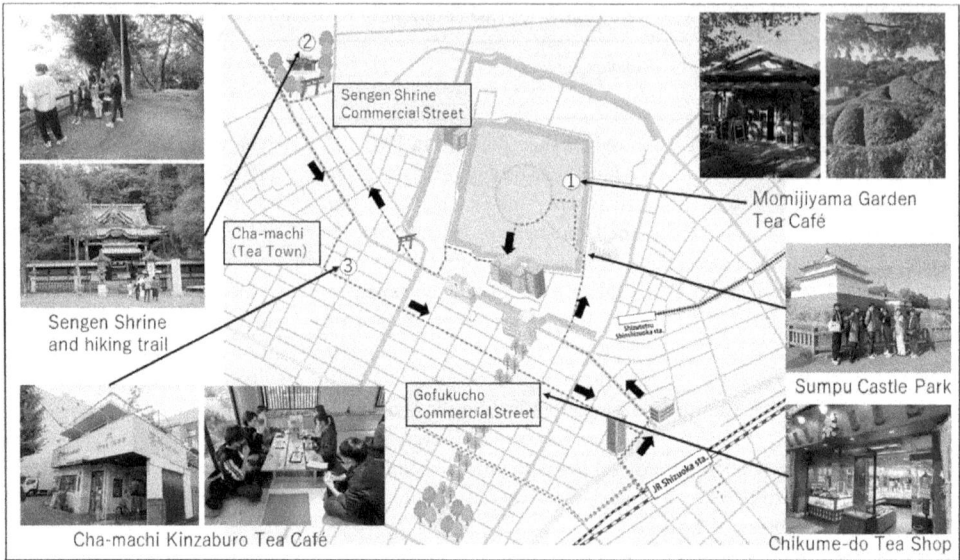

Figure 26.2   The community survey and linking urban tea cafés with nature

Source: Authors; modified from HYZ Studio's map, 2021

brew green tea. Then, Japanese black tea is sipped at Shizuoka City's Art Museum and traditional Japanese sweets are served with tea at T's Green Omachi Café. The programme ends with making tea wares at the Naoki Maeda Ceramic Studio.

Discussion

The three categories of cafés have different spatial characteristics based on their products and services and appear in both modern and traditional styles. Modern-style tea cafés are largely based on four functions (though traditional ones share some of these characteristics): (1) a space to communicate, (2) a place to relax, (3) a place to taste tea and (4) a space to learn about and exchange tea knowledge and culture.

As a space for communication, tea cafés are places for talking or working. They also play a significant role in promoting the region, such as providing tourism information and presenting the community's image and identity through tea products and services. Meanwhile, the relaxation and tasting of tea become valuable functions for tea cafés to promote health and wellness (Cheung *et al.* 2021). In terms of education and entertainment, tea cafés are global spaces for building people networks through sharing knowledge and culture related to tea and other interests.

These functions serve residents and indicate the potential for attracting both inbound and domestic tea tourists if the cafés are linked with community resources. However, tea cafés in Shizuoka City are not yet included in tea tourism routes because most of their consumers are locals. Thus, the creation of tea tourism routes linking tea cafés and community attractions can tremendously support tea cafés to diversify the community through tea tourism. Specifically, tea cafés create diversification of tourism products emerging from a wide range of sectors, including agriculture, food and beverage industry, trade, real estate, health, education, art and culture, and business services. This emergence of resources is the accumulation of community capital, including natural, financial, human, social and cultural, all of which lead to sustainable development (George *et al.* 2009; Khaokhrueamuang

2014). The diversity of tea-related tourism products provided in a tea café community increases tourists' satisfaction and their revisit intention owing to tea café values such as health and emotion (Cheung *et al.* 2021). The importance of diversifying the community by increasing the number of resources or strategies employed to improve the community and mitigate risk makes the communities more likely to overcome various shocks (Cochrane and Cafer 2017).

Although many tourists enjoy the city life, they equally admire the natural environment of a rural community (Nkemngu 2017). Thus, linking urban tea cafés with rural landscapes or tea-growing areas such as tea fields with tea tourism model courses is a practical approach towards community diversification. Tourism institutes such as destination marketing organisations and tour operators should consider this notion in promoting tea business and tourism. For this case study, Sofuto Kenkyu Shizu Ryoko Center, the local tour company, may create tea café tours combining urban tea cafés with the tea-producing communities in rural areas of Shizuoka.

Conclusion and implications

The review of the Shizuoka 'Ocha-Café' guidebook and a field survey of selected tea cafés and community resources in Shizuoka City clarify that tea cafés serve four functions for residents. They are a space to communicate, a place to relax, a place to taste tea and a space to learn about and exchange tea knowledge and culture. Through tea tourism model courses, these functions can diversify the tea café community by linking tea cafés with different community resources. Moreover, this chapter contributes to the literature on tea tourist typology that global tea cafés and tea tourism destinations can consider targeting relaxing and learning tea tourists, including creating tea tourism routes connecting tea cafés with community resources. Despite the interesting findings, this study lacks a survey on tea tourism demand from tea café perspectives. Future studies are recommended to examine the opinions and needs of both residents and tourists with regard to the proposed model courses.

Discussion questions

1   Why are tea tourists categorised by relaxation and learning motivations?
2   What are the functions of tea cafés in Shizuoka City?
3   How can tea cafés contribute to diversifying the community through tea tourism?

References

Cheung, L.M., Leung, S.K.W., Cheah, J-H., Koay, Y.K. and Hsu, C-Y.B., 2021. Key tea beverage values driving tourists' memorable experience: an empirical study in Hong Kong-style café memorable experience. *International Journal of Culture, Tourism and Hospitality Research*, 15(3), 355–370.
Cochrane, L. and Cafer, A., 2017. Does diversification enhance community resilience? A critical perspective. *Resilience*, 6(2), 129–143.
Fernando, P.I.N., Rajapaksha, R.M.D.K. and Kumari, K.W.S.N., 2016. Tea tourism as a marketing tool: a strategy to develop the image of Sri Lanka as an attractive tourism destination. *Kelaniya Journal of Management*, 5(2), 64–79.
George, E.W., Mair, H. and Reid, D.G., 2009. *Rural tourism development: localism and cultural change.* Cleveland: Channel View Publications.
Godfrey, K. and Clarke, J., 2000. *The tourism development handbook: a practical approach to planning and marketing.* London: Cassell.
Grey, H.P., 1970. *International travel: international trade.* Lexington, MA: Heath Lexington Books.
Jolliffe, L., ed., 2007. *Tea and tourism: tourists, traditions and transformations.* Cleveland: Channel View Publications.
Khaokhrueamuang, A., 2014. Sustainability of rural land use based on an integrated tourism model in Mae Kampong village, Chiang Mai Province, Thailand. *Geographical Review of Japan Series B*, 86(2), 157–173.

Khaokhrueamuang, A., 2021. International exchange in tea tourism: reconceptualizing Japanese green tourism for sustainable farming communities. *In*: Sharpley, R. and Kato, K., eds. *Tourism development in Japan theme, issues and challenges.* Oxon: Routledge, 140–159.

Khaokhrueamuang, A., Chueamchaitrakun, P., Kachendecha, W., Tamari, Y. and Nakakoji, K., 2021. Functioning tourism interpretation on consumer products at the tourist generating region through tea tourism. *International Journal of Culture, Tourism and Hospitality Research*, 15(3), 340–354.

Liu, Y., 2019. *Who is the potential tea tourist?* Master's thesis, Hospitality Management of California State Polytechnic University, Pomona.

Nkemngu, A.P., 2017. The urban-rural tourism mix: a partnership of convenience or sustainability imperative. *In*: Slocum, L. and Kline, C., eds. *Linking urban and rural tourism: strategies in sustainability.* Wallingford: Cabi, 115–127.

Plog, S., 1974. Why destination areas rise and fall in popularity. *Cornell Hotel and Restaurant Administration Quarterly*, 14(4), 55–58.

Tanaka, M. and Uchida, T., 2016. *Tokai area data book 2017.* Tokyo: Chunichi Shimbun-sya.

Vieux, F., Maillot, M., Rehm, C.D. and Drewnowski, A., 2019. Tea consumption patterns in relation to diet quality among children and adults in the United States: analyses of NHANES 2011–2016 data. *Nutrients*, 11(11), 2635. https://doi.org/10.3390/nu11112635

Yeap, J.A.L., Ara, H. and Said, M.F., 2021. Have coffee/tea, will travel: assessing the inclination towards sustainable coffee and tea tourism among the green generations. *International Journal of Culture, Tourism and Hospitality Research,* 15(3), 384–398.

Yoshida, Y., 2015. Popular in the health boom of America! Japanese tea exports are increasing rapidly (Kenkobumu no Amerika de ninki! Nihoncha no yusyutsu ga kyuzotyu). *President online (Purejident onrain)*, viewed 4 Jun 2021. Available from: https://search.yahoo.co.jp/amp/s/president.jp/articles/amp/16676%3Fpage%3D1%26usqp%3Dmq331AQQKAGYAYKGke6y7Yj4WrABIA%253D%253D

Young, C., 2018. Research on tourism commercialization utilizing tea resources in tea-producing regions (Chasanchi ni okeru ochashigen wo katsuyoushita kankousyouhinka ni kansuru kenkyu). *In:* Japan Marketing Academy, ed., *Japan Marketing Association conference proceedings (Nihon ma-keting gakkai kanfarensu puroshidingsu).* Tokyo, 162–174.

27

TEA TOURISM PROMOTION IN GLOBALLY IMPORTANT AGRICULTURAL HERITAGE SYSTEMS

Kyoko Ishigami and Amnaj Khaokhrueamuang

Introduction

Tea is thought to have originated in southwest China and southeast Asia. These areas are home to the wild varieties known as the Assam type (*Camellia sinensis* var. *assamica*) and the Chinese type (*C. sinensis* var. *sinensis*). These two varieties are grown commercially and are thus widely consumed all over the world. However, few tea-growing areas have been designated as world heritage sites. The preservation and transmission of local wisdom concerning traditional tea farming systems are now rare. Therefore, the UN Food and Agriculture Organization (FAO) has adopted this as its primary criterion for awarding world heritage status to tea-growing areas. Globally Important Agricultural Heritage Systems (GIAHS) is the label used by the FAO for this purpose (Khaokhrueamuang and Takehana 2018).

Characteristically, GIAHS are abundant in agricultural biodiversity and wildlife, are essential resources for indigenous culture, and create beautiful landscapes. They must therefore include (1) food and livelihood security; (2) agro-biodiversity; (3) local and traditional knowledge systems; (4) culture, value systems, and social organisation; and (5) remarkable landscapes and seascapes. As of July 2021, there are 62 GIAHS in 22 countries (FAO 2021). Four of these are traditional tea-farming systems. Two are based on agroforestry practices (the Pu'er Traditional Tea Agrosystem of the Yunnan Province in China and the Traditional Hadong Agrosystem in Hwagae-myeon, South Korea), while the other two are based on open landscapes (the Jasmine and Tea Culture System of Fuzhou city, Fujian Province, China, and the Traditional Tea-grass Integrated System in Japan's Shizuoka Prefecture; Khaokhrueamuang and Takehana 2018). The Anxi Tieguanyin Tea Culture System in China has been proposed as a new GIAHS (FAO 2021).

Usually, tea tourism is not among the aims of the GIAHS designation; however, its significance is gradually increasing (Khaokhrueamuang and Takehana 2018). This chapter, therefore, investigates tea tourism promotion in the Shizuoka Traditional Tea-grass Integrated System through the lens of the tea culture commodification framework proposed by Khaokhrueamuang (2021).

Tea tourism in GIAHS

Like UNESCO's World Heritage Site designation, the FAO's GIAHS certification can draw large numbers of tourists, although the use of local resources as tourist attractions is easier in World Heritage Sites than in GIAHS (Yotsumoto and Vafadari 2021). This is because one of the main objectives

DOI: 10.4324/9781003197041-32

of promoting GIAHS is conserving and managing traditional farming systems. In Japan, the main reasons for applying for GIAHS designation are the abandonment of traditional farming practices and farmlands due to a decreasing and ageing population (push factors) and biodiversity conservation (pull factor; Reyes *et al.* 2020). However, generating income and sustainably adding economic value to goods and services can be another important reason (Koohafkan and Altieri 2010). Tourism is an essential activity to achieve this goal.

Tourism in Japan's 11 GIAHS sites is based on the concept of Satoyama, an ideal model of agricultural landscape consisting of a mosaic of woodlands, orchards, grasslands, rice paddies, and irrigation ponds and canals, as well as human settlements (Khaokhrueamuang 2017). In these sites, green tourism, an activity aiming to support farmers through interaction between rural and urban communities, plays a significant role. One example is Noto's Satoyama and Satoumi in the Ishikawa and Minabe-Tanabe Ume (Japanese apricot) System of Wakayama Prefecture (Khaokhrueamuang 2021). The Traditional Tea-grass Integrated Farming System in Shizuoka Prefecture is one of Japan's GIAHS that has an action plan which includes green tourism (referred to as "agritourism"); (Association for Promotion of GIAHS "CHAGUSABA" in Shizuoka 2013).

In May 2013, the FAO recognized the Shizuoka Traditional Tea-grass Integrated System as world agricultural heritage in recognition of its diverse agricultural culture, which includes a good balance between high-quality tea production and biodiversity conservation, the culture of "Fukamushicha" (long-steamed tea) and hand-rolling, traditional events, the beautiful scenery of Satoyama, and the activities of the villagers who have inherited these traditions. This traditional system is known locally as "Chagusaba," meaning "semi-natural grasslands," where the grass is preserved for tea cultivation. This system remains only on 1% of the country's land.

The number of Chagusaba farmers has recently decreased due to the slump in the price of tea. The GIAHS designation, therefore, gives the region an economic opportunity (Kusumoto 2020). The Traditional Tea-grass Integrated System covers the areas of four cities (Kakegawa, Shimada, Makinohara and Kikugawa) and one town (Kawanehoncho). The grass in the Chagusaba zone (Japanese silver grass and Sasa broadleaf bamboo) is cut, dried in small heaps, and spread on the tea gardens from autumn to winter. This practice increases the effectiveness of fertiliser, controls weeds and enhances the soil's ability to retain water. These benefits invigorate the plants, improving the tea's flavour and aroma. The Chagusaba is also home to over 300 species of wild flora and fauna, including endangered wildflowers such as the Sasa lily and the Kikyo bellflower.

A conceptual framework for examining tea tourism promotion in GIAHS

Using customs (intangible culture) and objects (tangible culture) to add value to tourism can be seen as part of a process of "commodification." According to Khaokhrueamuang (2021), there are four key elements of the commodification of tea culture as a tourism product: tea spaces, tea communities, tea products and services, and tea-related activities.

Tea spaces linked to plantations can be commoditised for tourist activities such as tea plucking, festivals and folk music performances. Communities of tea producers exhibit both tangible and intangible culture that has the potential to be commodified. Tangible culture includes producers' farmhouses and tea factories, which can be commoditised as rental lodges or Japanese-style inns called "minshuku" or "nouka minpaku." Intangible culture includes customs, beliefs, livelihoods, and lifestyles such as tea-making methods, tea cuisine, and tea-related stories. As the commodification process progresses, tourism products and services are created, such as tea souvenirs, farmers' restaurants, and tea trails tours (Khaokhrueamuang 2021).

Research methodology

This chapter reviews the documents related to tea tourism promotion in the Shizuoka GIAHS-certified site. Both print materials (e.g., the GIAHS application, policy reports, action plans, promotional brochures, research articles) and digital materials (e.g., GIAHS-related websites and social media content) were used. The analysis was conducted based on Khaokhrueamuang's (2021) four key elements of cultural commodification discussed above.

Results

Tea spaces

Tea gardens in the Shizuoka GIAHS site are cultivated using the Traditional Tea-grass Integrated System. Grass (mainly silver grass, *Miscanthus sinensis*) is cut in autumn and winter, tied into bundles called "Kapposhi," dried, cut a second time, and laid between the rows of the tea gardens (in some areas, the dried grass is not cut). In the summer, the tea fields look like a patch of grass; the neatly cut grass fields dotted with Kapposhi can be seen in the autumn.

This method maintains good soil conditions in the sloped tea gardens; it also prevents run-off, which has a positive effect on the quality of the tea. *Miscanthus*, one of the typical grasses cut in the tea meadow, returns to the soil following a long period (10–20 years) after it has been spread between the rows of the tea garden. The decomposed soil is so soft that it crumbles when picking it up.

This system increases the carbon content of the soil thanks to the absorption of carbon dioxide by the tea plants through photosynthesis. More than 300 species of plants and animals, including many endemic species that are rarely seen elsewhere, have been found in the Satoyama grasslands maintained by tea farmers.

In the area where the tea-planting method is practised, the land is traditionally used in a clever way, taking advantage of its topographical features. The gentle slopes are lined with rows of tea plants, forming a beautiful landscape. When the tea plants are cut in the autumn and winter, the plantations are lined with tea bushes. These landscapes are an important tourist resource for the region (Figure 27.1).

As of the end of March 2021, 434 farmers were certified as practitioners of the Traditional Tea-grass Integrated System. The tea plantations amounted to 1047 ha and the plantations under management to 381 ha (Table 27.1).

Tea communities

The areas certified for the Chakusaba method are a representative production zone for Fukamushi-cha (long-steamed tea), which accounts for most of the tea distributed in Shizuoka Prefecture. In these tea communities, the silver grass used for Chagusaba has become part of daily life. For example, it is offered at the graves of ancestors. This reflects the link between tea cultivation and grass in the farming culture, which is passed down the generations as heritage. In the Chagusaba area, communities also share the management of tea factories through farmers' organisations.

Regarding the combination of the private sector and the tea communities, there are programmes and support services for residents and visitors. One example of tea tourism promotion is the tea training programme. At present, there are eight tea hand-rolling schools in the prefecture, each of which has its study group and strives to pass on its techniques. Furthermore, the hand-rolling preservation society participates in many tea-related events in the prefecture, offering practical lessons

Figure 27.1 The Chagusaba landscape of Mt. Awagatake

Source: Authors

Table 27.1 The number of tea farmers certified as practitioners of the Traditional Tea-grass Integrated System (Chagusaba) method

	2015	2016	2017	2018	2019	2020
Number of farmers	589	496	493	495	414	434
Tea glass field area (ha)	447	423	422	422	378	381
Tea plantation area (ha)	1234	1188	1185	1190	1014	1047

Source: Adapted from the Council for the Promotion of the World Agricultural Heritage "The Traditional Tea-grass Integrated System," 2021.

to the public and demonstrating the technique at kindergartens and elementary and junior high schools.

In the cities of Kakegawa and Kikugawa, a local taxi company operates the Chagusaba Taxi, which allows clients to go sightseeing the scenic tea meadows. Furthermore, the Council for the Promotion of the World Agricultural Heritage "Traditional Tea-grass Integrated System" (hereinafter the Council) has organised a business plan contest to create new businesses that use the area's resources for tourist purposes. In 2020, the Council received more than 30 business plans.

Tea products

In October 2013, a sticker indicating the degree of contribution to biodiversity conservation (commonly known as the "certification sticker") was launched to increase the awareness of the Chagusaba

Figure 27.2   The Chagusaba certification sticker on a tea product

Source: Authors

method. The sticker indicates that the products are made from tea leaves grown by certified practitioners of the Traditional Tea-grass Integrated System. Green tea packets, for instance, carry this sticker. The number of tea buds (between one and three) on the sticker signals the relative contribution to biodiversity conservation. Three tea buds mean that over 50% of the tea garden uses the Chagusaba method. Two buds indicate a percentage between 25% and 50%, while one bud is for gardens with 5%–25% of land cultivated according to Chagusaba (Figure 27.2).

At the end of March 2020, there were 151 registered sellers. These are found not only in Shizuoka Prefecture but also in cities all over Japan. Moreover, a tea producer in the Chagusaba zone that operates in 70% of the tea fields has used the GIAHS certification to market a community tea product. To support biodiversity conservation and return profits to the community, this ready-to-drink tea bottle is labelled with pictures of rare flora from Chagusaba areas.

Tea-related activities

The action plan of the Chagusaba certified system is based on the following items: the use of farm work volunteers, the promotion of exchanges between companies and villages, the fostering of citizens' awareness, the advancement of green tourism, and the strengthening of information dissemination.

To reduce the burden on elderly tea farmers, a volunteer system is being implemented. The Council has promoted the use of volunteers to help with the maintenance of the tea meadows, including subsidies to take on help in the certified areas. It has also put in place the necessary guidelines to ensure that the work is carried out safely and efficiently.

Every year, many employees of private companies and university students volunteer to support the Traditional Tea-grass Integrated System under the guidance of certified practitioners. This system reduces the labour for the farmers and spreads awareness of the Chagusaba farming method and the natural environment of Satoyama. In Kikugawa, the terraced rice paddy fields of Senggamachi create a beautiful landscape with a rich water environment that welcomes a variety of living creatures. A part of the terraces is used as a tea meadow. Every year, a learning programme is held here to raise awareness of the Chagusaba method and encourage the preservation of the waterfront's rich biodiversity.

Furthermore, in May 2018, the Council created a logo to support the Chagusaba farming method, which individuals and companies widely use to increase its exposure and raise awareness. The logo is used not only on tea products but also on items such as pamphlets, hats and signboards.

In 2020, to attract new tourists, model tours were conducted in Kakegawa and Kikugawa. In the former city, the potential of tourist resources such as local recipes and tour courses was explored in cooperation with accommodation venues around the certified areas. In Kikugawa, an online tour for children aged 10 to 15 was carried out to explain the value of the Traditional Tea-grass Integrated System and the proper way to drink tea easily.

Another important activity is tea pairing, that is, the combination of different kinds of tea with different dishes, somewhat like choosing the right wine. At one event, a variety of Chagusaba-grown teas were served with Teppanyaki dishes (food cooked on a hot iron plate) and French-inspired dishes. A certified Chagusaba practitioner was invited to talk about his passion for tea and his thoughts about the method. The event was attended by more than 240 people and was covered by the media. In 2018, to commemorate the fifth anniversary of the recognition of the Chagusaba method as a world agricultural heritage site, the Council held the "World Agricultural Heritage and Tea Pairing Party" from September to December in cooperation with high-class inns in the Izu peninsula and established restaurants in Shizuoka. These efforts to offer new ways of enjoying tea are helping to increase the demand for products grown according to the Traditional Tea-grass Integrated System.

According to the Council's website, the ecological practices of GIAHS dovetail with the current trends of ecotourism and heritage tourism. The landscapes of the tea fields and other historical interesting sights offer ample opportunities for this kind of tourism. One example of increasing popularity is taking the Oigawa steam railway to pass through splendid tea meadows. These activities entail the commodification of tea spaces as recreational sites of scenic beauty, where hikers and cyclists can practise their sports and learn about flora and fauna. A tea trail to Mt. Awagatake in the Higashiyama area of Kakegawa, where the Chagusaba method is preserved, is an excellent example of the commoditization of heritage tea spaces as ecotourism products. Higashiyama green tea has been promoted alongside the ecotourism activity of seeing the area's endangered wildflowers while climbing up the mountain, where the statue of priest Eisai, a father of Japanese green tea, is located.

According to the website of the Kakegawa Tourism Association, green tea tourism has attracted tourists not only from Japan but also from China, Korea and Thailand. "Chagusaba tourism" is an interactive tour offered on the website for visitors who are interested in experiencing Chagusaba with local farmers through walking, hiking and traditional hand-rolling. Adventure and sports activities are also offered in the Chagusaba area. For example, the Shincha (first flush tea) Marathon, which involves running in the tea fields, is held every year in April and attracts many visitors. A cycling map is also available for cyclists. As a result of these efforts, the number of tourists in the area has increased from 254 in 2016 to more than 90,000 in 2019.

Discussion

Tea tourism promotion in the Shizuoka GIAHS site involves the integration of different activities: green tourism, ecotourism, heritage tourism, gastronomic tourism, and sports tourism. These activities take advantage of the environmental components of local traditional tea farming.

Green tourism is found in the action plan of the GIAHS application in the form of agritourism plus certain activities run by the municipalities or the third sector (Ohe, as cited in Wijaya 2013). The green tourism concept is implemented by taking advantage of the landscape, culture and other resources to attract visitors and strengthen the ability to disseminate information through websites and social networking services. Ecotourism activities in the GIAHS-certified site include walking and hiking in the Chagusaba fields of Higashiyama. This can be defined as a form of responsible travel to communities located in traditional tea-growing areas that conserve the environment and culture. Ecotourism sustains the well-being of local people and enhances environmental education. As problems such as global warming become more serious, and as the need to achieve sustainable development goals increases, it is important to preserve ecological farming systems. Heritage tourism to GIAHS can help add value to such farming systems. Learning the Chagusaba farming method is a heritage tourism activity involving local tea culture. Using the steam railway to visit the tea gardens is another activity that enhances the value of the landscape. Gastronomic tourism is evident in the tea-pairing activities involving the local and the French cuisines. Even though the event in question was organised outside the GIAHS-certified area, the use of Chagusaba tea products supported the value of that tea culture. Sports tourism is also found in the area. On the Kakegawa tourism website, cyclists can download a map with relevant routes. The Shincha Marathon is an annual event that takes advantage of local tea spaces.

Tea tourism promotion interacts with various kinds of activities that involve learning and relaxation (Yeap *et al.* 2021). Therefore, developing and marketing tea tourism in the Chagusaba area should be based on educational and recreational activities. Considering the four elements of tea's culture commodification (Khaokhrueamuang 2021), the Chagusaba method can draw on Japan's emerging tourism trends.

Regarding tea spaces, the COVID-19 pandemic has created a need for social distancing that will impact future tourism. Educational and recreational activities that take place in tea fields, such as learning about tea culture and farming, walking or cycling, will become increasingly sought after. Online experiences such as tea tours can inspire tourists to visit the Chagusaba site and increase demand for its tea products. Due to an ageing society and depopulation, the renovation of abandoned traditional houses to accommodate tourists is a new trend in rural community revitalization in Japan (Murayama 2021). This occurs also in tea-growing communities. In the Chagusaba case, tea-pairing events and other food-related activities could take place in renovated community lodgings.

Concerning tea products, the certification sticker and the logo will help increase consumers' awareness. The logo may be used to brand the Chagusaba area as a tea tourism destination. As Ranasinghe *et al.* (2017) state, product-place co-branding is a powerful instrument to increase the popularity of tea products and tea-producing areas. The logo should convey the importance of Chagusaba and GIAHS in terms of tea tourism, particularly to promote the kind of inbound tourism that will make the traditional Japanese tea farming system known worldwide. To help achieve this goal, the sticker with the tea buds should contain information in English.

Implications and conclusions

Low tea prices, an ageing farming population and the lack of successors make it difficult to maintain and pass on Japan's traditional tea farming methods. Tea tourism can play a significant role in promoting sustainable farming systems. Both educational and recreational activities should be involved in this kind of tourism. The development of tea tourism entails the commodification of tea culture, which is the essential element in product-place co-branding. Focusing on this aspect may strengthen the Chagusaba traditional system and enhance tourists' intention to revisit.

Discussion questions

1 How has tea culture been commodified in the promotion of tea tourism in the Chagusaba area?
2 How can tea tourism promote sustainable tea farming systems as part of the GIAHS designation?

References

Association for Promotion of GIAHS "CHAGUSABA in Shizuoka", 2013. Biodiversity is conserved by farm management practices that produce high-quality tea-GIAHS Project Action Plan [online]. Available from: www.fao.org/3/bp800e/bp800e.pdf [Accessed 2 July 2021].

FAO, 2021. Agricultural heritage around the world [online]. Available from: www.fao.org/giahs/giahsaroundtheworld/en/ [Accessed 23 April 2021].

Khaokhrueamuang, A., 2017. Agricultural heritage systems of orchard based on the concept of Satoyama and sufficiency economy: green tourism perspectives for Japan and Thailand. *Journal of Thai Interdisciplinary Research*, 12(3), 38–49.

Khaokhrueamuang, A., 2021. International exchange in tea tourism: reconceptualizing Japanese green tourism for sustainable farming communities. *In*: Sharpley, R. and Kato, K., eds. *Tourism development in Japan theme, issues and challenges*. Oxon: Routledge, 140–159.

Khaokhrueamuang, A. and Takehana, K., 2018. Kankō ni okeru cha bunka no shōhin-ka dainanashō cha to heritējitsūrizumū Chūgoku, Kankoku, Nihon no sekai nōgyō isan no kēsu (Tea culture commodification in tourism, chapter 7: tea and heritage tourism: cases of agricultural world heritage in China, Korea and Japan). *The Tea*, 71(8), 6–11.

Koohafkan, P. and Altieri, M.A., 2010. Globally important agricultural heritage systems: a legacy for the future [online]. *Food and Agriculture Organization of the United Nations*. Available from: www.fao.org/3/i1979e/i1979e.pdf [Accessed 23 April 2021].

Kusumoto, Y., 2020. Biodiversity maintained by the traditional tea-grass integrated system in Shizuoka, Japan [online]. Available from: https://ap.fftc.org.tw/article/2492 [Accessed 23 April 2021].

Murayama, K., 2021. *Kanko saisei* (tourism revitalization). Tokyo: President.

Ranasinghe, W.T., Thaichon, P. and Ranasinghe, M., 2017. An analysis of product-place co-branding: the case of Ceylon Tea. *Asia Pacific Journal of Marketing and Logistics*, 29(1), 200–214.

Reyes, S.R.C., Miyazaki, A., Yiu, E. and Saito, O., 2020. Enhancing sustainability in traditional agriculture: Indicators for monitoring the conservation of Globally Important Agricultural Heritage Systems (GIAHS) in Japan. *Sustainability* [online], 12(14). Available from: https://doi.org/10.3390/su12145656 [Accessed 3 May 2021].

Wijaya, N., 2013. Contemporary problems in Japan's rural areas and opportunities for developing rural tourism: a case of Yamashiro district in Yamaguchi prefecture. *Journal of East Asian Studies*, 11, 59–72.

Yeap, J.A.L., Ara, H. and Said, M.F., 2021. Have coffee/tea, will travel: assessing the inclination towards sustainable coffee and tea tourism among the green generations. *International Journal of Culture, Tourism and Hospitality Research*, 15(3), 384–398.

Yotsumoto, Y. and Vafadari, K., 2021. Comparing cultural world heritage sites and globally important agricultural heritage systems and their potential for tourism. *Journal of Heritage Tourism*, 16(1), 43–61.

28

EXPLORING THE VALUE CREATION PROCESS IN THE JAPANESE BLACK TEA MARKET AND TOURISM

Risa Takano, Naoko Yamada and Daisuke Kanama

Introduction

In the mid-20th century, Japan exported more than 5000 tons of black tea, but this market decreased and finally disappeared. Although the black tea currently consumed in Japan is mostly imported, there has been a renewed interest in Japanese black tea. Almost all Japanese black tea farmers identified in this chapter also make a living as green tea farmers. As people's modern lifestyle has changed the primary tea consumption style from a pot to plastic bottles, because green tea leaves for pre-made beverages usually cost less than tea leaves intended for a pot, Japanese tea farmers have been affected.

In responding to this trend, some Japanese green tea farmers have decided to overcome these challenges by producing Japanese black tea. Some Japanese black tea products appear to have unique characteristics depending on the region of production and the farmer. Therefore, the number of customers interacting with local farmers and visiting producing areas is increasing.

Historically, there have been two difficulties in the Japanese black tea market. First, there were technological problems. The Japanese black tea industry declined after the liberalisation of black tea imports, and modern farmers did not have the expertise necessary for tea manufacturing. Although most Japanese black tea farmers used to produce poor-quality tea, the quality has drastically improved in the last few years, and the market has grown. Some Japanese black tea products have won accolades at international tea competitions.

Second, because various tea products, including imported black tea, had already spread throughout the tea market in Japan, farmers faced difficulties in differentiating their own products. They had to discover the value of their own products and succeeded in identifying a unique value. In 2018, more than 500 tea farmers produced Japanese black tea alongside green tea. This contributed to the development of new markets for Japanese black tea, not eroding the existing green tea market or the imported black tea market (Takano and Kanama 2019).

This study focuses on the second point where there are many unsolved elements compared with the first point. It discusses the customer value creation process from the perspective of authenticity in the Japanese black tea market considering tourism and the factors that have contributed to constructing authenticity in the tea market. This study aims to provide guidelines for improving the quality of value co-creation in the tea market and tourism.

This study makes three academic contributions. First, it focuses on the relationship between product innovation and tourism, while scholars (e.g., Hjalager 2010) have often focused on innovation activities in the tourism industry. Japanese black tea has experienced unique processes in which tourists are involved in the innovation process, especially in the early stages of product innovation. The examination of Japanese black tea should present the chemistry and cooperative effects between innovation activities and tourism, which have not been clarified yet.

Second, this study explores the synergy between innovation activities and tourism from the perspective of authenticity. Authenticity has been paid attention in the context of tourism and evolved around the perceptions of tourists about objects, places, experiences, and settings (e.g., Moore *et al.* 2021; Lovell and Bull 2018). Here, the concept of authenticity is applied in the interpretation of the innovation process. Additionally, the value co-creation activities between companies and customers, a black box in innovation research, is discussed.

Third, this study shows the possibility that innovation activities can be a form of tourism. While the early stages of the innovation process include high uncertainty, resulting in many entrepreneurs abandoning their activities, this research shows a new possibility to reduce the uncertainties and promote market expansion and also implies the potential of new opportunities for the tourism industry to further market growth.

Literature review

Innovation processes and value co-creation

In every stage of marketing activity, from the creation of new ideas to the establishment of technological superiority, commercialisation, market input, and profitability, there is uncertainty in the process of innovation. In particular, the higher the novelty and innovation, the more the process is filled with uncertainty, and the entrepreneurs must repeatedly make their decisions without clear economic rationality.

Prior studies explored the barriers that prevent commercialisation from being successful and the ways to overcome the barriers by analysing the case where innovation is realised by the companies (e.g., Cooper 2019; Salomo *et al.* 2003). A consistent result of the success factors revealed by these studies was the deep pursuit of customer orientation and a mechanism to acknowledge customers' needs. Past studies discovered cooperative and interactive acts between companies and customers (e.g., Cooper 2019; Takano and Kanama 2019). However, it is not easy to respond to such uncertainties and difficulties in innovative activities through value co-creation. While responding to the ever-changing environment, companies need to provide customers with authentic experiences that help the customers to recognise the customer value. In particular, in the early stages of the innovation process, the recognition of customers' authenticity has not been clear; thus, an entrepreneur is forced to search in the dark.

The demand for authenticity

In today's saturated food market, more is often preferred over affordable prices and good quality. As mass production has reduced prices, good-quality products are available at affordable prices in many countries. It is difficult for food products to compete with each other by focusing only on the value of goods. Accordingly, the economy has shifted from a product economy to a service economy and now from a service economy to an experience economy (Pine and Gilmore 1998). As the economy and society mature, the movement to seek authentic experiences has also expanded.

The desire for authenticity is realised by moving away from the living space filled with impurities, false images, fiction, and mass production. Through various authentic experiences, customers have become more willing to pay higher prices for consumption activities based on their recognition of the authenticity of a product, performance, location, and particular aspects of the producer (Carroll and Wheaton 2009).

Gentile *et al.* (2007) defined customer value created by such experiences as follows:

> The Customer Experience originates from a set of interactions between a customer and a product, a company, or part of its organisation, which provoke a reaction. This experience is strictly personal and implies the customer's involvement at different levels (rational, emotional, sensorial, physical, and spiritual).
>
> *(p. 397)*

Thus, experience value is derived from the interaction between companies and customers and emphasises consumers' activeness.

Production of authentic experiences

Since MacCannell (1973) argued the notion of staged authenticity, this debate also has been applied and discussed within the context of tourism. For example, Wang (1999) insisted that tourism experiences had three forms of authenticity: objective, constructive, and existential. Cohen and Cohen (2012) proposed hot and cool authentication processes. Authors have argued different forms of authenticity, such as iconic authenticity and indexical authenticity (e.g., Grayson and Martinec 2004). Although the interpretation of authenticity does not exist as the only one that has convinced everyone, there is agreement that authenticity describes a verification process and the evaluation of truth or fact (Newman and Dhar 2014).

Authenticity consists of interrelationships between the consumer, provider, and experience (Le *et al.* 2020). It is essential for companies to provide experiences that can be perceived as authentic by customers to increase customers' values. However, through this interaction, the process of construction, factors affecting the construction of authenticity and how it enhances experiences are still under discussion.

McIntosh (2004) discussed that the authentic experience of tourists is 'getting personally involved in the experience,' 'experiencing daily life that is more realistic and more natural,' and 'experiencing culture in its natural landscape.' Raffaelli (2017) found that the aspects of 'community,' 'curation,' and 'convening' are important to provide a personal and specialised customer experience.

As the results of the cited researchers or cases above show, there are a variety of constructs that influence the establishment of authenticity, but those studies show promoting authentic experiences are important to the establishment of authenticity. Other studies on innovation have shown the importance of adopting a customer orientation approach to creating customer value (Cooper 2019; Salomo *et al.* 2003). However, the value co-creation activities between companies and customers have been a black box. We discuss the process of creating customer value through cooperative and interactive acts between companies and customers by applying the concept of authenticity.

Method: field surveys and interviews with farmers and distributors

The method used for data collection in this study is as follows. The researchers visited five areas where Japanese black tea was harvested: Shizuoka, Ibaraki, Saga, Nara, and Fukuoka. They interviewed two distributors. One distributor runs a tea shop in Tokyo. The other distributor runs a tea shop and a café in Saga and distributes some Japanese black tea to retailers. The history of Japanese black tea production was also investigated, starting from its birth in the 19th century, using the multiple documents capture process. Additionally, materials were collected from academic conferences,

Table 28.1 The details of the interview survey and the questionnaire survey

Interview survey	
Form	Semi-structured interview form
Date	From June 2017 to February 2020
Interviewees	Five Japanese black tea farmers and two distributors
Questions	1. Japanese black tea production and manufacturing
	How do you manage the tea fields?
	What varieties of tea plants are cultivated, and for what purposes?
	What made you decide to produce the black tea?
	What are the technical difficulties of tea production, and how do you handle them?
	What are the characteristics and specialties of the black tea you produce?
	2. Japanese black tea business
	Where and how are your products sold?
	Who buys your products mainly?
	How have the sales of black tea changed?
	What is the profitability of the black tea business?
	3. Current issues and future approaches to the black tea business
	What do you need to expand your business?
	What are the current issues and future direction for your black tea business?
	What are your dreams and goals for the future?

Questionnaire survey

Form	Questionnaire survey
Date	From 1 June 2019 to 17 June 2019
Respondents	675 Japanese black tea farmers
Questions	1. The area where you produce
	2. The year when the black tea production is started
	3. The reason for started the black tea production
	4. The amount of black tea production in 2018
	5. The black tea production rate of total tea production
	6. The methods of tea cultivation
	7. The season to pick the tea leaves for the black tea
	8. The varieties of tea plants for the black tea
	9. The plans of the black tea production in the next year
	10. The selling channel of the black tea
	11. The price range of the black tea
	12. The impact of black tea sales on green tea sales
	13. The status of intellectual property acquisition
	14. The topics to be learned and discuss for the black tea business
	15. The technical difficulties in black tea production
	16. Your dreams and goals for the black tea business

Notes: In the interview survey, distributors were not asked question 1. In the questionnaire survey, questions 2 to 14 are selecting questions, and questions 1, 15, and 16 are descriptive questions.

lectures, and information on the internet regarding recent trends not covered by books. A questionnaire survey was also conducted with 675 Japanese black tea farmers, with the cooperation of the Jikoucha Institute, making up the largest network of Japanese black tea farmers. The questionnaires included 16 items related to management, production, sales, and distribution (Table 28.1). Depending on the question, a description or multiple-choice type answer was required. This survey was conducted in June 2019, and the final responses were received from 219 farmers.

Case study 28.1   Japanese black tea farmer

As a case study of Japanese black tea farmers, one producer was selected from the Sashima District in Ibaraki. In this case study, it was aimed at clarifying the specific initiatives and activities towards the value co-creation made by a Japanese black tea farmer in Sashima. This area has been producing green tea under the brand name of 'Sashima-cha' for a long time. Among 30 tea farmers in this area, 12 farmers produce black tea.

One of the farmers is Mr. Masahiro Yoshida. Yoshida Tea Farm produces high-quality Japanese black tea, winning many awards in various tea contests. His farm cultivates a rare cultivar of tea named 'Izumi' for green tea, and he noticed that this would be suitable for black tea. When he first began to produce his Japanese black tea, he sold it alongside his green tea at his online and offline stores of the tea farm. Gradually, in addition to his existing sales channels, he actively tried to find newer channels to sell his products. One of the channels he found is a nationwide event called the 'National Locally Grown Tea Convention.'

Since this event is organised in various areas every year, it could be a good opportunity for black tea farmers to showcase their products to customers that they never would have had the chance to meet. This opportunity also proved beneficial for them to develop new tea products after receiving customer feedback from tastings and talking. For example, these opportunities led Mr. Yoshida to understand that customers wanted to enjoy the differences in the taste and flavour of the Japanese black tea. He began to produce types of Japanese black tea that placed emphasis on the flavour characteristic of 'Izumi'. Currently, his tea is evaluated and enjoyed by customers for its fruity and flowery flavours.

On the other hand, the process took him a long time to realise his ideal tastes and flavours. There was initially very little information to produce high-quality black tea in Japan, so he sought out the production technologies by himself. To gain knowledge about new production technologies, he actively contacted outside mentors, and improved his production skills by collaborating with mentors who were knowledgeable about tea production and could provide guidance. Gradually, his black tea began to be evaluated in contests.

Nowadays, many tourists from far away visit Sashima and purchase his Japanese black tea at his own stores near his tea farm. He now recognises it as an opportunity for co-creation and tries to connect with more customers, and just opened a new café for tourists. In this way, he has actively engaged in a value co-creation process, as described above. The direction of these acts can be categorised into two approaches. One approach is to increasingly get in touch with customers. The other approach is from the perspective of production technology. As can be seen from this case study, without the development of technologies, even if he could understand a customer orientation approach through the value co-creation process, he would not have been able to incorporate it into the creation of his products. Also, without contacts with customers in events, he would not have been able to understand the concept of customer orientation in depth and would not have been able to create products that were evaluated by so many people. It can be seen that it is important to take both approaches.

Result and discussion: value co-creation in an innovation process: Japanese black tea

History of the revival of Japanese black tea and estimation of the market scale in recent years

The 19th century marked the birth of the Japanese black tea industry. The Japanese government encouraged black tea manufacturers to engage in foreign trade and promoted the tea industry. In 1954, the export of black tea reached its highest recorded levels since World War II (Figure 28.1).

As the economy grew rapidly, the cost of Japanese tea production increased due to labour shortage and low productivity (Shimizu 1978). Additionally, the liberalisation of tea imports began in 1970. Subsequently, farmers rarely produced Japanese black tea. Almost all domestic tea production changed from black tea to green tea (Shimizu 1978). However, when the green tea market shrank, some green tea farmers responded positively to their difficult circumstances. Black tea production has been growing steadily since the 2010s. Between 2008 and 2016, the number of tea farmers increased from 100 to 600 (Arai and Nagata 2017).

The production of Japanese black tea has been expanding, but the scale of this market is still unclear. Based on the questionnaires, the researchers found that 309.3 tons of Japanese black tea were produced, and the estimated market size was approximately 2.7–4.1 billion yen in 2019.

Innovation activities and tourism from the perspective of authenticity

Innovation activities of Japanese black tea farmers

Food has traditionally been developed in various parts of the world, suited to the climate of each region. With this reason, 'local' related to the place is considered essential to the original characteristics of food. Arguably, 'local' influences the authenticity expected by customers, contributing

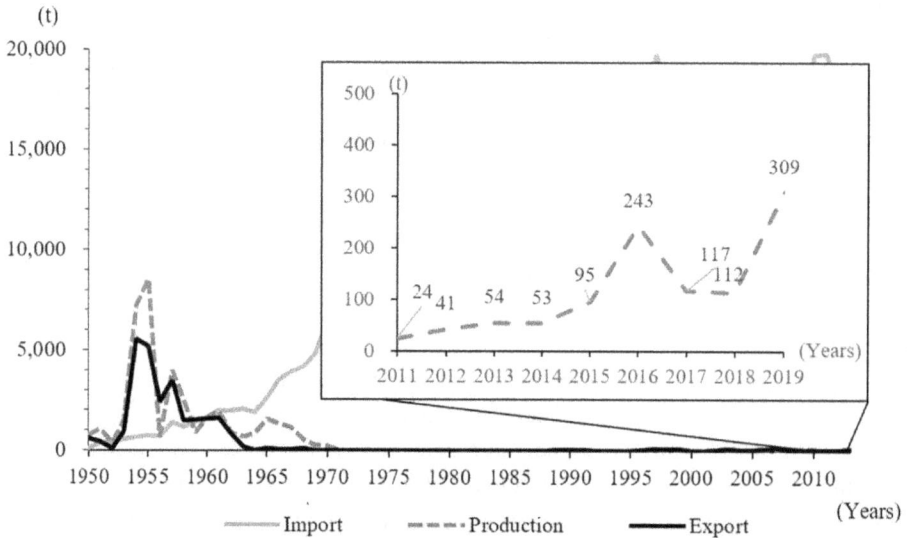

Figure 28.1 The production, import, and export of Japanese black tea in Japan

Source: Modified from Shimizu 1978; Nihon Koucha Kyokai 1990, 2014; and Japanese Association of Tea Production 2019

to the value of the customer experience (Autio *et al.* 2013). Food specifically includes the characteristics of the food itself, such as flavour, colour, taste, and stories about the produce and the production methods. As a result, the characteristics unique to the region and the natural environment are regarded as elements that create the characteristics and the story of the food itself. These reasons suggest focusing on the unique resources and natural environment of the region for Japanese black tea.

First, the unique resources of the region are noticed in the variety. One of the varieties of Japanese black tea is native. The native-variety tea trees were grown from seeds. When grown from seeds, the surviving tea trees were selected as suitable for the natural environment in a particular region so that only native varieties were cultivated. This point attracted consumers. However, breed varieties are clones with the same characteristics in different natural environments. Although uniformity can generate large product quantities, it makes it difficult to provide differentiation.

Second, the method of picking tea leaves reflects the natural environment. Tea leaves are currently picked by machines in Japan. However, some farmers have tea farms where the land is not flat and cannot use machines. They pick tea leaves manually as a traditional way. This results in the fact that the tea manufacturing processes of fermentation are consistent, and the products are of good quality. Consumers' response to this narrative may increase their purchase intention.

Additionally, the geographical characteristics and season of tea leaf picking reflect the natural environment. The components of tea are known to differ based on exposure to sunlight and temperature. The taste and flavour of Japanese black tea are influenced by geographical characteristics and the time of harvest (Anan and Nakagawa 1974).

According to Petrini (2013), customers are increasingly demanding that food be 'good,' 'clean,' and 'fair.' These are emerging movements that contrast with globalised and homogenised food, such as fast food. Japanese black tea has been produced in such a way that it can be regarded as a traditional and local food, and the results of this study are consistent with this understanding.

Relationship between Japanese black tea and tourism

For all these reasons discussed, farmers spend a long time considering various elements such as climate, varieties, cultivation method, picking season, and geographical characteristics to pursue the ideal black tea. These processes result in varied tastes and flavours, and these differences are understood as the original characteristics of locally produced black tea. Some of the farmer-led activities have helped construct the authenticity of Japanese black tea.

First, a pioneer of these activities is the National Locally Grown Tea Convention. It was formed in 2002 by tea farmers, and every year they organise a presentation and sale of the year's Japanese black tea in tea-producing areas. From the early stages of innovation, Japanese tea farmers have taken it upon themselves to organise events that bring customers to production areas. In this event, customers can see the natural environment and tea gardens in the region, sense the differences in flavour created by the different tea gardens and varieties, and feel the enthusiasm and dedication of the farmers. This experience helps customers feel that Japanese black tea embodies both the original and local. Originality and localness are one of the components of perceived authenticity to tourists (Skinner *et al.* 2020).

Second is the establishment of an academic society, Jikoucha Institute, which holds conventions for researchers, tea tasters, blenders, and farmers. It has led to the creation of a community among people involved in Japanese black tea. Recently, some communities openly disclose these activities to online customers, allowing them to see the daily lives of farmers and tea tasters. Through such real backstage access, usually unseen, customers view it as a story about the production of Japanese black tea without any inaccuracy. Seeing the backstage at a destination contributes to authenticity perceived by visitors (Bryon 2012; Daugstad and Kirchengast 2013).

Third is the organisation of contests. There are multiple Japanese black tea contests held in Japan. In these contests, tea experts with a deep knowledge of Japanese black tea evaluate the quality of teas by identifying any shortages in the manufacturing process in terms of the colour, flavour, and taste. Accordingly, the experts evaluated the unique Japanese black tea produced in each region, which results in discovering the characteristics of Japanese teas. This sort of 'award-winning' certification represents what Wang (1999) described as objective authenticity or what Beverland *et al.* (2008) illustrated as indexical authenticity by which people believe an object as official, authorised, or validated.

Impact on tourism

As a result of these activities, consumers enjoy Japanese black tea with various product characteristics and seek to understand the nature and differences of Japanese black tea through their experiences, such as visits to the producing areas and communication with local farmers. The National Locally Grown Tea Convention is a representative example, attracting more than 6000 tea fans from all over the country to tea-producing areas. Additionally, an increasing number of customers visit stores run by tea farmers in the tea production area to meet them or participate in tours to experience producing Japanese black tea with the farmers. These experiences helped customers to construct their own perceptions of authenticity, which contributed to their authentic tea experiences. The experiences highlighted the appeal of Japanese black tea as a tourism resource and attracted people who look for more authentic experiences.

These activities represent a movement towards the need for authentic experiences that enhances the promotion of Japanese black tea tourism. Since 2019, Japanese black tea farmers in Sashima have accelerated the movement to open a café, targeting tourists. Their aim is not only to attract people looking for an authentic experience, or gain income, but also to explore the value of co-creation of Japanese black tea with consumers through tourism.

Conclusion: elements of authenticity and the influence of authentic experiences in Japanese black tea

In this study, the relationship between innovation activities and tourism is explored from the perspective of authenticity, using the case of Japanese black tea. While past studies have focused on innovation activities in the tourism industry, few studies focused on the relationship between product innovation and tourism. Therefore, the concept of authenticity was applied in the innovation research to deepen the understanding of value co-creation. This study used information and data obtained through multiple interviews and questionnaires with farmers and distributors.

According to the case of Japanese black tea, producing this tea has claimed that it has regional value as a local food from the early stages of innovation. The differences in flavour, colour, and quality of Japanese black tea between regions have been accepted as its original characteristics. Various Japanese black teas attract people who look for an authentic experience. They seek to understand the nature and differences of Japanese black tea through their experiences, such as visiting producing areas and communicating with local farmers. These movements in the search for authentic experiences enhance the promotion of Japanese black tea tourism. This research implies that providing customers with authentic experience in innovation activities would consequently reduce the uncertainties and difficulties in creating customer values, and furthermore, lead to expanding the tourism market.

Farmers have provided customers with authentic experiences since the early stages of innovation. The four practices played a key role in constructing authenticity by customers in the case of Japanese black tea: curation of people who share the same interests, community building, convening real and rare conversations, and evaluation through contests. These excellent authentic experiences highlighted the rarity, difficulty of imitation, recursiveness, uniqueness, and regionality of Japanese black tea and attracted people looking for authentic experiences. However, the farmers' aim is not only to provide a more authentic experience. Through tourism, farmers try to explore the value co-creation of Japanese black tea with consumers. It indicates the potential of a new possibility of tourism to help promote market expansion.

Discussion questions

1   What kind of communication channels, ways, and/or process should tea farmers offer for value co-creation with customers who will not visit a farm?
2   What is the most effective approach to make customers aware of authenticity?
3   How could Japanese black tea farmers embrace tourism in the future?

References

Anan, T. and Nakagawa, M., 1974. Chaba no Kagaku seibun ganryo ni oyobosu hikari no eikyo. (Effect of light on chemical constituents in the tea leaves). *Journal of the Agricultural Chemical Society of Japan*, 48(2), 91–96.
Arai, S. and Nagata, J., 2017. Kokusan kocha seisanshagun no keieiteki gijutsuteki seikaku: Okinawaken wo jirei ni. (Management and Technological adoption of black tea producers emerging in Japan: Case study in Okinawa). *E-journal GEO*, 12(2), 192–208.
Autio, M., Collins, R., Wahlen, S. and Anttila, M., 2013. Consuming nostalgia? The appreciation of authenticity in local food production. *International Journal of Consumer Studies*, 37(5), 564–568.
Beverland, M.B., Lindgreen, A. and Vink, M.W., 2008. Projecting authenticity through advertising: consumer judgments of advertisers' claims. *Journal of Advertising*, 37(1), 5–15.
Bryon, J., 2012. Tour guides as storytellers – from selling to sharing. *Scandinavian Journal of Hospitality and Tourism*, 12(1), 27–43.
Carroll, G.R. and Wheaton, D.R., 2009. The organisational construction of authenticity: an examination of contemporary food and dining in the US. *Research in Organisational Behavior*, 29, 255–282.
Cohen, E. and Cohen, S.A., 2012. Authentication: hot and cool. *Annals of Tourism Research*, 39(3), 1295–1314.
Cooper, R.G., 2019. The drivers of success in new-product development. *Industrial Marketing Management*, 76, 36–47.
Daugstad, K. and Kirchengast, C., 2013. Authenticity and the pseudo-backstage of agri-tourism. *Annals of Tourism Research*, 43, 170–191.
Gentile, C., Spiller, N. and Noci, G., 2007. How to sustain the customer experience: an overview of experience components that co-create value with the customer. *European Management Journal*, 25(5), 395–410.
Grayson, K. and Martinec, R., 2004. Consumer perceptions of iconicity and indexicality and their influence on assessments of authentic market offerings. *Journal of Consumer Research*, 31(2), 296–312.
Hjalager, A.M., 2010. A review of innovation research in tourism. *Tourism Management*, 31(1), 1–12.
Japanese Association of Tea Production, 2019. Tea production results in Japan. Available from: www.zennoh. or.jp/bu/nousan/tea/dekiru03.htm [Accessed 10 February 2019].
Le, T.H., Arcodia, C., Novais, M.A. and Kralj, A., 2020. Producing authenticity in restaurant experiences: interrelationships between the consumer, the provider, and the experience. *Tourism Recreation Research*, 1–13.
Lovell, J. and Bull, C., 2018. *Authentic and inauthentic places in tourism: from heritage sites to theme parks*. London: Routledge.
MacCannell, D., 1973. Staged authenticity: arrangements of social space in tourist settings. *American Journal of Sociology*, 79(3), 589–603.
McIntosh, A.J., 2004. Tourists' appreciation of Maori culture in New Zealand. *Tourism Management*, 25(1), 1–15.
Moore, K., Buchmann, A., Månsson, M. and Fisher, D., 2021. Authenticity in tourism theory and experience. Practically indispensable and theoretically mischievous? *Annals of Tourism Research*, 89, 103208.
Newman, G.E. and Dhar, R., 2014. Authenticity is contagious: brand essence and the original source of production. *Journal of Marketing Research*, 51(3), 371–386.
Nihon Koucha Kyokai, 1990. *Nihon no kocha shijo no gaiyo* (Overview of the Japanese black tea market). Tokyo: Nihon Kocha Kyokai.
Nihon Koucha Kyokai, 2014. *Kocha tokei*. (Tea statistics). Tokyo: Nihon Kocha Kyokai.
Petrini, C., 2013. *Slow food nation: Why our food should be good, clean, and fair*. New York: Rizzoli Publications.
Pine, B.J. and Gilmore, J.H., 1998. Welcome to the experience economy. *Harvard Business Review*, 97–105.

Raffaelli, R., 2017. Reframing collective identity in response to multiple technological discontinuities: The novel resurgence of independent bookstores. *Harvard Business School*, 1–9.

Salomo, S., Steinhoff, F. and Trommsdorff, V., 2003. Customer orientation in innovation projects and new product development success-the moderating effect of product innovativeness. *International Journal of Technology Management*, 26(5–6), 442–463.

Shimizu, G., 1978. *Kocha nyumon* (Introduction to black tea). Osaka: Hoikusha.

Skinner, H., Chatzopoulou, E. and Gorton, M., 2020. Perceptions of localness and authenticity regarding restaurant choice in tourism settings. *Journal of Travel & Tourism Marketing*, 37(2).

Takano, R. and Kanama, D., 2019. The growth of the Japanese black tea market: How technological innovation affects the development of a new market. *Journal of Economic Structures*, 8(1), 13.

Wang, N., 1999. Rethinking authenticity in tourism experience. *Annals of Tourism Research*, 26(2), 349–370.

29

TEA TOURISM AND TEA TOUR GUIDING

The case of Iran

Hamira Zamani-Farahani

Introduction

Iran is one of the tea-loving countries, with tea cultivation covering an area of about 28,000 ha with 50,000 tea growers (Mehdizadeh 2020). Iran started cultivating and producing tea from the late 19th century in the north of the country, on the shores of the Caspian Sea. The households of tea growers in the country are scattered in the two provinces of Gilan (90%) and Mazandaran (10%). In 2020, Iran was the seventh-largest producer of tea in the world, producing 160,000 tons (World Atlas 2020).

However, tea tourism activity has started only in recent years in the country, and there is a lot of room for development. In order to hold a successful tea tourism programme or tour, in addition to the tea attractions, facilities and tour planning, the presence of a professional and qualified tour guide is essential. A tour guide is the most effective link between tourism attractions and the visitor. Tour guides are the ones working on the front line to enhance tourists' experience and organisational success (Houge Mackenzie and Raymond 2020; Ap and Wong 2001). Although researching the influence of tour guide performance on visitor or tourist sustainable behaviour is considered as an important feature in tourism, tour guiding in special interest tourism has received little academic research.

Tour guide training has an important impact on sustainable tourism development and tour management in developing countries (Weiler and Ham 2010). Professional qualifications (completing relevant training) can help raise this task's quality of services and standards. This chapter investigates specifications and characteristics of tea tourism and for tea tour guides using Iran as case study. The chapter begins with an overview on tea tourism in the Middle East, examines such tourism in Iran and reviews tour guiding in tourism and tea tour guide training and developing in Iran. Finally, a brief discussion and conclusion is presented. Discussion questions are included at the end of the chapter.

Literature review

Tea tourism in the Middle East

The Middle East is the transcontinental area between Western Asia and Egypt and includes countries that share common factors like ethnic groups, geographic features, religious beliefs and political

DOI: 10.4324/9781003197041-34

history. The region is generally understood to include Bahrain, Cyprus, Egypt, Iran, Iraq, Israel, Jordan, Kuwait, Lebanon, Oman, Palestine, Qatar, Saudi Arabia, the Syrian Arab Republic, Turkey, the United Arab Emirates and Yemen. The Middle East is therefore a large and diverse geopolitical and geographical area located in southwest Asia and northeast Africa. Its physical geography is varied. Mountains, greenspace and deserts divide the Middle East region into six geographically distinct zones that have influenced the development and maintenance of cultural traditions through much of the history of the region.

The region is located in a strategic and geopolitical position and has great natural and cultural attractions, as well as the potential for tourism (Zamani-Farahani *et al.* 2019). The dynamics of tourism destinations have remarkably differed across the region. The main development of tourism in the country has been more in the form of religious-pilgrimage tourism, cultural-heritage tourism, recreational tourism and adventure tourism. Special-interest tourism has received less attention, for example including rose and flower tourism (Iran); saffron, barberry and jujube tourism (Iran); lavender tourism (Turkey); health and thermal tourism (Iran, Turkey, Jordan); desert adventure and safari tourism (Iran, United Arab Emirates); as well as gastronomy tourism, geotourism, photography tourism, wellness tourism, and finally more specifically "tea tourism".

Among the countries in the Middle East, there are only the two countries of Turkey and Iran, with their tea plantations and processing facilities, that have the potential to host tea tourism. Turkey holds a significant place among the world's largest tea producers and ranks sixth in the world production area of tea (Worldatlas 2020). But the primary goal of tea cultivation in Turkey is to meet the domestic consumer demand and consumption (Seyis *et al.* 2018), and this trend has continued to date. Turkey has developed limited tea tourism and tea farms, and tea products are known as an important agricultural product rather than a tourist attraction, although in the tea capital city of Rize, there have been some efforts at creating attractions for tea tourism opportunities such as the Çaykur Tea Museum (Turkey from the Inside 2021) and construction of the 30-metre-high, seven-floor building shaped as the largest tea glass in the world, intended to create a tea museum with tea houses inside (Dailysabah 2021). Consequently, amongst the countries in the Middle East, only Iran has taken steps to develop tea tourism.

Tour guiding in tourism

Historically tourist guiding has developed for decades as a distinct profession within the tourism industry alongside other complementary professions that contribute to the visitors' overall experience (FEG 2020; Brito and Farrugia 2020). A tour guide is defined as a "Person who guides visitors in the language of their choice and interprets the cultural and natural heritage of an area, which person normally possesses an area-specific qualification usually issued and/ or recognised by the appropriate authorities" (WFTGA 2021). The tour guides need to be aware of the different interactions between tourist and host destination, and the moderation of any negative impacts (Noroozi 2021). The gaps between what local guides know and can do, and what they need to know and do, define their training needs (Weiler and Ham 2002). This statement demonstrates that tour guiding standards and professionalism is required through proper education and training programmes. Therefore, obtaining proper education, an appropriate certificate and being adequately prepared are essential to become a qualified tourist guide (Lovrentjev 2015) and to improve self-awareness and skills. Professional certifications can assist in improving tour guide performance, raising, and maintaining guiding standards (Black and Ham 2005), especially in natural-based settings. Licensed tour guiding means quality and can be a contributing factor in building the destination brand (Brito 2020); this feature can be preserved and enhanced through a continuous professional training programmes.

Besides knowledge of the tour area, a tourist guide should require a certain level of talents like good language skills, timeliness and punctuality, flexibility and patience, interest, charismatic personality, problem-solving skills, great organisational and leadership skills, responsibility, strong empathy and understanding, creativity, ability to communicate effectively, outgoing, energetic, keen ability to improvise and adapt, interpretation, focuses on building rapport, engaging storyteller and actor, good sense of humour, eagerness to learn, willingness to teach others and professionalism (Zammit 2020; El-Menshawy 2016; Lovrentjev 2015; Kruczek 2013). Alternatively, tour guide performance was found to have a significant direct effect on tourist satisfaction with guiding service, quality experience and visitor sustainable behaviour (Hwang *et al.* 2020; Alazaizeh *et al.* 2019; Baeti 2019; Kuo *et al.* 2018; Hwang and Lee 2018; Chang 2014); in turn, tourist satisfaction results in greater intention to repurchase or return and recommend the tour.

Methodology

Findings in this chapter are based on analysis formulated through local tea plantations research by observation, discussion by local and key people involved, and photographing which was done a few years ago in order to write a book titled *An Introduction to Tea Tourism (Iran): Practical Guide*, published in Persian in 2019. Research was done on theoretical and practical courses of tea tourism guiding, as well as analysis of published data including academic studies, government documents, current media sources, and statistical databases. The author has presented the finding of these researches as a case study that is part of Iran's overall tea tourism picture.

Results

Tea tourism in Iran

Tea tourism could be defined as the tourism experience of the tea plantation and production, history, culture and etiquette of the cultivation, preparation and consumption of tea. Iran is located in the eastern portion of the Northern Hemisphere, in the southwest of Asia serving as a bridge between central and western Asia, which links Asia to continental Europe. The Iranian plateau is one of the oldest centres of ancient civilisation in the world. The coastline of the Caspian Sea extends over a length of 700 km and comprises the green and fertile plateau. The climate of this region is warm and humid, and the average annual precipitation level is 1000–2000 mm. According to the author's research, Iran's tea production areas are located near the coastal plains of the Caspian Sea mainly in the two northern provinces of Gilan and Mazandaran. After rice, tea is the most important agricultural product of Gilan Province. Today, tea consumption is an integral part of ordinary life in Iran and also plays a significant role in common culture. Before planting tea in Iran, the tea was first exported through the Silk Road by the Chinese; later, when cultivation and production gradually flourished in India, tea was imported to Iran through southern trade routes.

Tea cultures in Iran and those of Russia and Turkey have some commonalities. Nevertheless, there are particular features regarding Iranian tea culture in Iran that are not found in other places. Tea is a significant part of Iranian life. Iranians drink tea when they get up in the morning, through the whole day, afternoon chats, reception of guests, ceremonies, special occasions and parties, after meals or while watching television (Gholami 2019). Traditionally, tea is served from a samovar, a heating vessel originally imported into Persia from Russia. Samovars are made from copper, brass, silver or gold and are still used throughout Russia, central Asia and Iran. Literally meaning "self-boiler", the samovar is used to keep water hot for prolonged periods of time through a fuel-filled pipe in the middle of the structure that heats the contents surrounding it, and ornate versions from the Qajar dynasty may still be found in use (Iran Tourism News 2019).

In 1958, to support tea producers and consumers, an organisation called Tea Organisation of the Country was established in the city of Lahijan (in the north of Iran, on the shores of the Caspian

Sea and the southern slope of the Alborz mountain range) in Gilan Province, to manage the affairs related to tea cultivation and industry. Later, to solve existing bottlenecks and provide appropriate technical and scientific services to tea growers, the Tea Research Institute was established in 1984 in the same city. This institute carries out research, laboratory and services related to the tea industry. There is also the Union of Producers, Traders and Tea Packaging Industries of the Country which opened and registered in 1975.

In comparison with China and India and other similar countries, holding tea tourism tours in Iran is at the beginning stages of its development. Attention to the importance of tea tourism occurred in the last few years, when tea tourism tours have been periodically organised by travel agencies (i.e., Iran Tourism Center). The Tea Tasting Tour in Iran or Tea Tourism Tours are generally designed from one to five days, depending on the programme, and are performed by air and land, mainly departing from Tehran, the capital of the country to Gilan Province. During guided walks, tourists have the opportunity to visit the tea gardens and farms, picking fresh tea leaves, learning about soil, insects and weeds that are different in a stable ecosystem, becoming familiar with the local farming and rural lifestyle through the tasting of tea and experiencing private farm tours. Tours typically also include visits to tea factories (seeing the processing steps that process the green leaf into a final tea product), the tomb of the father of Iranian tea and the Iran National Tea Museum. On multi-day tours, visitors see other tourist attractions such as the city of Rasht (the capital of Gilan Province, UNESCO Creative City of local gastronomy); Rudbar (olive gardens); Manjil (famous for its wind turbines and as the land of wander winds); and Gilan Rural Heritage Museum (to see real life style and culture of Gilan region), tasting wonderful local foods and other tourism attractions (Iran Tourism Center 2018). One of the suitable accommodations for tea tourism is the first farm hotel in the country named Moein, a five-star hotel, which is surrounded by tea gardens in Gilan Province. This hotel with a beautiful view overlooking the Masouleh Mountains and tea gardens was opened in 2015. Near this hotel, there are some tourist attractions such as the historical village/town of Masouleh and Qaleh Rudkhan Castle (a brick and stone medieval castle).

Development process of tea use and production

Iran's history of tea planting begins in the late 19th century. Some say that the history of tea consumption in Iran, according to the travelogues of orientalists, dates back to the 17th century. Nevertheless, according to authentic historical documents, Iranian scientists and pharmacists were the first to know this plant and its medicinal and therapeutic properties, and used it as a medicinal plant alone or in combination with other herbal remedies. Therefore, they first became familiar with tea as a medicinal substance and listed its name in medical and pharmacological books.

The wide drinking of tea in Iran dates back to the Safavid period (1501–1736). At this time, in the capital city of Isfahan in addition to coffee shops, there were also teahouses that served customers with "Khatai tea" (tea imported from Khotan China), as well as Indian tea. The tea-drinking utensils such as samovars and teapots and cups for tea were imported from Russia to Iran about 100 years before tea was cultivated in the country and gradually spread among different segments of the population. After the mass production of tea in Iran, samovars and tea utensils replaced coffee utensils. From then on, tea was offered in coffee shops instead of coffee.

In the early days, the activity of drinking tea belonged to the aristocracy. After the development of tea cultivation and mass production of this plant, tea entered all markets of Iran and became available to the public at a cheaper price than imported tea. The first Pahlavi government (1925–1941) established the first tea factory and the first tea research centre in Lahijan region. The development of tea cultivation is considered one of the main successes of the government's modernisation policy at that time.

The first promoter of planting and production of tea in Iran

Around 1883–1884, a person named Haj Mohammad Hossein Isfahani brought tea seeds from China to Iran. Therefore, the first tea seeds came to Iran in the same way that the tea plant was once imported as a medicine. The first tea farms were formed in the country's northern provinces because of its climate suitability. However, because of sociopolitical factors his attempt was unsuccessful. Afterward, Mohammad Mirza, Kashif al-Saltanah, Consul General of Iran in India, with the encouragement of the king at the time and according to his mission, after about one and a half years of effort and work on tea plantations in India brought several thousand tea seedlings and several boxes of tea seeds from India to Iran. In 1902, he established the first tea gardens in Lahijan region and around Tonekabon area, in Gilan and Mazandaran Provinces. The outcome of his efforts was successful. Then he wrote a book, *Tea Cultivation Instructions*, and taught the workers how to cultivate and produce tea. In 1929, when returning from China and Japan, while gaining new experiences in this field, he was killed in a car accident near Bushehr (port city and capital of Bushehr Province, southwestern Iran). His body was transported to Lahijan, and according to his will, he was buried in his tea-growing hills of Lahijan (Yousefdehi 2020). Since he was the first person in Iran to process tea, he became known as the father of Iranian tea. His mausoleum is currently one of the attractions of Lahijan and next to it is the Museum of Iranian Tea History, established in 1996 (Zamani-Farahani 2019). This museum in Lahijan is the only specialised tea museum in the country that introduces the Iranian tea industry from the beginning of its entry into its current state. The tomb of Kashif al-Saltanah and the Museum of Iranian Tea History (both are tea tourism attractions) belong to the contemporary period of tea history and have a national registration number (Figure 29.1).

Tea production

As previously mentioned, the area under tea cultivation is located in the northern regions of the country in the plains and foothills, with about 30% of the gardens located in the plains and the rest

Figure 29.1   The tomb of the father of Iranian tea and the Iran National Tea Museum in Lahijan region
Source: Author

in the foothills. Tea is mostly cultivated in Lahijan, Fooman, Langrud, Amlash, Siahkal, Rudsar, Ashrafieh, Soomehsara, Shaft and Rasht regions in Gilan Province and Tonekabon and Ramsar District in Mazandaran Province. The green tea leaves are harvested in three crops: spring, summer and autumn. Spring harvesting begins in early May, summer in the second half of June and autumn harvesting in September and ends in October.

One of the remarkable points is that in the cultivation and production of Iranian tea, no chemical pesticides are used. The harvested product of tea growers is processed annually in 182 active factories with 38 tea brands (Mehdizadeh 2020). Despite local daily and weekly (markets in the region, unfortunately there is still no market for the supply and sale of domestically produced tea. Most of the tea produced in Iran is black, green and white, but white and green tea is produced less than black tea due to its low profitability. In Iran, it is customary to brew tea with aromatic plants such as cardamom, mint, cinnamon stick, ginger, citrus aurantium blossom and damask rose. In addition to their properties, these compounds give a pleasant taste to this drink.

According to Iran's tea organisation, the most important challenges facing the Iranian tea industry can be listed in three sectors: production, processing and trade. Needs to renovate tea gardens, dated factories and production line automation, improper packaging and incompatibility of the capacity of tea factories with green leaves are among the most important problems facing tea processing. Many of the country's tea-growing fields have changed their use, and some of these lands have become residential settlements due to their proximity to cities. Sometimes winter frosts and summer droughts have also caused serious damage to the Iranian tea industry. Many lands have been abandoned by tea growers due to a lack of encouraging policies. Some northern tea growers have sold their fields to make a living and so on. Considering the issues stated, the development of tea tourism is one of the ways to help the development and diversification of Iran's tea industry and slowly improve the living conditions of local residents.

Developing tea tourism guides

In order to hold a successful tea tourism program or tour, in addition to the tea attractions, facilities and tour planning, the presence of a professional and qualified tour guide is essential, requiring specifications for and characteristics of a tea tourism tour guide. Tour guides, while having excellent moral and personality characteristics, might also have a very wide range of general and specialised knowledge, which is discussed in the previous section. Tourist guides are key players in the tea tourism experience. As far as the author's research shows, there are not yet specific training programmes for tea tourism tour guides in countries such as China, India and Sri Lanka, where tea tourism is promoted; therefore, there is the need to develop such an educational framework.

In Iran, to hold a tour there must be a trained tour guide with a tour leader card (identification card). The travel agency is not allowed to conduct a tour with an untrained tour guide without an identification card. Tour guide training is carried out in training centres that have been licensed by the Ministry of Tourism of Iran (the Ministry of Cultural Heritage, Tourism and Handicrafts [MCTH]) and its list is informed and updated through this ministry and its representatives in the provinces of the country. The issuing authority for licences and identification cards for available types of tourist guides (cultural tour guide, eco-tour including tea tourism guide, geo-tour guide) is also MCTH and its provincial branches. General conditions are required for applicants to participate in these courses; these are including citizenship of the Islamic Republic of Iran, belief in the religion of Islam or a minority religion recognised by law of the Islamic Republic of Iran, being at least 22 years of age (maximum age is not considered), have no criminal record and no drug addiction, have at least a bachelor's degree, have a conscription card or exemption for male applicants, have physical health appropriate to the subject of activity and mental health, and have fluency in one of the foreign languages (for guiding international tours).

The next step is to register and participate in the tour guide course (in person or online) and pass the related exams; this process will take about 9 to 11 months. After taking the interview test in an international language of choice (English, German, French, Italian, Spanish, Russian, Turkish, Chinese, Armenian, Japanese, Arabic, etc.), in which the level of proficiency in foreign languages should be at an advanced level, and then having success in the course exam in the training centre and participation in the comprehensive exam of the MCTH (the whole test usually lasts 140 minutes) and success in all the above examinations (passing minimum score of 12 out of 20 for each lesson), and in order to obtain a GPA certificate, a minimum of 12 will be required). This comprehensive exam is usually held twice a year (summer and winter).

During all these steps, the applicant should pay the related costs. Those who have a bachelor's degree or higher in related fields such as tourism management, geography and tourism planning and tourism marketing do not need to go through a training course. After registering and participating in the language interview test and then the comprehensive test, applicants will receive their certificate. To get a tour guide identification card, applicants must submit the general documents as already stated, a certificate of success in the exams and participation in the interview of the technical card issuance committee. The issuance and receipt of this card takes about three months, and the validity of the tour guide identification card will be one year for the first time and for three years after the extension for the second time.

Training courses in the field of tour leadership and specialised guide are various and include Iran tour guide (cultural), nature tour guide (eco-tour leadership), mountain guide, birdwatching guide, child tourism guide (age limit 22–55 years), specialised guide for bicycle tours (age limit 20–40 years), safari guide, agricultural tourism guide and so forth. Tea tourism guides fall under the category of eco-tour leadership and agricultural tourism guides. If tea tourism in the future will have a significant development, a special training course should be implemented for this category of tour guiding.

In general, the tour guide courses include theoretical topics and several practical field trips. There are about 22 to 33 theoretical and practical training courses for different tour guide programs. Tea tourism tours were previously guided by ecotourism and agricultural tourism guides, but the finalisation of the course for tea tourism guides must be done. The suggested thematic titles of theoretical and practical courses of tea tourism guide courses and training framework are presented in Table 29.1 (Astiaj Tourism 2019).

These above theoretical courses will be completed with related seminar(s), educational videos and trips including familiarity with tea production areas in the country, tea farms, tea production process and tea production centres (traditional and industrial factories and packaging centres), tea educational-research centres, tea museums, tea bazar(s), local tea-making customs and so on. Travelling also helps tour guides to keep up to date with what is happening abroad, to learn, appreciate and to be inspired by what others are doing successfully (Zammit 2020). The training course of the tea tourism tour guide by considering general topics, should be prepared, compiled and implemented, taking into account needs, facilities and conditions of a tourist destination. These courses should be changed, adapted and modernised through the years to provide the best education and preparation to future tourist guides (Zammit 2020).

Discussion and conclusion

The current chapter investigated specifications and characteristics of tea tourism and tea tourism tour guiding in Iran as a case study. According to the evidence uncovered, Iran is the only country in the Middle East and West Asia that has taken steps to develop its tea tourism. However, as far as the author's research shows, there still are not specific training programs for tea tour management and guiding. Since Iran's tea tourism is at the beginning of its development, it requires more planning

Table 29.1 Proposed curriculum for tea tour guide training

1. Introduction to tourism	2. Ecotourism and natural tourism (specialised and basic terms)
3. Familiarity with local ecotourism and conservation	4. Familiarity with local tourism geography and climatic diversity
5. Introduction to special interest tourism	6. Familiarity with local plant diversity and plantation of tea, tea preparation
7. Introduction to tea garden (planting and harvesting), properties, tea processing, types	8. Vocabulary and terms of natural tourism/tea tourism
9. Recognition of local museums	10. Introduction to touring skills
11. Nature settlement equipment, techniques and map reading and working with GPS	12. Application of information technology in travel
13. Familiarity with how to plan and execute a tea tourism tour	14. Familiarity with the rules and regulations of tour management
15. Ethics of tourism profession and behavioral patterns	16. Behavioural patterns and social etiquette in tourism
17. Familiarity with local communities and the principles of facilitation	18. Familiarity with the attitudes and behaviour of nations
19. Principles and foundations of interpretation of cultural and natural heritage	20. Familiarity with the local culture and ethnic groups
21. Psychology of tour participants	22. Photography in nature
23. Familiarity with handicrafts and traditional arts	24. Contemporary political and social history
25. Rhetoric	26. Travel hygiene and first aid
27. Crisis management in travel, rescue and medical emergencies	28. Related seminars, workshops, and study tours/ field trips

and development attention. More attractions such as visiting related organisations, tea research centres, tea cultivation treasury, traditional and modern tea-processing centres, local tea making, related handicraft production and display centres, and so on must be considered in planning related tour itineraries. In addition, the creation of a special tea market and creativity, enhancement and standardisation of the tea museum, and holding local and regional festivals with a focus on tea, along with the introduction of local culture, foods, products and handicrafts in the region, and visiting and drinking tea in local teahouses should all be included as part of the local tea tours. Beside the potential positive socioeconomic impact on the local community, the purpose of holding tea tours can be to pay more attention to maintaining tea gardens and reviving abandoned tea factories in the country.

Proper guidance for tea tourism is indispensable. In order to meet the expectations of tourists and organised tours, tour guiding standards and professionalism are required through proper education and training programs. Iran is a country in which education, training and a licence are required for the tourist guide to meet certain criteria in order to provide a high-quality guiding product. However, these training courses should be specialised and regularly updated. Tea tourism in Iran for further development requires better planning, effective marketing strategies, holding specific training courses, and job evaluation or improvement programs of the quality and reputation of the tourist guides. Continuous training over the years and how to deal appropriately with tourists for locals such as gardeners and other village activists, as well as tour guides and planners of these type of tours, are essential. The tour guiding process needs a good monitoring system to ensure high standards of service performance by the tour guides.

This chapter calls attention to fulfilling the gap of tea tourism and tea tour guiding development in Iran and other global tea regions. It is suggested that to meet the required qualification and raise the standards of the tea tourism guiding profession, each country should design its own appropriate standard and training course program due to its own environmental and sociocultural characteristics. Therefore, it is recommended that future research should identify and investigate the efficacy of tea tourism tour guide performance and related training programs within a particular country or context.

Discussion questions

1 What are the characteristics of tea tourism in the Middle East region?
2 What are the features of tea tourism in Iran?
3 What are the characteristics of a perfect tea tourism tour guide?
4 What specific training programs and courses needed to be considered for the tea tour guide?

References

Alazaizeh, M.M., Jamaliah, M.M., Mgonja, J.T. and Ababneh, A., 2019. Tour guide performance and sustainable visitor behavior at cultural heritage sites. *Journal of Sustainable Tourism*, 27(11), 1708–1724.

Ap, J. and Wong, K.K.F., 2001. Case study on tour guiding: professionalism, issues and problems. *Tourism Management*, 22(5), 551–563.

Astiaj Tourism, 2019. *Tea tourism in Iran: an unpublished comprehensive research project*. Tehran: Astiaj Tourism Research and Consultancy Center.

Baeti, A.N., 2019. *An analyzing tour guide knowledge of tourist satisfaction*. University Muhammadiyah of Purwokerto [online]. Available from: www.researchgate.net/publication/338401598_TOURISM_GUIDING/link/5e12a809299bf10bc3928b74/download [Accessed 28 March 2021].

Black, R. and Ham, S., 2005. Improving the quality of tour guiding: towards a model for tour guide certification. *Journal of Ecotourism*, 4(3), 178–195.

Brito, L.M., 2020. The consequences of guiding profession deregulation for the status and training of tourist guides: a Portuguese overview. *International Journal of Tour Guiding Research*, 1(1), 34–44.

Brito, L.M. and Farrugia, G., 2020. On tourist guiding: reflecting on a centuries-old profession and proposing future challenges. *International Journal of Tour Guiding Research*, 1(1), 4–12. Available from: https://arrow.tudublin.ie/ijtgr/vol1/iss1/3 [Accessed 2 April 2012.].

Chang, K.C., 2014. Examining the effect of tour guide performance, tourist trust, tourist satisfaction, and flow experience on tourists' shopping behavior. *Asia Pacific Journal of Tourism Research*, 19(2), 219–247.

Dailysabah, 2021. Turkey's tea capital Rize aims high with glass-shaped building [online]. Available from: www.dailysabah.com/turkey/turkeys-tea-capital-rize-aims-high-with-glass-shaped-building/news [Accessed 2 April 2021].

El-Menshawy, S., 2016. Effective rapport in tourist guiding (interpretation of themes). *Journal of Socialomics*, 5(3), 1–5.

FEG, 2020. Definition of a tourist guide. *European Federation of Tourist Guide Associations (FEG)* [Online]. Available from: www.feg-touristguides.com/post.php?i=why-definitions-matter [Accessed 14 May 2021].

Gholami, M., 2019. Cornerstone of Iranian culture: tea [Online]. Available from: www.visitouriran.com/blog/cornerstone-of-iranian-culture-tea/ [Accessed 15 March 2021].

Houge Mackenzie, S. and Raymond, E., 2020. A conceptual model of adventure tour guide well-being. *Annals of Tourism Research*, 84, 102977. https://doi.org/10.1016/j.annals.2020.102977.

Hwang, J., Kim, J.J., Soo-Hee Lee, J. and Sahito, N., 2020. How to form wellbeing perception and its outcomes in the context of elderly tourism: moderating role of tour guide services. *International Journal of Health*, 17(1029), 2–16.

Hwang, J. and Lee, J.H., 2018. Relationships among senior tourists' perceptions of tour guides' professional competencies, rapport, satisfaction with the guide service, tour satisfaction, and word of mouth. *Journal of Travel Research*, 58(8), 1331–1346.

Iran Tourism Center, 2018. Iran tea tasting tour [Online]. Available from: https://irantourismcenter.com/portfolio-item/iran-tea-tasting-tour// [Accessed 22 April 2021].

Iran Tourism News, 2019. The beauty of Iran tea production and tea culture [Online]. Available from: https://irantourismnews.com/the-beauty-of-iran-tea-production-and-tea-culture/ [Accessed 20 April 2021].

Kruczek, Z., 2013. The role of tourist guides and tour leaders in the shaping of the quality of regional tourist products. *In:* Marak J. and Wyrzykowski, J., eds. *Regional tourism product – theory and practice*. Wrocław: University of Business in Wrocław (UBW), IV. 47–56. Available from: www.researchgate.net/publication/258244104 [Accessed 22 April 2021].

Kuo, N.T., Cheng, Y.S., Chang, K.C. and Chuang, L.Y., 2018. The asymmetric effect of tour guide service quality on tourist satisfaction. *Journal of Quality Assurance in Hospitality and Tourism*, 19(4), 521–542.

Lovrentjev, S., 2015. Education of tourist guides: case of Croatia. *Procedia Economics and Finance*, 23, 555–562.

Mehdizadeh, M., 2020. Tea production and trade information in 1398(2020). *Iranian Tea Organization* [Online]. Available from: www.irantea.org/fa/Details/7708/ [Accessed 30 February 2021].

Noroozi, H., 2021. The tangible and intangible heritages of Iranian Nomads: the touristic potential of pastoral nomadism. *International Journal of Tour Guiding Research*, 2(1), 63–77. Available from: https://arrow.tudublin.ie/ijtgr/vol2/iss1/7 [Accessed 22 April 2021].

Seyis, F., Yurteri, E., Ozcan, A. and Savsatli, Y., 2018. Organic tea production and tea breeding in Turkey: challenges and possibilities. *Ekin Journal*, 4(1), 60–69.

Turkey from the Inside, 2021. RİZE [Online]. Available from: www.turkeyfromtheinside.com/places-to-go/r/227-rze.html [Accessed 18 July 2021].

Weiler, B. and Ham, S.H., 2002. Tour guide training a model for sustainable capacity building in developing countries. *Journal of Sustainable Tourism*, 10(1), 52–69.

Weiler, B. and Ham, S.H., 2010. Development of a research instrument for evaluating the visitor outcomes of face-to-face interpretation. *Visitor Studies*, 13(2), 187–205.

WFTGA, 2021. *What is a tourist guide?* World Federation of Tourist Guide Association (WFTGA) [Online]. Available from: https://wftga.org/about-us/what-is-a-tourist-guide/ [Accessed 22 May 2021].

Worldatlas, 2020. The world's top tea-producing countries [Online]. Available from: www.worldatlas.com/articles/the-worlds-top-10-tea-producing-nations.html [Accessed 5 June 2021].

Yousefdehi, H., 2020. *Kashif Al-Saltanah: Mirza Mohammad Kashif Chaikar.* Rasht: Farhangeilia (in Persian).

Zammit, V., 2020. Roles and responsibilities of a tourist guide and their trainers: reflections and recommendations. *International Journal of Tour Guiding Research*, 1(1), 18–22.

Zamani-Farahani, H., 2019. *An introduction to tea tourism Iran: practical guide.* Tehran: Mahkameh Pub (in Persian).

Zamani-Farahani, H., Carboni, M., Perelli, C. and Torabi Farsani, N., 2019. Islamic tourism in the Middle East. *In:* Timothy, D.J., ed. *Routledge handbook on tourism in the Middle East and North Africa.* London: Routledge, 125–136.

PART V

Resilience in tea tourism

30
RESILIENCE THROUGH TEA TOURISM
A tea region case from India

Sujama Roy

Introduction

Can resilience contribute to the development of tourism within a tea-producing region? In the context of this question, this chapter examines resilience through tea tourism in a tea region of India, the Dooars in the West Bengal state of India. In the field of tourism studies, resilience is viewed as a more community-focused alternative to the traditional sustainable development paradigm (Lew 2014). Resilience thinking opens up new ways of thinking in order to find solutions to otherwise complex problems, using participatory, collaborative and interdisciplinary solutions (Grove 2018). Resilience-based analysis therefore provides a useful approach to understanding how a specific shock impacts destinations, their economic development and local communities (Walker and Salt 2012). Such analysis tends to provide a wider and more coherent vision taking into account the of uncertainty and unexpected changes.

The tea-producing Dooars region has no political identity; it has only a geographical existence as an alluvial floodplain of the Himalayan foothills in north-eastern India. Historically, the region was controlled by the kingdom of Bhutan until the British annexed it in 1865 after the Anglo-Bhutan war (Sunder 1895). Today, there are about 150 tea gardens existing in this region of West Bengal. Many of these tea gardens are located in the lap of mountains and beside rivers such as the Rydak, Toorsha, Dima, Kaljani, Nonai, Jainti and Sankosh. Some of the tea gardens are contiguous to protected forests, such as Jaldapara National Park which offers the Jaldapara Tea and Forest Resort to visitors.

Tea plantation history and living realism of area gardens

Tea plantations in North Bengal, as elsewhere in India, share colonial roots, which can broadly be read as the history of labour migration. It is significant that the colonial capitalists were interested not in individual but in family migration, as this would ensure a reproduction of cheap labour (Sarkar 2019). Plantation labour has historically been put under oppressive labour regimes that were armed with substantial powers, both legal and extra-legal, to discipline the migrant labour in such a manner that their mobility and even their out-garden sociability could be thoroughly restricted. Hence the migrant labour force turned into a labour "held in bondage in free market", according to Das Gupta (1992, 66).

The isolated territorial location of the plantations, coupled with the restricted mobility of the workers throughout the colonial period, continue to influence labour relations to this day (Xaxa

DOI: 10.4324/9781003197041-36

1985). While government laws (such as the Plantation Labour Act 1951 and other relevant acts) were passed and some degree of security for the labour force was introduced, workers still remained at the margins. In contemporary times, this marginalization has taken new shapes and given birth to new problem areas.

Given the structural specificities of the common methods of producing tea in North Bengal, the livelihood of the workers is solely dependent upon garden management. Accordingly, the employment, social security and politico-legal entitlements – the essential components of "decent work conditions" – of the workers are destined to be denied as and when the management locks out or abandons a garden. Several studies have shown that the tea estates of North Bengal, barring a few exceptions, have failed to provide "decent work conditions" to the huge number of workers they employed, the majority of them belonging to marginalized and resource-less tribal communities (Prasanneswari 1984, Bhowmik 2009).

Both global economic factors and local constraints have affected the tea industry of the Dooars. Chakraborty (2013) observed that many tea estates failed to manage the shock they witnessed in production and worker benefit dilapidation. Cachar Plantation Enquiry Committee explained, "Economic deterioration and sickness of the tea industry had begun in the colonial time; it was accelerated under the ownership of Indian proprietors whose motive is not only to manufacture tea but in the speculative value of land" (cited in Thakur 1995). With regards to the state of the tea gardens, the Tea Board of India (1995) surveyed 115 tea gardens reporting natural causes (cold climate, flood, drought, heavy rain and hail) as responsible for this state of the tea industry in northeast India. More than a thousand deaths were reported to have occurred due to starvation and malnutrition in several shut-down tea estates of the Dooars (e.g., the estates or Kathalguri, Red Bank, Bundapani, Bharnobari, Dheklapara, Lankapara and Surendra Nagar) during the last decade. According to Sen (2015), the administration, management and some trade unions associated with tea gardens do not acknowledge this starvation phenomenon but are concerned about the issue of malnutrition.

Confronting these transformations taking place in the overall tea plantation industry, the grossly underpaid plantation labour has to bear the impact of a flexible labour regime that characterises the plantations of North Bengal in contemporary terms, which forces its workers, half of whom are women, to migrate outside in search of livelihood, often ending up falling prey to traffickers (Biswas *et al.* 2005; Chakraborty 2013; Ghosh 2015). Thus, analysing the vulnerability of labourers of the closed down, abandoned and sick tea gardens of North Bengal, it becomes imperative to think of the livelihood options for these plantation workers (Roy and Biswas 2018).

Tea tourism in the Dooars

The Dooars region of North Bengal is considered as the second-leading tea production region in India. Tea gardens of the Dooars are basically surrounded by rich flora and fauna and have majestic views of the eastern Himalayan foothills, various orchids, cultural heritage, rituals and festivals. The tea gardens have always provided a rich visual rhetoric in any tourism promotional discourse of the Dooars. Though tea tourism is not an unknown concept of Dooars tourism, the region is yet to witness any major development in this area. There is also a general lack of related study on tea tourism here in the academic literature. However, Datta (2018) identified the most attractive pull factor for tourism in the Dooars region is natural beauty, followed by wildlife and tea gardens, which altogether could be experiential in the frame of tea tourism. As highlighted by Jolliffe (2007), tea tourism can offer visitors a variety of experiences that might include following tea trails, participating in tea traditions, visiting tea gardens and staying at historic tea estate bungalows repurposed for tourism. All of these experiential product attributes are present all over the tea region of the Dooars.

Chakraborty and Islam (2020) stated that tea gardens of the Dooars are directly related to the entire socioeconomic and sociocultural features of the people of the Dooars. In earlier times, tea tourism was limited only to the managers and owners' families. Later, tea tourism of the Dooars has been lifted up with their facilities like adventure, nature, experiences of culture, tradition and customs to appeal to a broader audience. This includes sightseeing tours, tea harvesting, tea tasting, local food and experience of ethnic culture. There are consequently a number of factors which are supportive to any tea tourism projects of the region.

Firstly, many Dooars tea gardens lie along the National Highway (NH31), which makes such gardens accessible with the development of feeder roads connecting tea gardens and the national highway. Secondly, there is scope for watching wildlife, as most tea gardens share boundaries with national parks of the Dooars. Thirdly, most of the workers are tribal people from the contiguous areas of Jharkhand, Chattisgarh and Orissa. These include the Oraons (who form about half of the tribal population), the Mundas, Kharias and Santhals, among others. There are also nontribal artisan communities from the same area of origin, such as Mahali, Chik Baraik, Ghasi and Turi. These groups collectively form a common identity that can be distinguished from the other people in the region (e.g., Bengalis, Nepalis) on an ethnic basis. The rich cultural fabric brings potential for ethnic tourism within the region.

Despite these factors, the gardens in the region tend to lack basic tourist facilities, amenities and infrastructure, government information centres and proper tour guides. Security issues too could prevail in touring the gardens. Chakraborty and Islam (2020) have pointed out reasons that are undermining the overall tea tourism development in the region: (1) lack of coordination between local tea garden authorities and Government Forest and Tourism Departments and (2) lack of systematic planning and proper organizational structure. Apart from these, the tea tourism area poses certain threats like human-wildlife conflict (especially human-leopard conflict) in the garden areas, and above all, the impoverished condition of the closed garden labours makes any tea tourism project tasks challenging.

The global tourism industry has emerged as a more demanding space where travellers are more informed and seeking value for money for their experiences. This has inevitable impacts on regional tourism. Thus, tour operators are also viewing increased partnerships and collaborations as well as staff training and management as essential in developing the value-added products needed for tea tourism to develop. To support this growth, more tea-related attractions and activities need to be developed, providing authentic experiences in natural settings, while also offering modern facilities in terms of accommodations and provisions such as local culinary experiences. Travellers are also demanding Wi-Fi connectivity at the tea estates that are often in remote locations. Examining the challenges in provisions for tea tourism in the Dooars noted above and observing the lack of dedicated marketing initiatives, tea tourism here could be considered to still be at a nascent stage. However, this stage of development provides an opportunity to organize the process and arrangements to ensure products and experiences are available for future visitors.

Table 30.1 Number of tea gardens in the Dooars (1876–2013)

Year	Number of tea gardens	Year	Number of tea gardens
1876	13	1941	189
1881	55	1951	158
1892	182	1961	155
1901	235	1971	151
1921	131	1990	163
1931	151	2013	150

Source: Tea Board of India (1995), Government of West Bengal (2013), Xaxa (1985)

Discussion

The operational challenges faced by the tea estates have led some to diversify their business, venturing into tea tourism. This would allow the tea industry here through diversification to create extra income streams while providing some alternative employment to the local tea estate workforce. However, this growth in tea tourism might be restricted to the larger tea estates and not as available to the many smallholders in the region.

It is important to note that the first state Tea Tourism Policy (Government of West Bengal 2013) limited the land meant for tea tourism to 5 acres for any garden, and the area for civil construction was limited to 1 to 1.5 acres. It was stated in the policy that the tea garden company should have a majority share in the joint venture company and that "[no] outside entity will be allowed" (Government of West Bengal 2013, 3). The latest Tea Tourism Policy (Government of West Bengal 2019) has been renamed the Tea Tourism and Allied Business Policy 2019, wherein the tea gardens will be allowed to utilize 15% of the total grant area subject to a maximum of 150 acres for tea tourism and allied business activities.

According to the first Tea Tourism Policy:

> Out of the allowable area a maximum of 40% can be used for construction activities in conformity with extant Rules and Regulations and provided the proposed activity is in harmony with the ecology and the environment. . . . The civil construction will be eco-friendly and buildings so constructed will be limited to two stories. The project should be self-contained in terms of water and sanitation and should not result in additional loads on the environment.
> *(Government of West Bengal 2013, 2)*

The latest policy declares that:

> The allowable business activities shall include Tea Tourism, Plantation, Animal Husbandry, Hydro Power Non-Conventional Energy Resources, Social infrastructures and services. An illustrative list of activities under above categories may include tourism resorts, wellness centres, schools, colleges, universities, medical/nursing colleges, hospitals, cultural/ Recreational & Exhibition Centres, Horticulture, Floriculture, Medicinal Plants, Food Processing Units, Packaging Units etc.
> *(Government of West Bengal 2019, 1)*

It has also claimed that the policy

> is aimed at generating enhanced investment and employment opportunities for sustainable and inclusive economic development by way of effective utilization of vacant/surplus land in Tea Gardens without any curtailment/compromise in area under tea plantation. . . . There shall be no reduction in the area under tea plantation and no retrenchment of existing labour force engaged in Tea Garden. Implementing a new project should in no way harm the existing ecology of the area and shall have to be in strict compliance of environmental regulations. . . . Employment from new project must be generated in a manner that 80% of local people get opportunity and are absorbed. Preference shall be given to the wards of the workers of the concerned tea gardens.
> *(Government of West Bengal 2019, 3)*

What is clear from these statements is that the policy is seeking business resilience, and tea tourism here is one of the many components for that. Thus, it tends to evade the actual nuances that tea tourism can offer. It is noted that the latest policy involves many different sectors under the umbrella term "Tea Tourism and Allied Businesses" whose motives may or may not be aligned with that of only tea tourism. Whereas the tea gardens in the Dooars suffer from the frequent change in management, this multi-industry strategy could bring more confusion and constraints in future. The present chapter also argues that the Tea Tourism Policy of West Bengal is a paradox where ecological factors are promised to be maintained and, at the same time it is finding an economic way forward through a more lucrative but makeshift process overlooking the actual problem of sickness that the gardens are having since long ago. In so doing, the policy actually tries to bring a more evasive mode of garden economy which is far from real community or social resilience.

In the question of community resilience through tourism in a broad sense, it is necessary to look at the overall tourism development of the Dooars region that had taken place in the past two decades. During these decades the Department of Forest and Tourism Department of West Bengal Government have taken several initiatives to develop rural tourism in the Dooars region. The author was present as training coordinator in one of such initiatives in 2014, where nearly 100 youths from the forest villages and tea garden adjacent areas were trained in a Homestay Hospitality and Rural Tourism Entrepreneurship programme, followed by which many of them have come up as rural entrepreneurs, by setting up their homestays and operating into the rural tourism market. However, it was found that despite being in the vicinity of the training venue, there was no participation from the youths of the tea gardens of the Dooars. The point here is to signify that the gross absence of the tea garden population is not merely an accident. The reason demands further probing.

Historically, the colonial plantation population of the Dooars was a captive labour force. Many restrictions continue to influence labour relations with this population to this day. The ethnic factor has played a role, in that the workers as migrants to the area are accorded low social status. Hence, though this work force has contributed to building the wealth of the state through their labour in the tea industry, there are barriers to their entry into the regular economy. While such communities are numerically large in the region, they have never been politically effective. Non-worker, nontribal trade union leaders (mainly belonging to the dominant Bengali community) lead the trade unions and make decisions on behalf of the tribal workers.

As a result, despite being a part of the formal sector and having a high degree of unionization, tea garden workers in the Dooars region of West Bengal are not totally as free as those in other industries. The fact that the plantation workers always remain in enclaves hinders their growth and adaptability socially. These workers remain vulnerable and may be subject to exploitation by the companies where they can be made to work for low wages and under exploitative conditions. In fact, early labour conditions on plantations were created so that the workers remained captive. Later developments did not help workers to become emancipated from the historic constraints of low-paid work.

Rural tourism in the Dooars is mainly operational as an informal and unorganised sector. Tourism here often requires little start-up capital, can raise demand for local products and services, and create investment and entrepreneurial opportunities while improving transportation, infrastructure and utilities. Thus, tourism has the potential to help local populations break the poverty cycle through formal and informal employment, entrepreneurship, training and community betterment. All of the factors discussed in this section are indisputable in the rural tourism of the Dooars region.

Recommendations

Following the observations above about the development of tea tourism in the Dooars region, it is suggested that there could be a number of actions taken to nurture tea-related tourism. Firstly, it needs a fundamental change in the tea sector, where tourism as an ally could be legally more permissible to operate following the community-based tourism (CBT) model. Tea tourism as an alternative source of employment will increase workers' bargaining power and will also reduce the isolation of the plantations.

Secondly, it is argued earlier that the Plantation Labour Act (PLA; Government of India 1951) needs to be seriously implemented. The act provides for all inputs needed for uplift of the workers' situation. These include improvement of living conditions by standard housing, provision for drinking water, sanitation, recreational facilities and education. Had the provisions of this act been fully implemented in the past, plantation labour would not be experiencing its present conditions (Bhowmik 2011). Tourism is known for the potential of uplifting the social condition of marginal communities. Tea tourism in the Dooars thus needs to align more truthfully with the PLA (1951) in being a more responsible and conscientious form of tourism.

Thirdly, in rural tea-producing areas the development of small-scale homestay accommodation can be an important tool as a form of equitable growth offering tea workers alternative sources of livelihood. Homestay accommodations can also contribute to the preservation and development of skills and protection of the local natural environments. In the Dooars the development of homestay accommodation could thus contribute to developing responsible forms of tourism enhancing community capacity building and developing resilience.

Fourthly, the lack of proper education for children of tea garden workers has affected their occupational mobility. Like their parents, these children are destined to work in low-paid positions, often as casual labour, with little or no possibility of moving into improved or skilled occupations. It is thus necessary to establish centres for technical training (including hospitality and tourism) for the new generation so that they can have other opportunities. Finally, the current West Bengal State Tourism Policy (2019) needs to be more attentive to tea tourism where a multilateral cooperation could be developed between all actors including the Tourism Department, Forest Department, Tea Garden Management, Tea Associations and Tea Board.

The present Tea Tourism Policy of West Bengal Government (2019) was formulated on a flat assumption focusing only on high-end tourism. It is evident that a more people-centric, responsible tourism approach needs to be incorporated. A revised Tea Tourism Policy needs to be therefore capable of delving deep into the historical fault lines that occurred across the region and mitigating these divisions by utilizing tourism as a source of resilience. As noted by Mondal and Samaddar (2021) in their assessment of tea tourism in India, integrated planning involving all stakeholders is required for tea tourism to move forward in a sustainable manner within the tea-producing regions of India. Planners must therefore find sustainable methods of working to implement tea tourism, with a view to preserving the natural and cultural heritage resources for such tourism.

Limitation and future research

The chapter was developed at a time when India is hit by the second wave of the COVID-19 pandemic (April–August 2021, in which the author became affected by COVID-19) and in an even worse situation than the first wave (March–September 2020). All interstate movements were apparently closed and/or heavily restricted. There was a nationwide lockdown/restriction of all kinds of mobilities. In such a situation, thus, the present research draws heavily upon secondary sources. Several discourses on the subject have been focused critically and had to rely more on retrospection of the author's previous engagement within the field as the rural tourism project coordinator (2012–2014). However, due to the same reason perhaps, the chapter could potentially be able to probe into the root of the problems which was otherwise not graspable.

There remains albeit a basic lack of providing the research with necessary site-specific details and inventory checks (accommodation, management capacity, number of gardens with potential for developing tea tourism in future). Further studies are required for validation of the current assumption and detailed empirical evidence and observations are needed from the field to bring more accuracy in the study. Since tea tourism is seen as a measure for resilience therefore it is recommended to incorporate the UN Sustainable Development Goals (SDG) at the very beginning of the process, and for that the Tea Tourism Policy must be revised. The term "sustainable development" is used rather blatantly without addressing it more critically; however, in the case of Dooars tea tourism, the labour condition and redemption is intimately connected with the SDGs.

So far, tea tourism tends to be apolitical and focused only on colonial legacy. Tea tourism should also be seen as an industrial heritage, like heritage railways, and more action research is required. The subverted history of the gardens needs to be retrieved to build a more responsible and ethical form of tea tourism.

Conclusion

For tea tourism, plantation culture has always been seen through the colonial frame of reference where labour sources of history have remained silent or not been so celebrated. The labour condition has remained the centre of attention for discourses of political science or economics. However, it is the call for greater engagement of tourism studies to take these accounts into the disciplines of tourism too. Most of the researchers stressed upon the public-private partnership (PPP) model, but before devising any such model, the historical drawbacks need to be re-evaluated. It seems any developmental process or policy decisions are merely overriding the region without addressing its historical fault lines. It is believed that without getting to the root of the causes it would always remain unsuccessful to build any sustainable tea tourism project, and the aspiration of resilience through tea tourism would remain off limits both in discourse and in material practices.

Discussion questions

1   What are the historically underlying fault lines in the development of tea-related tourism in the Dooars?
2   What are the causes of the non-involvement of the tea plantation workers in tourism?
3   To what extent is the Tea Tourism Policy effective in developing tea tourism in the region?
4.   In what way is the Tea Tourism Policy actually overriding the region's tea tourism potential by overlooking its historical fault lines?

References

Bhowmik, S.K., 2009. Politics of tea in Dooars. *Economic and Political Weekly*, 44(9), 21–22.

Bhowmik, S.K., 2011. Ethnicity and isolation: marginalization of tea plantation workers. *Race/Ethnicity: Multidisciplinary Global Contexts*, 4(2), 235–253.

Biswas, S., Chakraborty, D. and Talwar, A., 2005. *Study on closed and reopened tea gardens in North Bengal*. Kolkata: Paschim Banga Khet Majdoor Samity (PBKMS) and International Union of Food, Agriculture, Hotel, Restaurant, Catering, Tobacco, Plantation and Allied Workers' Association (IUF).

Chakraborty, A. and Islam, S.S., 2020. Impact of tea tourism in Dooars, North Bengal: an overview. *Mukt Shabd Journal*, 9(6), 5798–5804.

Chakraborty, S., 2013. Tea, tragedy and child trafficking in the Terai Dooars. *Economic and Political Weekly*, 17–19.

Das Gupta, R., 1992. *Economy, society and politics in Bengal: Jalpaiguri 1869–1947*. Delhi: Oxford University Press.

Datta, C., 2018. Future prospective of Tea-Tourism along with existing forest tourism in Duars, West Bengal, India. *Asian Review of Social Sciences*, 7(2), 33–36.

Ghosh, B., 2015. Post-reform closure and sickness of plantation industry marginalisation of workers and vulnerability of women and children in Jalpaiguri, West Bengal. *In:* Panda, R. and Meher, R., eds. *Trend, magnitude and dimensions of inequality in post-reform India*. New Delhi: Concept Publishing Company, 260–292.

Government of India, 1951. Plantation labour act. Available from: https://labour.gov.in/sites/default/files/The-Plantation-Labour-Act-1951.pdf [Accessed 28 February 2022].

Government of West Bengal, 2013. Tea tourism policy. Available from: https://wbxpress.com/files/2013/10/Tea-Tourism-Policy.pdf [Accessed 15 March 2021].

Government of West Bengal, 2019. Tea tourism and allied business policy (Kolkata Gazette). Available from: www.wbtourismgov.in/home/download/pdf/3816_tea_tourism_and_allied_business_policy_2019.pdf [Accessed 16 March 2021].

Government of West Bengal, 2019. West Bengal Tourism Policy 2019 (Kolkata Gazette). Available from: https://wbtourism.gov.in/home/download/pdf/west_bengal_tourism_policy_2019.pdf [Accessed 16 March 2021].

Grove, K., 2018. *Resilience*. London: Routledge.

Jolliffe, L., ed., 2007. *Tea and tourism: Tourists, traditions and transformations*. Clevedon: Channel View Publications.

Lew, A.A., 2014. Scale, change and resilience in community tourism planning. *Tourism Geographies*, 16, 14–22.

Mondal, S. and Samaddar, K., 2021. Exploring the current issues, challenges and opportunities in tea tourism: a morphological analysis. *International Journal of Culture Tourism and Hospitality Research*, 15(3), 312–317.

Prasanneswari, 1984. Industrial relation in tea plantations: the Dooars scene. *Economic and Political Weekly*, 19, 956–960.

Roy, N.C. and Biswas, D., 2018. Closed tea estates: a case study of the Dooars Region of West Bengal, India. *Vision*, 22(3), 329–334.

Sarkar, S., 2019. Labour migration in the tea plantations: colonial and neo-liberal trajectories of plantation labour in the Dooars tea belt of West Bengal. *Journal of Migration Affairs*, 2(1), 25–43.

Sen, R., 2015. Tea workers – distressed in the organised industry in North Bengal. *The Indian Journal of Industrial Relations*, 535–549.

Sunder, D.H.E., 1895. *Survey and settlement of Western Dooars in the district of Jalpaiguri 1889–1895*. Calcutta: Bengal Secretariat Press.

Tea Board of India, 1995. *Techno economic survey of Dooars Tea Industry*. Calcutta: Tea Board of India.

Thakur, U., 1995. "Sick" tea plantations in Assam and Bengal. *Labour, Capital and Society/Travail, capital et société*, 44–66.

Walker, B. and Salt, D. 2012. *Resilience practice: building capacity to absorb disturbance and maintain function*. Washington, DC: Island Press.

Xaxa, V., 1985. Colonial capitalism and underdevelopment in North Bengal. *Economic and Political Weekly*, 1659–1665.

31

HUMAN-WILDLIFE INTERACTIONS IN TEA TOURISM

The Dooars in India

Chandan Datta

Introduction

Tea tourism is an interest that is motivated by the history of tea, the traditional beliefs and the style of tea consumption. Tea tourism is about picking tea leaves in the tea garden, visiting tea factories, visiting tea stalls and getting acquainted with the local tea culture (Jolliffe 2007), as well as the wild animals of the Dooars. All the tea garden areas, adjacent to the protected area, have become second homes to wild animals of the Dooars. As a result, human-wildlife interaction in most parts of this region has become one of the major obstacles to sustainable tea tourism development. The human-wildlife interaction in the tea garden area in this region plays a dual role in the development of tea tourism. The positive interaction between man and wild animals in tea tourism in this region will help tourists to experience both the tea garden and surrounding wildlife of the area. A unique nature-based tourism will appear in this landscape. The revenue from tea tourism can be shared between the government and the locals to make the environment of human-wildlife interaction friendly. The governments have shown their interest towards tea tourism, and some tourists are coming to take the test of existing tea tourism in the Dooars. Interest of the tourist is the main key to large-scale development of tea tourism, but a large number of tourists show less interest towards it due to listening and seeing news about negative human-wildlife interaction in tea garden areas in electronic and print media.

Literature review

Tea trees were planted in Assam in 1834 under the supervision of Governor William Bentick. Tea cultivation began in 1853 at Wayanad in the Nilgiri Hills of southern India. In 1990, there were 163 tea estates in the Dooars of West Bengal. Currently, the number of tea gardens in the region has multiplied. The tea industry in West Bengal is 200 years old. In 1835, Mr. Gordon brought tea tree seeds from China and planted them in the Darjeeling hills. In 1862, Mr. Haughton planted the first tea garden in Gazoldoba, Dooars (Mitra 2010). Su *et al.* (2019) emphasized the need to improve community livelihoods through the integration of tea and tourism. They used a sustainable livelihood approach in this regard.

Su and Zhang (2020) expressed the view that how tourists find physical and mental well-being through drinking tea and how drinking tea has been socialized by tourists. Lin and Wen (2018) highlighted how the development of tourism has changed the *ethnic marriage and labour division*. Jolliffe and Aslam (2009) highlighted how the bungalows, plantation, culture in the tea garden area and the natural environment in tea production has developed *tea heritage tourism* in Sri Lanka and how it

DOI: 10.4324/9781003197041-37

has improved tea community livelihood. Feng *et al.* (2012) put emphasis on the willingness to pay the tourists for protecting tea tourism resources. They also highlight the total economic value of the tea tourism resources. Cheng *et al.* (2012) discussed the contribution of stakeholders in the development of tea tourism. Datta (2018) highlighted the strengths, opportunities, weaknesses and threats about tea tourism of the Dooars. According to Cheng *et al.* (2010), emphasis has been placed on the attitude and perception of tourists in matters related to tea and tea tourism.

The nature of tourists' interaction with wildlife is an important and less discussed element of sustainable tourism (Moscardo 1998; Moscardo and Saltzer 2005). This proves that the opportunity to see wildlife is one of the most important factors in the travel decisions and travel destinations of most international and domestic tourists. The opportunity for tourists to interact or visit wildlife helps to create a more positive attitude among tourists towards wildlife conservation (Moscardo *et al.* 2001).

Crisis is a situation when life, health and infrastructure are adversely affected. Many have described the negative impact of an event on individuals and property as a crisis. Crisis is an event that creates instability and insecurity situations for an individual, group, community or society (McCool 2012). This study highlights the causes, effects and problems of human-wildlife interaction in tea tourism and how to make human-wildlife interaction an important part of tourism. More specifically, this work undertakes an assessment of perception of the respondents regarding human-wildlife interaction.

Although all literature reviews discuss various aspects of tea tourism, they do not discuss human-wildlife interaction in the tea garden area. Human-wildlife interaction is a major problem in various countries in South Asia, especially in countries such as India where tea gardens share a boundary with national parks. Not much research has been done on tea tourism in India yet.

Methods

The qualitative research method has been applied in this study. A random sampling survey has been conducted for data collection throughout the region. A 5-point Likert scale has been used to measure respondents' perception regarding certain questions for getting detailed analysis and comparison. The analytical hierarchy process method has been used to assess the preferences of the respondents regarding reasons for incorporating human-wildlife interaction with tea tourism in the Dooars. The survey was conducted from 2017 to 2021. It was possible to conduct surveys in the months of January to March only in 2020 and 2021 due to the COVID-19 pandemic. A total of 355 respondents were interviewed, out of which 212 were men and 143 were women. The experiences and perceptions regarding human-wildlife interaction in tea tourism have thus been collected from respondents.

Case study 31.1   Human-wildlife interactions in Dooars tea tourism

The Dooars is a geographical region located at the foot of the mountains between the Teesta and Sankosh rivers with a number of passages through which one can reach Bhutan and which connect the plains of India and the hilly region of Bhutan. There are many tea gardens in this region where tea tourism can be developed. The main problem in the development of tea tourism in the region is the occurrence of negative human-wildlife related incidents in the tea garden region. However, several tea gardens have shown interest in developing tea tourism in the region. The Dooars' geographical location is an important factor in itself for promoting tea tourism in this region.

The Dooars is a paradise of tea gardens, forests and wildlife that attracts tourists. The region has a long history of human-wildlife interaction. Tourists have been visiting this region since

---

time immemorial. For this reason, human-wildlife interaction in the tea garden area of the Dooars is becoming increasingly important in the development of tea tourism.

Table 31.1 Respondents' profile

Types of respondents	Frequency (n)	Percentage
Tea tourist	149	37
Want to be a tea tourist	133	42
Local people	73	21
Gender	Frequency (n)	Percentage
Male	212	60
Female	143	40

Source: Author

Among the respondents, 37% are tea tourists (i.e., those who enjoy tea, tea gardens, tea garden wildlife, natural beauty and tea culture); 42% are ordinary tourists who want to be tea tourists and want to enjoy the taste of tea tourism in the future; and 21% are local people who work in tea gardens, live in tea garden workers' quarters or are tea garden managers. All these local people want the current economic situation of the tea garden to change with the help of tea tourism (Table 31.1).

The study highlights the respondents' perceptions of human-wildlife interactions in tea gardens. In answer to the question of what kind of wildlife can be seen roaming in the tea garden, the respondents report that 63% elephant (n-223), 44% leopard (n-156), 36% bison (n-128), 54% deer (n-193), 60% peacock (n-213), 34% python (n-120), 47% monkey (n-168) are noticed in the tea garden.

In response to the question of at what time of the day the maximum number of wild animals are seen in the tea garden area, the respondents indicate 53% early morning (n-187), 44% morning (n-156), 37% noon (n-132), 50% afternoon (n-178), 65% evening (n-232) and 78% night (n-277). A variety of negative human-wildlife interactions are observed in this region. Common negative interactions include wildlife attacking humans, humans attacking wildlife, house damage by wildlife, livestock loss by wild animals, tea garden damage, disruption of tea production process and car damage by wildlife.

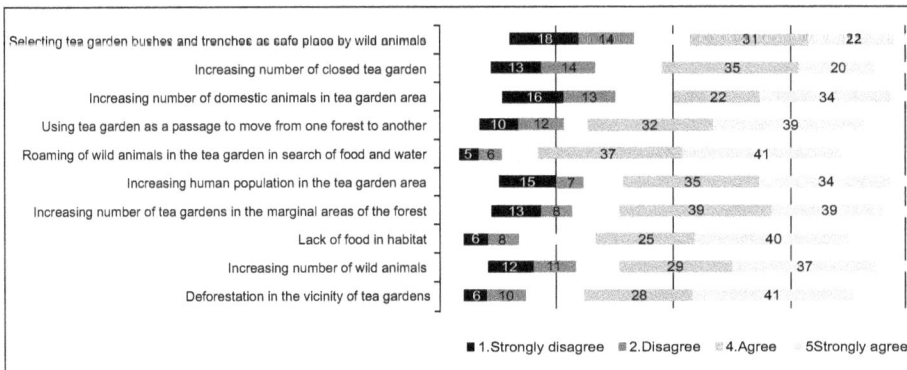

Figure 31.1 Percentage distributions of responses on causes of human-wildlife interaction in tea garden area. 'Neutral' responses (3) are represented by the blank space in each stacked bar. N = 355.

Source: Field survey, 2017–2021

The reasons for the negative interaction of wildlife with humans in the tea garden are presented (Figure 31.1). The reasons why the respondents have strongly agreed are the use of tea gardens as a passage to move from one forest to another, roaming in the tea gardens in search of wildlife food and water, lack of food in the wildlife's own habitat, and deforestation in the vicinity of the tea gardens. Other reasons are the selection of tea garden bushes and trenches as a safe place by wildlife, closed tea gardens, increase in the number of domestic animals in the tea garden area and so forth.

There are positive, negative and neutral opinions among respondents on how human-wildlife interactions will help develop future tea tourism in the Dooars tea gardens. There are positive opinions about tea gardens near forests, especially near national parks; there are negative opinions in tea gardens adjacent to small forests; and there are neutral opinions in closed tea gardens and tea gardens away from forests (Figure 31.2). Tea gardens located east and west of Garumara National Park and Buxa Tiger Reserve have the highest number of positive responses. Figure 31.3 shows the distribution of intensity of human-wildlife negative interactions in the tea gardens of the region. It is observed that the negative human-wildlife interactions more frequently occurred in between Jaldapara National Park and Buxa Tiger Reserve and to the east and west of Garumara National Park.

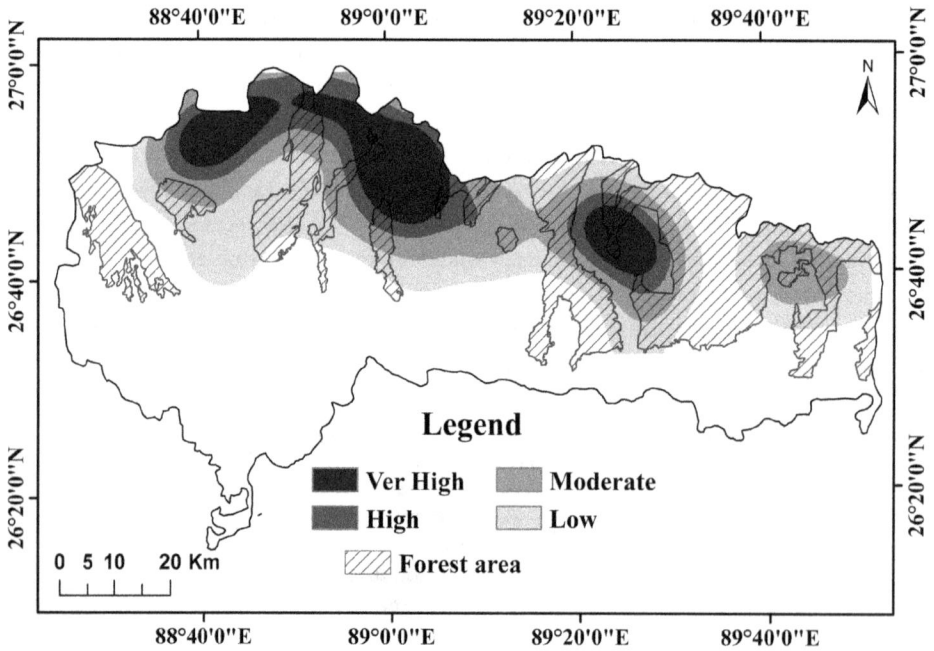

Figure 31.2 Distribution of tea garden that have positive, negative and neutral views on human-wildlife interaction in tea tourism for the future

Source: Field survey, 2017–2021

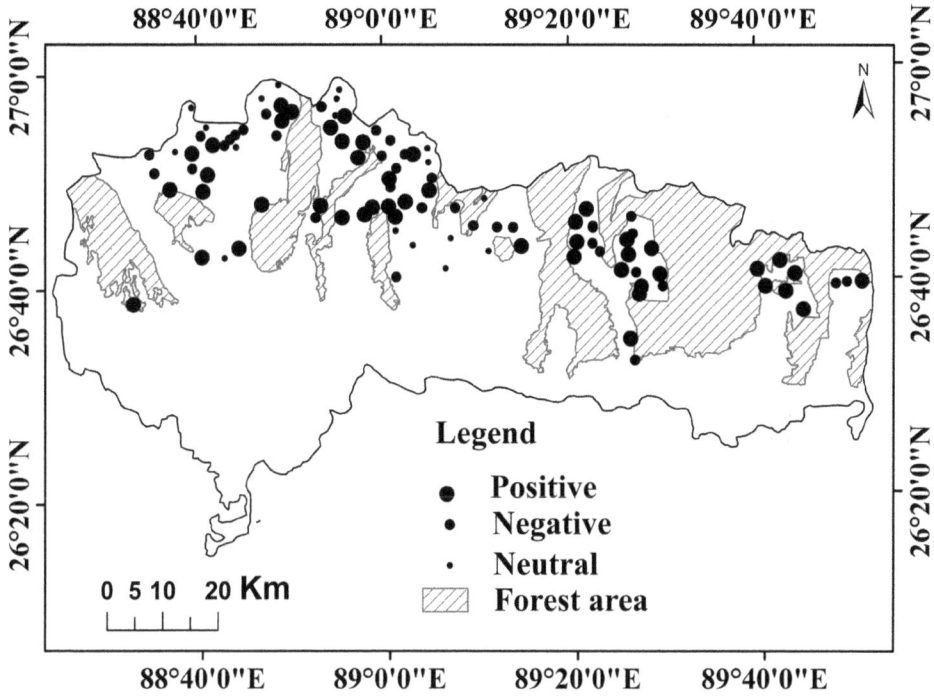

Figure 31.3 Frequency of negative human-wildlife interaction in tea garden area of the Dooars

Source: Field survey, 2017–2021

Table 31.2 Pairwise comparison matrix showing the respondent's relative preferences of selecting a reason for incorporating the human-wildlife interaction with tea tourism

	R1	R2	R3	R4	R5	R6	R7
R1: Seeing wildlife in their own habitat	–	5.00	0.33	0.11	0.14	0.20	0.33
R2: Being able to see animals not previously seen	0.20	–	5.00	0.14	0.11	3.00	5.00
R3: Opportunity to see wildlife closely	3.00	0.20	–	0.11	0.14	5.00	5.00
R4: Pleasant environment	9.00	7.00	9.00	–	1.00	7.00	5.00
R5: Excitement and adventure	7.00	9.00	7.00	1.00	–	5.00	7.00
R6: Being able to know about the behaviour of wildlife	0.20	0.33	0.20	0.14	0.20	–	7.00
R7: Telling others about experience	0.33	0.20	0.20	0.20	0.14	0.14	–
Total	20.73	22.73	22.73	2.71	2.74	21.34	30.33

Source: Author

Respondents were asked to comment on how to incorporate wildlife interactions into the development of tea tourism in the region. Respondents expressed their preference by giving relative rankings for seven reasons. As a result, most respondents reported preferring pleasant environments (31.78%) and excitement and adventure (32%) to positively engage in the human-wildlife interaction with tea tourism in the region.

311

According to them, the green beauty of the tea garden as well as the opportunity to see the wildlife will bring a special status to the Dooars as a tourism destination. Other reasons respondents like their interaction here are the opportunity to see wildlife closely (9.83%), being able to see animals not previously seen (9.58%), being able to know about the behaviour of wildlife (6.21%) and seeing wildlife in their own habitat (5.63%; Table 31.3). Findings suggests that respondents have given seven times more preference to R5 and R6 than R7. Similarly, R7 has a 0.31 times lower preference than R1 (Table 31.2). For this reason, the pairwise comparison matrix provides detailed relative choices scores for each factor that helps respondents select the preferred score.

Respondents have expressed that they strongly agree and agree with views on a number of steps to reduce the number and impact of negative human-wildlife interactions in the region.

Table 31.3 Normalized matrix showing the respondent's preferences of selecting a reason for incorporating the human-wildlife interaction with tea tourism

	R1	R2	R3	R4	R5	R6	R7	Average
R1: Seeing wildlife in their own habitat	0.05	0.22	0.01	0.04	0.05	0.01	0.01	0.056381
R2: Being able to see animal not previously seen	0.01	0.04	0.22	0.05	0.04	0.14	0.16	0.095837
R3: Opportunity to see wildlife closely	0.14	0.01	0.04	0.04	0.05	0.23	0.16	0.098336
R4: Pleasant environment	0.43	0.31	0.40	0.37	0.36	0.33	0.16	0.337845
R5: Excitement and adventure	0.34	0.40	0.31	0.37	0.36	0.23	0.23	0.320093
R6: Being able to know about the behaviour of wildlife	0.01	0.01	0.01	0.05	0.07	0.05	0.23	0.062182
R7: Telling others about experience	0.02	0.01	0.01	0.07	0.05	0.01	0.03	0.028278
Total	1.00	1.00	1.00	1.00	1.00	1.00	1.00	1.00

Source: Author

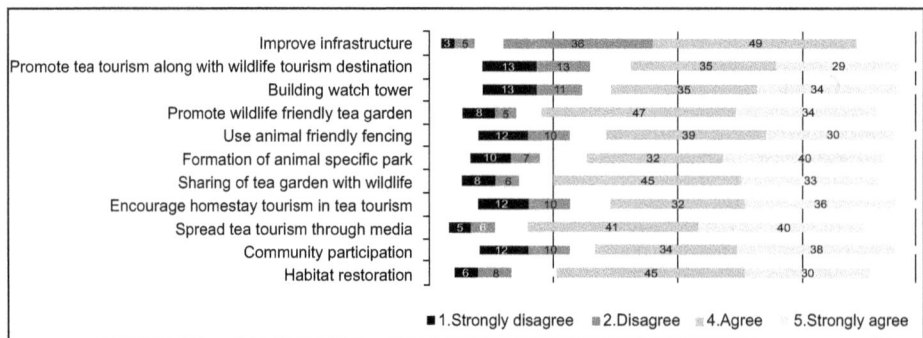

Figure 31.4 Percentage distributions of responses on initiative to be taken to minimize the negative human-wildlife interaction and promote tea tourism. 'Neutral' responses (3) are represented by the blank space in each stacked bar. N = 355.

Source: Field Survey, 2017–2021

The steps that respondents strongly agree and agree with are improving infrastructure (49%), formation of animal-specific park (40%), spreading tea tourism through the media (40%), promoting a wildlife-friendly tea garden (47%), sharing of tea garden with wildlife (45%) and habitat restoration (45%). More or less, disagree and strongly disagree opinions have been observed among the respondents for each step. If all these steps are taken, the Dooars will also be considered as an important tourist attraction point for tea tourism in the future (Figure 31.4).

Analysis

Tea tourism is not only a product but also a cultural feature that is conveyed through cultural adaptation (Jolliffe 2007). Interaction between humans and wildlife is critical in many ways, especially its impact on human health and well-being (positive and negative), human attitudes and behaviours towards wildlife, and the benefits and risks of wildlife (Soga and Gaston 2020). The discussion of human-wildlife interaction in tea tourism is of growing interest for current researchers. There are many main reasons for this, such as that human-wildlife interaction can provide a range of emotional pleasures to people in addition to the beauty of the tea garden, human fear and awareness of the negative consequences of some human-wildlife interactions (attacks on humans by large carnivores, poisoning by poisonous animal bites, wildlife and vehicle collisions, etc.) and human-wildlife interactions to assist in the conservation of wildlife in any region. Many times people deliberately try to interact with wildlife. The human-wildlife interaction will take place only when the tourists come to the tea garden to see the wildlife around the tea garden, take pictures and feel the excitement about the presence of wildlife. The negative impact of human-wildlife interaction depends on how physically close tourists are staying close to the wildlife and how consciously tourists are interacting with the wildlife.

The negative impact of human-wildlife interaction depends on how physically close tourists are staying to the wildlife and how consciously tourists are interacting with the wildlife. And negative perceptions develop when they know they are likely to be attacked by wildlife. Human-wildlife interactions vary from tea garden to garden and from time to time. The reasons for this change are the distance from the forest to the tea garden, the presence of wildlife, accessibility, the presence of tea tourism, convenience of home-stay, the willingness of tourists to go, the local culture and the socioeconomic situation. However, the dense settlement of tea gardens may hamper the possibility of positive wildlife interaction with tea tourism in the region.

There are a number of things that are important to incorporate wildlife interactions with tea tourism in the region, such as the opportunity to see wildlife and the views of the government, tea garden owners, local people and tourists. Undoubtedly, the scope of human-wildlife interaction varies from place to place and from time to time. It is not static. Wildlife interactions with people in and around the tea garden area are changing depending on global climate change, human distribution and how people want to interact.

The incidence, frequency and type of wildlife interactions with humans vary from time to time (daily, seasonally and annually). Wildlife such as elephants, leopards, bison, deer, peacocks, monkeys and pythons are found in this area. Elephants come to the tea garden at night and early in the morning, but sometimes the elephant herds get stuck in the tea garden if they cannot return to the forest before sunrise. This can be a factor in attracting tourists. Leopards have chosen the tea garden bushes and trenches as ideal places to give birth to their young. This is why leopard cubs are often found in the tea gardens of the Dooars. Leopards are usually seen in the evening and in the afternoon. When there is a shortage of food inside the forest in summer, pythons move to the adjacent tea garden area in search of food. Bison move mainly to the tea gardens around the forest in search of water in summer and winter. The diverse geographical features, culture and socioeconomic situation determine what the human-wildlife interaction will look like in the development of tea tourism.

However, many people are injured and killed every year due to the negative human-wildlife interaction in the tea gardens in this region. In the last few years, numbers have multiplied due to the increase in the number of wildlife, the loss of human ecological knowledge and the increase in the number of people. Many tourists have been injured while taking pictures.

A capacity building plan needs to be developed so that local community people can participate in the development of tea tourism and enjoy its benefits. Local people need to be made aware of modern tourism policies and aspects. Education should be provided to the locals on how to develop tourism jointly with private companies, tourism market demand, skills and increasing the quality of tourist services. Negative human-wildlife interaction is a kind of crisis in the development of tea tourism. Although various measures have been taken for crisis management, those steps can be successful by following the crisis management phase below given by Pearson and Mitroff in 1993.

A. Early warning and signal detection by which to understand the systems and clues of the impending crisis.
B. Preparation and prevention means taking steps to alleviate the crisis when it is impossible to avoid the crisis completely.
C. Damage containment refers to actions taken to reduce the spread and impact of a crisis.
D. Business recovery refers to the methods, procedures and functions of short-term and long-term development to bring normal business back to normal.
E. Organizational learning means understanding and reflecting on the later stages of the crisis. Identify the positive and negative aspects of crisis response.

However, there are a number of barriers to crisis preparedness in these areas, such as a negative attitude towards crisis planning, lack of disaster management, limited financial resources, lack of crisis knowledge, low risk perception, small tea gardens and the existence of private ownership in tea gardens (Wang and Ritchie 2012).

Conclusion

Tourism development, especially tea tourism, has recently become a priority in India. India is a fast-growing emerging economy, and the number of domestic tourists in India is increasing as a result of economic development. Tea tourism is a nature-based platform with positive human-wildlife interaction. Currently determining tourist destinations in terms of wildlife passion and psychology is an emerging tourism demand. To popularize wildlife interaction with tea tourism, both wild animals and human interactions with wildlife need to be managed. In order to save the wildlife, the local people of the area have to be saved. Tea tourism can save the existence of local people in all these areas by bringing about economic prosperity. The image of the tourism destination can be enhanced in front of the tourists by adopting such a crisis management strategy.

Discussion questions

1 Why, when and at what level can human-wildlife interaction be observed in a tea garden?
2 How do human-wildlife interactions affect tea production, the tea labour community and tourists?
3 How can human-wildlife interaction become an important part of tea tourism?
4 How does human-wildlife interaction play a vital role in choosing a tourist destination?

References

Cheng, S., Hu, J., Fox, D. and Zhang, Y., 2012. Tea tourism development in Xinyang, China: stakeholders' view. *Tourism Management Perspectives*, 2, 28–34.

Cheng, S.W., Xu, F.F., Zhang, J. and Zhang, Y.T., 2010. Tourists' attitudes toward tea tourism: a case study in Xinyang, China. *Journal of Travel and Tourism Marketing*, 27(2), 211–220.

Datta, C., 2018. Future prospective of tea-tourism along with existing forest-tourism in Duars, West Bengal, India. *Asian Review of Social Sciences*, 7(2), 33–36.

Feng, W., Wang, Y., Tanui, J.K. and Li, X., 2012. Evaluation on the economic benefits of tea cultural tourism resources for Biluochun tea in Suzhou Dongting Mountain. *Journal of Tea Science*, 32(4), 353–361.

Jolliffe, L., ed., 2007. *Tea and tourism: tourists, traditions and transformations*. Clevedon: Channel View Publications.

Jolliffe, L. and Aslam, M.S.M., 2009. Tea heritage tourism: evidence from Sri Lanka. *Journal of Heritage Tourism*, 4(4), 311–344.

Lin, Q. and Wen, J.J., 2018. Tea tourism and its impacts on ethnic marriage and labor division. *Journal of China Tourism Research*, 14(4), 461–483.

McCool, B.N., 2012. The need to be prepared: disaster management in the hospitality industry. *Journal of Business and Hotel Management*, 1(2), 1–5.

Mitra, D., 2010. *Globalization and industrial relations in tea plantation-a study on Dooars region of West Bengal*. Delhi: Abhijeet Publication, 13–31.

Moscardo, G., 1998. Interpretation and sustainable tourism: functions, examples and principles. *Journal of Tourism Studies*, 9(1) 2–13.

Moscardo, G. and Saltzer, R., 2005. *Understanding tourism wildlife interactions: visitor market analysis* (Technical Report). Townsville: CRC Reef Research Centre.

Moscardo, G., Wood, B. and Greenwood, T., 2001. *Understanding visitor perspectives on wildlife tourism* (Wildlife Tourism Research Report Series: No. 2). Townsville: CRC Reef Research Centre.

Pearson, C.M. and Mitroff, I.I., 1993. From crisis prone to crisis prepared: a framework for crisis management. *The Executive*, 7(1), 48–59.

Soga, M. and Gaston, K.J., 2020. The ecology of human–nature interactions. *Proceedings of the Royal Society*, B287, 20191882.

Su, M.M., Wall, G. and Wang, Y., 2019. Integrating tea and tourism: a sustainable livelihoods approach. *Journal of Sustainable Tourism*, 27(10), 1591–1608.

Su, X. and Zhang, H., 2020. Tea drinking and the tastescapes of wellbeing in tourism. *Tourism Geographies*, 1–21.

Wang, J. and Ritchie, B.W., 2012. Understanding accommodation managers' crisis planning intention: an application of the theory of planned behavior. *Tourism Management*, 31(5), 1057–1067.

32

A RESILIENT TEA DESTINATION

The Azores case

Jose Soares de Albergaria Ferreira Pinto

Introduction

Azores tea spans across generations, the three sectors of the economy connecting the present to the past and the heritage of these nine islands. Tea contributes to tourism before, during and after visitors stay and is part of the humanised scenery, gastronomy, health, well-being and diversified offers, and is part of families' local traditions: tea is seen as medicinal and a ritual. Perhaps tea availability outshines its uniqueness and value amid other things the Azores has to offer to visitors. This chapter explains how resilience is encrypted in the culture of local people and how tea contributes to tourism under the most challenging times through its role in co-creating and enhancing experiences that satisfy and prompt tourists to recommend and return to the Azores.

Literature review

The new possibilities the enlightenment brought to the arts, architecture and science with the new agro-industrial transformation are outstanding and co-existed with the late baroque movement. The chinoiserie movement, the imitation or interpretation of the Chinese and other Asian cultures and traditions, had a powerful influence in late 18th-century European societies. This influence lasted until the beginning of the 20th century in arts, tapestry, painting, utilitarian and decorative porcelain, furniture, architecture and others, including the habit of drinking tea (Impey 1977).

Amongst other influences, the tea ceremony brought to light for the English noble society by Catarina de Braganza influenced the habits of the wealthy English (bourgeoisie) and the rest of the society. Tea was in demand in Europe, and it became an aspirational product associated with rituals of social gathering to the most sophisticated circles of that time. It represented networking with the most reputed elements of society of that time (Quitério 1987).

At the same time, the Azores archipelago was in a race against the clock. During the last quarter of the 19th century, the inhabitants of this island had witnessed an unprecedented change. The declining production of most crops, mainly oranges, represented the majority of the output from the soil (Dias 1995). Jose do Canto stood out as a leader to recover from this crisis. He was focused on changing the existing landscape of agriculture practices and suggested a shift towards experimenting with newly introduced crops like pineapple, tobacco and tea (Riley 2001).

The rupture with previous agriculture practices in the Azores and, consequently, the election of tea, tobacco and pineapple was a long process of effort and resilience. First, there was an attempt to

DOI: 10.4324/9781003197041-38

absorb the shock of the initial commercial challenges with the orange plantation and harvesting in the Azores. Second were the diminishing quality of the St. São Miguel Islands orange, the increasing costs of production and competitive prices from Spain and Italy. Third, a natural difficulty to get the crops in harsh weather conditions out of São Miguel Island created market availability issues, disrupting the value chain and, ultimately, eroding trust (Moura 2019).

In an attempt to recover from this situation, resources were deployed aiming at scaling orange production and coping with the immediate demand from English orange merchants in quantity and quality. This included the fight against diseases, yield per acre and profit margin in the increasing competitiveness of the orange market value chain. Tea finally stood out, and the tea plantation and production took place in the Azores archipelago (Dias 1995). A new era for São Miguel Island's agriculture was about to start.

Tea introduction at São Miguel Island was preceded by circumstantial attempts in continental Portugal. Visconde Vilarinho de São Romão; the king of Portugal, his highness Dom Fernando II (at Sintra); and later Pereira de Castro are a few examples. Other authors claim the precious plant seems to be connected to Brazil tea introductions and the redundancy of efforts in different areas of continental Portugal, making it difficult to ascertain which attempt is responsible for the tea plantation and production for commercialisation in the 19th and 20th centuries. Jose do Canto's investment, the experimental nature of his work, the network of scientists and naturalists, and the ownership of an extensive area of arable land at São Miguel Island (also known historically as St. Michael Island) seems to be connected to tea commercialization and the increase in the number of tea factories across the island. Edmond Goeze, a prominent botanist who worked at Kew Gardens, in London, is credited to have advised Jose do Canto's tea experimentation efforts. Under such circumstances, it is fair to acknowledge Jose do Canto botanic collection was perennial to the study of different tea varieties. This research paved the way to understand which variety would better adapt to the climate and soil at São Miguel Island (Moura 2019).

A genuine concern was underlying Jose do Canto's approach. He aimed at selecting the seeds that would be able to compete with the growing market availability of the British production at Assam, the Chinese (Darjeeling), the tea from Sri Lanka (Ceylon) and the Dutch production in Java (Indonesia). Under such a scenario, the endeavour to export tea was directed towards continental Portugal, given the international pressure from other tea-producing areas. Even so, historical records show that the competition was fierce, including tea from Mozambique. In addition, tea blends made at the point of sale would downgrade the quality of the tea produced at São Miguel Island.

São Miguel's tea represented 2% of the market share in 1898, 22% in 1921 and, surprisingly, 51% in 1936. From then onwards, the trend in market share changed and led to the closure of most of the tea factories in 1966. Only two locations remain working and are currently operating simultaneously as tourist attractions (Moura 2019).

Tea plantation and production happens only at São Miguel Island. Those areas are Gorreana and Porto Formoso. Gorreana produces 38 tons on average per year; most of the production is sold locally and some in France and Germany. Porto Formoso produces 12–14 tons of black tea, wherein most of the production is sold at the factory shop to tourists as a souvenir. This information is consistent with the current national account on tea production for 2018, 51 tons and 2019, 53 tons (LUSA 2013; INE 2021). These two locations work simultaneously as a tea plantation, manufacturing, commercial post to sell tea locally and abroad. In addition, these locations work as a tourism attraction connecting nature and health (*RTP Play* 2017). Moreover, the Gorreana tea factory owner owns a travel agency on the island; thus, a large percentage of the tourists who visit São Miguel Island end up visiting and experiencing tea there (personal communication, Madalena Motta, 20 July 2022).

Tourism marketing campaigns, the availability of accommodation, and the opening of low-cost air connections have accelerated the pace of this transition. Such changes are influential to the family business of producing and commercialising tea. The typical visit to the tea factory became a nature walk, a learning process about tea plant growth and harvesting, and involved the tasting of São Miguel tea as a nutraceutical (Baptista *et al.* 2012, 2014; Paiva *et al.* 2021) and matching the needs and the wants of health-conscious visitors and locals (IPDT 2015).

Azores and COVID-19 pandemic impact

The Azores' Observatory for Tourism (OTA) provides the analysis, disclosure and follow-up on tourism activities. It has a dedicated section about COVID-19 pandemic follow-up in the Azores archipelago. According to this institution, more than half of those surveyed admit they would travel to another island in the archipelago to stay with friends and relatives. This domestic travelling intention was fostered by government incentives (Azores Government 2020a), sustaining enough demand to keep the tourism sector running without significant changes or lay-offs. Notwithstanding the strenuous effort, local authorities have set upon fighting the pandemic; the tourism sector suffered a drastic reduction in the number of passengers landing in the archipelago, peaking in April 2020 (99% reduction) and May 2020 (98% reduction). Surprisingly, there was an increase in the number of accommodations, 2874 in December 2020, and 2771 in December 2019 (supply, 2019–2020), most likely due to ongoing investments which started before the pandemic (OTA 2021; SREA 2021).

Resilience in islands a natural liability?

Resilience in islands seems to be associated with the size of the land, population density, a higher frequency of natural disasters and an unfortunate proximity of terrestrial, coastal and marine ecosystems. However, what seemed to be a vulnerability associated with insularity has proved to have built up the resilience of islanders by several factors: first, on the belief in their capacity; second, the familiarity with the environment; and third, the understanding of what forms of adaptation are needed for a particular context or situation. In addition, some authors refer to an ability to anticipate change and prepare a diverse array of strategies and self-awareness of human impact on the environment. These particularities of islanders are usually associated with a robust social capital: supporting livelihood diversity changes associated with fostering island networks and support systems. This complex array of drivers includes traditional and local environmental knowledge on early warning systems, innovative livelihood diversification initiatives, women empowerment, creative food and water security systems, place-based adaptation and community engagement (Souza *et al.* 2015).

Researchers and practitioners working for international organisations support the idea that resilience can be understood by the interplay of absorption, adaptation and transformation capabilities before and after a disruptive and unexpected event. Their rationale is that the resilience construct was redefined many times in the literature, creating confusion (Béné *et al.* 2016). This study uses absorption, adaptation and transformation as anchor words that connect capacities to outcomes (see Figure 32.1). The constructs can be summarised as follows: (1) absorption, the ability to prepare for and mitigate the negative impacts of an event by using existing systems and resources (coping and persisting); (2) adaptation, the ability to adjust, modify, or change characteristics or actions to reduce damage, maximise opportunities and remain without significant changes in function and identity; and (3) transformation, the ability to create a new system generate new abilities, so known disruptive events have limited impact in the future. Béné *et al.* (2014) explain that resilience is a confluence of all of these three capacities, each of them leading to different outcomes: persistence (absorption), adjustment (adaptation) and change (transformation). This framework orients the study and, consequently, the structured interview that follows. Some contradictory views on tourism stakeholders' collaboration in a critical situation like a pandemic guides this study. It seems that notwithstanding their awareness of the potential damage of natural disasters to the destination, they fail to develop

An ability to prepare for and mitigate the negative impacts of an event by anticipating, using existing systems and resources to cope and persist.

Absorption

An ability to adjust characteristics to reduce damage, maximize opportunities, and remain without major changes in function and identity.

Resilience

An ability to create a new system, and generate new abilities, so known disruptive events have limited impact in the future.

Adaptation

Transformation

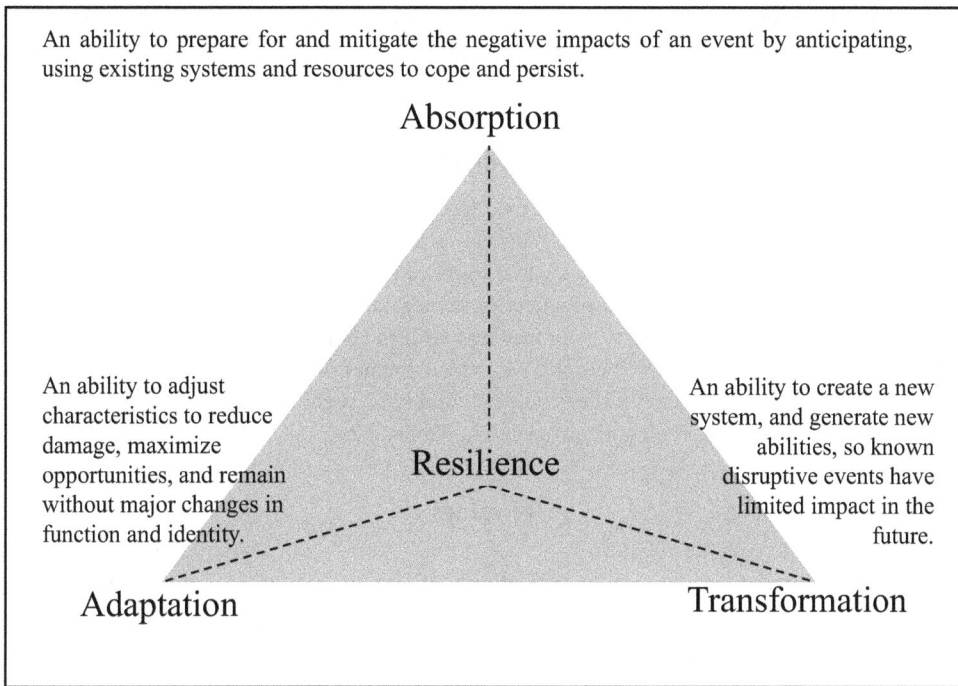

Figure 32.1 Resilience framework as a capacity

Source: Adapted from Béné *et al.*, 2014

effective measures to build a destination-wide and organisational resilience response (Filimonau and Coteau 2020).

Method

Research design

This study is exploratory and focuses on identifying underlying ideas about tea and the recovery after the pandemic in an archipelago with nine islands. A semi-structured interview process to look at tourism and tea as an element of recovery within a case study approach was based on the information collected interviewing the tea producers and owners of the tea factories in the Azores. The author deploys a deductive approach to this interview using as a reference the framework proposed in Figure 32.1; therefore, case study research is deemed adequate by the exploratory nature of the research, and the type of questions (e.g., "How do tea and tourism support the Azores recovery?"). See Table 32.1 on the questions and Table 32.2 about the codes for the interview.

Data collection

As a standard practice, both interviewees were briefed about the research purpose, anonymity and academic use of transcripts. The interview guide and questions were designed based on the work of Béné *et al.* (2014, 601) and adapted to the study research question: "How is tea supporting the recovery after the COVID-19 pandemic?" (Table 32.3). These semi-structured interviews were conducted in Portuguese and took approximately 45–90 minutes each. Recordings and notes were taken during the interview process and translated into English. In addition, the author used rephrasing, paraphrasing and prompts to the questions in Table 32.3 due to the exploratory nature of this study.

Data analysis

The audio recordings and notes were transcribed immediately after each interview and later translated to English, excluding pauses, status, filler words and grammar editing. The transcripts were uploaded to Delve, an online qualitative analysis software (Twenty to Nine 2021). Transcripts were coded TPO1 and TPO2 (TPO – tea producer or owner), analysed using a deductive coding procedure wherein anticipated responses were matched with the existing codes. As we identified new themes, we took them as inductive insights (emergent themes) and later checked with the audio recording and script to understand if they could be considered within the scope of the codes created or not. The coding and sub-coding are presented in Table 32.2.

Findings

Overall context

In hindsight, the recovery process was described by all interviewees as surprising. The struggle and fear they felt at the beginning of the pandemic soon translated into new business challenges and opportunities. All interviewees described the government-enacted recovery measure enforced by the Azores government since 16 June 2020 (see Table 32.3; Azores Government 2020b). Overall, the Azores portrays a remarkable recovery, and herd immunity appears to have been attained in five of the nine islands. From 31 December 2020 to 11 October 2021, 173,743 have been vaccinated

Table 32.1 Questions used in the interview process

Theme: Recovery after COVID-19: tea and tourism contribution
Research question: How do tea and tourism support the Azores recovery?
Q1. In what ways has COVID-19 impacted the place where you live?
Probe 1.1 How did your company endure it?
Probe 1.2 How did your company adjust?
Probe 1.3 How did your company change?
Q2. What are the most important factors for COVID-19 recovery?
Probe 2.1 What other factors can be considered?
Probe 2.2 How may tea support tourism after the COVID-19 pandemic?
Probe 2.3 How may tourism support tea after the COVID-19 pandemic?

Table 32.2 Coding and definitions for the interviews

Coding and definitions
Absorption: ability to prepare for and mitigate the negative impacts of an event by using existing systems and resources
Adaptation: the ability to adjust or change characteristics or actions to reduce damage, maximise opportunities and remain without major changes in function and identity
Transformation: ability to create a new system to generate new abilities so disruptive events have limited impact

Table 32.3 Azores archipelago inter-island incentive scheme (Azores Government 2020b)

Azores Government Incentive Scheme (2020)
Scope: Azores archipelago residents aged 2 and above
Budget allocation: 1.75 million euros
Travelling by air between Azores islands
   Incentive: 50% of the travel package cost, or up to 150 euros
Extra 25 euros if using car rental services
   Conditions to enjoy the incentive:
   Boat round trip from their place of residency to another island
   Book three nights in a local accommodation
   Three meals per resident of at least 15 euros each meal/resident
   One tourism activity of at least 30 euros each per resident
Exemptions: for children aged 2–12, the incentive is scored at 100 euros
Travelling by sea between Azores islands
   Incentive: 50% of the travel package cost or up to 100 euros
Extra 25 euros if using car rental services
   Conditions to enjoy the incentive:
   Boat round trip from their place of residency to another island
   Book three nights in a local accommodation
   Three meals per resident of at least 15 euros each meal/resident
   One tourism activity of at least 30 euros each per resident
Exemptions: for children aged 2–12, the incentive is scored at 70 euros

with the first dose (73.4%), and 192,750 completed the vaccination (81.4%) under the regional vaccination plan (Azores Government 2021). All interviewers were confident of the future, given that most of the tourist attractions are outdoor experiences thus perceived as safe. The recent Earth Check certification highlights these destination brand characteristics by taking nature to underscore the natural balance (Earth Chek 2019).

Adapting to the new normal

While some businesses relying on direct sales suffered from the onset of the restrictions to travel due to the COVID-19 pandemic, others started selling products online. This change was significant for tea and tea products (spa treatments, cookies, bulk sell for kombucha). From a hospitality point of view, adaptations were aimed at the (new) inter-island tourist profile who have different expectations from a foreigner and, more importantly, less disposable income.

Business development and trust

While the critical economic concern around business survival and revival was assured by a government-sponsored inter-island program and the additional incentive plans to the tourism sector, interesting actions are reported having in mind health recommendations (capacity, mask use, distancing and hygiene) all aimed at building the trust and perception of it on the safeness of the destination, as reflected by the following comments made by our interviewees.

> *TPO1*: "We followed government guidelines and instructions: the factory was closed for a long time. My trusted clients buying tea from France and Germany ensured sales continued throughout the COVID-19 pandemic by email and other online systems. Visitation has not been my primary source of revenue; Tea is."

TPO2: "We were closed for three months (April until May/June 2020). According to the local authorities, customer capacity should be reduced to 50 per cent, mandatory use of masks and all visits to the tea factory were cancelled."

The fulfilment of the World Health Organization and local jurisdiction health recommendations regarding the pandemic set the agenda across the world as a priority for the hotel industry (WHO 2021). The aspect of safety is also described as a reassurance to reduce the fear of infection and the quality of the experience. It seems reassurance, quickness, intimacy and proximity are the key drivers of customer-intended outcomes as a recovery from the pandemic (Bonfanti *et al.* 2021) and are aligned primarily with providing the evidence (tangible and intangible) of the reduction, elimination and mitigation of the risk of infection by COVID-19. Once the new or revamped hygiene agenda became a standard procedure, business owners looked at persuading and politely reminding tourists to follow the recommended measures. The best possible combination of factors to prevent or treat COVID-19 infection entails preventive and prophylactic measures, vaccines and medicines that proved effective in attenuating COVID-19, amidst other solutions. Tea presents itself as a nutraceutical that adds value to the existing well-being services and medical and thermal tourism (Quintela 2004). These characteristics sustain the feeling of reassurance, building trust. This trust is of utmost value because tourists take more precautions for destination choice, wherein hygiene and the enforcement of social distance seem to drive decision-making for nature-related destinations (OECD 2020).

Alterations

Changes in the context of the current COVID-19 pandemic occur when the ability to adjust to the new normal continues to push the boundaries of the existing beliefs, methods, ways of working and how hospitality professionals were taught to believe tourists would like to experience a destination. Though some of the interviewees recognise that this COVID-19 pandemic had a terrifying effect and intensified the need to enforce existing standards in hospitality, the support provided by the government somehow made these alterations mild or irrelevant. The interviewees believe that Azores people are resilient by default and have endured challenging times before. The above highlights something contradictory: Azores' ability to endure through times of difficulty diminishes the perception of the need to transform; hence, the learning process drives communities to focus on absorption and adaptation rather than creating disruptive solutions that may open new opportunities and overcome challenges (transformations).

TPO1: "Our company has been in the tea business for 140 years. We have endured through several changes. The worst ones were the political ones, not the pandemic, and indeed not this one."

Emotional solidarity and social distance are described within the context of domestic tourism, focused on the involvement of residents in activities that are a priori designed for tourists and find seclusion over residents' willingness to visit (Joo *et al.* 2018).

Visitor experience and visitor profile changes

The Azores archipelago has a diverse array of visitors. The type of tourist ranges from the younger, more independent traveller health and nature conscious to the more sophisticated visitor eager to experience and share their own experience at all times on all social media platforms. On the other end of the spectrum, visitors above retirement age from northern European countries come in with enough disposable income to invest in experiences connecting them to nature and local culture. Overall, they contribute more significantly to visitation to the Azores archipelago and stay longer than the younger cybernauts. However, this trend changes with the digital nomads who decide to live temporarily embedded with the local culture and the way of living. Overall, the introduction of low-cost carriers facilitated a better geographical distribution and flattened the visitation seasonality peaks in the Azores (SREA 2021).

Tourism saturation, over-tourism?

A common theme underlying all interviews is that the pandemic came in when tourists' presence became annoying and felt like an impediment to residents' daily routines. This was particularly true for São Miguel Island, because it is the first stop for almost two-thirds of tourists who travel to the Azores. Concurrently, other concerns addressed referred to preserving nature, traditions, heritage and legacy. Some interviewees refer to this as a consequence of the delay in applying the territory development guidelines and their articulation with the marketing campaigns.

> TPO1: "I trust that preserving what nature is giving us is a priority. We sell this natural beauty to those who come to the archipelago to visit us, and we wish to continue to do so."
>
> TPO2: "I fear the avalanche of people coming in July and August. I do not think this over-tourism is beneficial to the visitor's experience."

The transformative possibilities of recovery after COVID-19 are delayed by the complexity of interrelated issues in tourism: the biodiversity conservation and climate change agenda and the destination ability to enhance well-being whilst promoting an equitable development of tourism across different geographies (Hall *et al.* 2020). The economic and social reset opposing the perpetual growth model opens a window to reimagine tourism as it is: people live in more remote areas and find sustainable ways to remain in business (Everingham and Chassagne 2020). In tandem, the Azores observatory for tourism is now leading initiatives to expand the idea of the creation of new tourism activities and services (Richards *et al.* 2021) and the development of green areas across the archipelago (OTA 2020). Surprisingly, and to the extent of our knowledge, none of those programs addresses tea and tea production in other areas aside from the existing ones: Gorreana and Porto Formoso.

Tea as a surprise, tea as an experience and tea as a memory

The research identified three central ideas. First, tea is an attraction in the context of so many other things you can do when visiting the Azores. Tea is about extending and tying up memories, connecting recent and remote experiences across the Azores archipelago in hospitality settings. Moreover, tea is used to co-create and sustain the living memory of how people lived in ancient times before tourism came to be in the archipelago: a collective memory of people in the 19th and 20th centuries.

> TPO1: "We trust we are contributing to tourism by sharing the story of tea on this island which is also a bit of the history of my family across several generations. Our factory was one of the first places to welcome tourists, and those remain loyal clients to our tea until today. That loyalty is unprecedented, given it goes across generations. Additionally, I trust tea helps to sell the Azores as a destination too. Tea aligns with the dominant narrative: history and health."
>
> TPO2: "Tea is a surprise for the tourists: as they discover tea, the history, how it is made and the taste of it. We connect that experience with the appreciation of tea, the teacup, and the tea biscuits. Tourists are invited to enjoy an outdoor walk through the plantations and later learn about the manufacturing process. It is a learning experience that culminates with tourists

brewing and tasting tea. Ultimately, they buy some tea as a reminder of their visit to the Azores. They do so to remind themselves and tell others about this beautiful place called the Azores."

The role of souvenirs is described as influential on tourism post-purchase behaviour. The quality of those souvenirs also shapes their perception of the destination (Suhartanto 2018; Sthapit and Björk 2019). In descriptions provided, tea is a representative "memoir" of visitors' stay at the Azores: it brings health and taste in one cup together with pleasant lasting memories. These memories are co-created with tourists participating every year in the annual harvesting of the tea on the first Saturday in May, reinforcing tourists' connection with the local community.

> TPO2: "We are aware of the traditions involved in harvesting tea in the Azores and have been working with ethnographers to recreate every year what would be the harvesting of Tea in the old days. This is an initiative of the local tea brotherhood, which I am proud to say is celebrating the first decade since its inception."

As presented, tea can be a driver of transformation and the co-creation of alternatives for visitors to experience, participate, enjoy and taste tea and recall their visit to Azores (Campos *et al.* 2018). The tourist is involved and participates in harvesting tea leaves and, in doing so, being part of the tea as a way of life: the combination of nature, scenery, history, tea preparation and tasting through the best combination of senses in search for long-lasting memory (Bosse *et al.* 2008; Tung *et al.* 2017).

Conclusion

This study approached this exploratory research with the following research question: "How do tea and tourism support the Azores recovery?" Using as guidance a modified resilience framework, this study identified that the intensity of change associated with short-term transactional costs is determinant to elicit absorptive, adaptive and transformational capacities. Findings confirm previous authors work on the resilience of island communities (Souza *et al.* 2015).

In addition, this study highlights that government incentives of financial nature created a sense of psychological safety at an earlier stage of the pandemic. This sense of safety brought an optimistic perspective towards the future and recovery but delayed somehow the adaptation on the reassurance COVID-19 was a fad. The incentives were welcomed but had limited effects in creating adjustments or transformations. The Azores is an archipelago situated 1500 miles from continental Portugal, so the initial perception was that this pandemic would be a continental problem. When people realised how detrimental COVID-19 was, a state of emergency was declared, and from March until May 2020, the Azoreans felt vulnerable and waited for government intervention.

This study supports the co-existence of the three abilities: absorption, adaptation and transformation interact as capacities yet only manifest within a specific context (Béné *et al.* 2014). Therefore, persistence, incremental adjustment and transformational responses that lead to alterations depend on at what stage support is provided and, if done at the initial stages, transformational responses are limited. Consequently, the study highlights that limited transformational responses occur if there is an established sense of accomplishment. Ultimately, the confluence of resilient traits (the three capacities, Figure 32.1) passed on through generations and learned through exposure to previous challenging times of scarcity and starvation seem to be embedded in contextual cues.

Historical crises create a trilogy of disruption. The onset of tea at the Azores emerged at the end of the orange crisis: starvation, a shortage of resources where usually there is abundance, pressure, the immediate and relentless pursuit for a resolution out of the scarcity and perspective shift; new ways of thinking about the problem thus resulted in a scientific approach to agriculture and the novelty of new crops with good yield.

In response to an agriculture crisis, tea came to be in the Azores in a long process of adoption that had its apex under Jose do Canto, a highly educated farmer and a scientist in this endeavour.

Tea at the Azores beyond the turn of the century was found in all sectors of the economy, tightly connected to tourism, prompting long-lasting memories and co-creating unforgettable experiences for tourists to enjoy, return and recommend the Azores archipelago. Somehow, tea remains and persists as the testimony of how the Azores people have endured and succeeded in difficult times.

Discussion questions

1   What role does tea play on the co-creation of the tea experience, making it unique and appealing to visit the Azores?
2   What was unique about the onset of the three resilient traits in the process of recovery?
3   How can tea plantations, production and commercialization contribute to visitors reflections about the balance between tourism and the preservation of nature?
4   What connects the recovery from the COVID-19 pandemic to tea production in the Azores?
5   How can souvenirs help in recalling the visitor's tea experience, thus extending it to others who may not have visited? Can tea as a souvenir make such a connection? Would this influence their future travel plan decisions?

Acknowledgements

I want to acknowledge the support given by friends Miguel Quental, Cristina Tavares Quental, Luis Pessanha and Gonçalo Oom, and also my wife, Carolina Ferreira de Sousa. Their insight, conversations and distinct views helped me to develop my narrative around resilience, tea and tourism in island communities.

References

Azores Government, 2020a. Resolução do Conselho do Governo n.o 236/2020 de 4 de setembro de 2020. RCM 236/2020.

Azores Government, 2020b. Resolução do Conselho do Governo n.o 168/2020 de 16 de junho de 2020. RCM 168/2020.

Azores Government, G. dos A., 2021. Safe Destination Azores [online]. *Safe Destination Azores*. Available from: https://destinoseguro.azores.gov.pt/ [Accessed 31 July 2021].

Baptista, J., Lima, E., Paiva, L., Andrade, A.L. and Alves, M.G., 2012. Comparison of Azorean tea theanine to teas from other origins by HPLC/DAD/FD. Effect of fermentation, drying temperature, drying time and shoot maturity. *Food Chemistry*, 132(4), 2181–2187.

Baptista, J., Lima, E., Paiva, L. and Castro, A.R., 2014. Value of off-season fresh *Camellia sinensis* leaves. Anti-radical activity, total phenolics content and catechin profiles. *LWT – Food Science and Technology*, 59(2), 1152–1158.

Béné, C., Al-Hassan, R.M., Amarasinghe, O., Fong, P., Ocran, J., Onumah, E., Ratuniata, R., Tuyen, T.V., McGregor, J.A. and Mills, D.J., 2016. Is resilience socially constructed? Empirical evidence from Fiji, Ghana, Sri Lanka, and Vietnam. *Global Environmental Change*, 38, 153–170.

Béné, C., Newsham, A., Davies, M., Ulrichs, M. and Godfrey-Wood, R., 2014. Review article: resilience, poverty and development. *Journal of International Development*, 26(5), 598–623.

Bonfanti, A., Vigolo, V. and Yfantidou, G., 2021. The impact of the Covid-19 pandemic on customer experience design: the hotel managers' perspective. *International Journal of Hospitality Management*, 94, 102871.

Bosse, T., Jonker, C.M. and Treur, J., 2008. Formalisation of Damasio's theory of emotion, feeling and core consciousness. *Consciousness and Cognition*, 17(1), 94–113.

Campos, A.C., Mendes, J., Valle, P.O. and Scott, N., 2018. Co-creation of tourist experiences: a literature review. *Current Issues in Tourism*, 2(4), 369–400.

Dias, F.S., 1995. A importância da 'economia da laranja' no Arquipélago dos Açores durante o século XIX. *ARQUIPÉLAGO – Revista da Universidade dos Açores*, 189–240.

Earth Chek, 2019. The Azores – the world's first certified archipelago [online]. Available from: https://earthcheck.org/news/2019/december/the-azores-the-worlds-first-certified-archipelago/ [Accessed 30 July 2021].

Everingham, P. and Chassagne, N., 2020. Post COVID-19 ecological and social reset: moving away from capitalist growth models towards tourism as Buen Vivir. *Tourism Geographies*, 22(3), 555–566.

Filimonau, V. and Coteau, D.D., 2020. Tourism resilience in the context of integrated destination and disaster management (DM2). *International Journal of Tourism Research*, 22(2), 202–222.

Hall, C.M., Scott, D. and Gössling, S., 2020. Pandemics, transformations and tourism: be careful what you wish for. *Tourism Geographies*, 22(3), 577–598.

Impey, O., 1977. *Chinoiserie: the impact of oriental styles on Western art and decoration*. London: Oxford University Press.

INE, 2021. *Agricultural Statistics – 2020*. Brief Analysis Report.

IPDT, 2015. *Plano estratégico e de marketing do turismo dos Açores*. Portugal: O IPDT – Turismo e Consultoria é reconhecido pela consultoria em turismo, sendo especialista no desenvolvimento de planos estratégicos e de marketing turístico, estratégias de comunicação e promoção turística, processos de certificação de sustentabilidade para o turismo e produção de informação turística. É Membro Afiliado da Organização Mundial de Turismo (OMT) e do Global Sustainable Tourism Council (GSTC) e integra uma rede mundial de organizações públicas e privadas do setor. Research.

Joo, D., Tasci, A.D.A., Woosnam, K.M., Maruyama, N.U., Hollas, C.R. and Aleshinloye, K.D., 2018. Residents' attitude towards domestic tourists explained by contact, emotional solidarity and social distance. *Tourism Management*, 64, 245–257.

LUSA, 2013. Plantações de chá dos Açores produzem 50 toneladas anuais. *Açoriano Oriental*, 13 May.

Moura, M.F. de O., 2019. Introdução da cultura do chá na ilha de São Miguel no século XIX (subsídios históricos). Doctoral Thesis. Universidade dos Açores, São Miguel, Azores.

OECD, 2020. Tourism Policy Responses to the coronavirus (COVID-19) [online]. *OECD*. Available from: www.oecd.org/coronavirus/policy-responses/tourism-policy-responses-to-the-coronavirus-covid-19-6466aa20/ [Accessed 16 July 2020].

OTA, A.T.O., 2020. Projeto I&D Green GA [online]. *Observatório do Turismo dos Açores*. Available from: https://www.otacores.com/greenga/ [Accessed 9 Aug 2021].

OTA, O. do T. dos A., 2021. COVID-19: evolução e impactos [online]. *Observatório do Turismo dos Açores*. Available from: https://otacores.com/covid-19-evolucao-e-impactos/ [Accessed 31 July 2021].

Paiva, L., Rego, C., Lima, E., Marcone, M. and Baptista, J., 2021. Comparative analysis of the polyphenols, caffeine, and antioxidant activities of Green Tea, White Tea, and Flowers from Azorean *Camellia sinensis* varieties affected by different harvested and processing conditions. *Antioxidants*, 10(2), 183.

Quintela, M.M., 2004. Cura termal: entre as práticas 'populares' e os saberes 'científicos'. Presented at the VIII congresso Luso-Afro-Brasileiro de Ciências Sociais, Coimbra, 22.

Quitério, J., 1987. *Livro de bem comer: crónicas de gastronomia portuguesa*. Lisboa: Assírio & Alvim.

Richards, G., Qu, M., Huhmarniemi, M., Baptista, M., Henriques, N., Ormond, M., Hawes, F.-M. and Duxbury, N., 2021. CREATOUR International Webinar – Highlighting distinctiveness: Connecting travel to community and sense of place.

Riley, C.G., 2001. José do Canto, um gentleman farmer açoriano. *Análise Social*, 36(160), 685–709.

RTP Play, 2017. Episode 1, Fabrico Nacional Episódio 1 – de 18 Mai 2017 – RTP Play – RTP. Video, Radio Teledifusão Portuguesa. May 18.

Souza, R.D., Henly-Shepard, S., McNamara, K. and Fernando, N., 2015. Re-framing island nations as champions of resilience in the face of climate change and disaster risk. UNU-EHS Working Paper series.

SREA, 2021. Azores Tourism Dashboard [online]. Available from: https://app.powerbi.com/view?r=eyJrIjoiODAyNTRkZGQtNmRmOC00OWM5LWE4ZjctZGRjNjZkY2UwMTdiIiwidCI6IjE0YWI3NzE4LTNlNzEtNDAxOS04OTBhLTU0ZWQ5YjkyZjk4YSIsImMiOjh9 [Accessed 31 July 2021].

Sthapit, E. and Björk, P., 2019. Relative contributions of souvenirs on memorability of a trip experience and revisit intention: a study of visitors to Rovaniemi, Finland. *Scandinavian Journal of Hospitality and Tourism*, 19(1), 1–26.

Suhartanto, D., 2018. Tourist satisfaction with souvenir shopping: evidence from Indonesian domestic tourists. *Current Issues in Tourism*, 21(6), 663–679.

Tung, V.W.S., Lin, P., Qiu Zhang, H. and Zhao, A., 2017. A framework of memory management and tourism experiences. *Journal of Travel & Tourism Marketing*, 34(7), 853–866.

WHO, 2021. Coronavirus disease (COVID-19) – World Health Organization [online]. Available from: www.who.int/emergencies/diseases/novel-coronavirus-2019 [Accessed 8 August 2021].

EPILOGUE

Li-Hsin Chen and Amnaj Khaokhrueamuang

Tea has been appreciated as one of the most venerable and prevalent beverages for generations. The way people make and consume tea, along with its aesthetic values, have created refined and profound tea traditions and cultures globally. For example, "tea time" is an essential feature in the British way of life. In Iran, almost every neighbourhood has a tea house where residents discuss their thoughts, problems, or business; it is said that the country's independence began with tea. In Japan, the art of the centuries-old tea ceremony (Chado or Sado) carries many social values and spiritual meanings. In tea-farming regions such as Sri Lanka, India, and Kenya, the tea industry embraces many tangible and intangible heritage and cultural aspects while providing livelihoods for millions of workers. Tea is the most popular beverage in the day-to-day life of people and is a drink that many people cannot live without. Although the long history implied that tea drinking is a particular activity for seniors, the United States has recently promoted iced tea to international markets and made tea popular among younger generations (Fact MR 2021). Also, Taiwanese bubble tea (also called "black pearl tea" or "boba tea") creates a global addiction and makes Millennials and Generation Z big fans of this tea beverage (Purswani 2021). The newly established luxury hotel, The Place Taipei, embodies local Pouchong tea tradition and the tea production process to make the interior design more unique and attractive for cultural tourists (Meccano 2022). The diverse rituals and cultures related to tea drinking demonstrate that this age-old beverage is inseparable from the sociocultural and economic being of human society.

By and large, tea is not just a daily beverage; instead, tea drinking is a social experience that allows people to learn about another culture and explore new knowledge. To some extent, tourism and tea share the same features: both are significant sources of social and cultural capital for many regions. However, in terms of economic values, tourism undeniably can provide alternative income sources and training opportunities to tea farmers. On the other hand, tea can be integrated into touristic and hospitality activities and assist tour operators or destination management organisations (DMOs) in creating unique selling points for tourists. Therefore, tea tourism, just like coffee or wine tourism, should not be viewed merely as a combination of tea and tourism. The multiplier effects of tea tourism should be valued because it can be a driving force for preserving culture and heritage, conserving the ecological system, and improving the living conditions for workers connected to the tea industry (Chen *et al.* 2021a).

Despite developing tea tourism being more accessible and cost-effective than other economic activities, such as establishing the manufacturing industry in rural areas, DMOs, tea tour operators, and farmers encounter many challenges and problems when promoting such tourism products. For example, Mondal and Samaddar (2021) identified several issues related to tea tourism in India, such as unsuitable planning and marketing efforts, inadequate collaboration among stakeholders, lack of involvement of residents, socioeconomic inequality between farmers and owners, and consumer's attitudes. Doctor's (2020) report about Darjeeling in India demonstrates the competition for land usage between tea tourism and tea farming. As the state government permitted a part of tea plantations for touristic purposes, many doubts have been raised about the proper use of tea gardens as merely the backdrop to tourism. These challenges become more severe when the COVID-19 lockdown hits tea-farming areas. Although some destinations attempt to use tea tourism to compensate for the loss of profit and the decline of the tea business caused by the pandemic, we must ask if developing tea tourism has any meaning without the improvement of the quality of tea livelihoods and the well-being of tea farmers or workers.

On top of these operational issues, tea tourism as a research subject has not been explored abundantly yet compared to other special-interest tourism, such as wine tourism or culinary tourism. Given such an investigation is at its beginning stage, Chen *et al.* (2021) pointed out that most of the existing research focuses on a particular place or single stakeholder without cross-case comparisons. Furthermore, most tea tourism studies tend to be cross-sectional and descriptive without a theoretical justification. As a result, the existing body of knowledge is relatively heterogeneous. Thus, they call for more holistic, interdisciplinary, and longitudinal research to advance the understanding of tea tourism.

Being aware of this gap, this *Handbook* was intended to provide contemporary viewpoints and a wide range of perspectives on the various features of tea tourism. Although most chapters presented in this *Handbook* did not use complicated research methods, they have covered topics and cases that are not yet well investigated in the literature. Furthermore, the multidisciplinary scopes adopted by many authors have also offered fresh thinking and innovative theoretical lenses to enrich the understanding and study of tea tourism. Besides, this *Handbook* was designed not just for academics but also for practitioners. Reading through chapters, the readers will be able to learn of the opportunities and challenges brought by tea tourism globally. Case studies across five continents enable tea business owners, marketers, and policymakers to compare the strengths and weaknesses among various examples and to find a model as the benchmark for their interests. Despite the initial intention of this volume being to comprehensively and systematically examine the diverse dimensions of tea tourism, some research themes regarding the intersections of tea and tourism have not been fully covered as the global situation changes rapidly. Thus, this concluding chapter provides suggestions for areas that need further research attention and sets future agendas for policy and practice. Specifically, six themes are identified and discussed: (1) theoretical foundations; (2) nature disasters and crisis management; (3) responsible destination management and marketing; (4) the use of information and communications technology in tea tourism; (5) emerging markets and innovative products in tea tourism; and (6) tea tourism and peace.

Theoretical foundations

The theoretical foundations of tea tourism are mainly based on theories from the tourism field as applied to this form of touristic activity. The knowledge and theories about tea, especially in terms of history, development, production, consumption, and use, provide an enhancement to these foundations. This volume has also introduced the concept of tea tourism being a part of the formalised study of tea, as reflected by the knowledge system of teaics, a logical, objective, and structured framework of knowledge relating to tea (see Chapter 8 by Brian Park).

Researchers focusing on tea tourism in the last few decades have taken a number of theoretical approaches to their work, employing various quantitative and qualitative methods in forms of independent study and research that do not seem to represent a systematic approach. More recently, some of this knowledge has been reviewed in a more in-depth manner, as with a systematic review of the literature (Chen *et al.* 2021) and a paper calling for improved methodological approaches (Mondal and Samaddar 2021). While this volume has added to the knowledge base and employed various theoretical approaches, it has perhaps not taken a systematic approach, due to the number of topics and chapter authors involved in the endeavour. Also, while the call for contributions document took a systematic approach to topic inclusion, within the response it was not possible to have complete coverage of all of the areas initially identified. There is thus an opportunity in the future for the proposal of a more systematic classification of both what is known about tea tourism and the methods and theories that can be employed to implement research-based on such a taxonomy. In conclusion, within this volume theoretical issues regarding tea tourism are introduced and discussed from a variety of perspectives, while working towards a more comprehensive theoretical and methodological foundation for this form of tourism.

Natural disasters and crisis management

Natural disasters such as earthquakes, tsunamis, hurricanes, tornadoes, floods, and landslides are created by the forces of nature and may result in serious damages and deaths. The tea regions of the world are located in natural areas including both mountainous, hilly, and coastal regions, often in tropical or sub-tropical climatic zones that may be subject to natural hazards and disasters. The tea plant (*Camellia sinensis*) as part of agriculture is also affected by pests, diseases, and frost. Natural disasters occurring in the tea-producing areas of the world can thus possibly threaten or even halt the development of tea tourism in affected regions. Such disasters can also affect the global supply of tea, threatening both communities in producing countries and impacting the availability of tea for hospitality operations and experiential tea settings in consuming countries. In these contexts, practical research into crisis management strategies in tea tourism settings that will be useful for all stakeholders is needed. In particular, planning for post-disaster recovery requires a thorough investigation and analysis to ensure the efficient and effective sustainable tourism practice in tea regions. Crisis management is also important in other situations, such as in the COVID-19 pandemic period when many tea tourism entities were forced to pivot their operations, creating new services and products. This trend was facilitated by technology as operators offered virtual tea tastings, tours and classes.

Responsible destination management and marketing

The economic contributions and social benefits of tea tourism were discussed in many chapters in this *Handbook*. However, the pro-profit mindset of DMOs in tea tourism destinations has been challenged recently. We can observe that some owners of tea gardens try to use tourism to increase their revenues. Still, a few of the tour guides are either not from the local community or not appropriately trained to provide meaningful and in-depth learning experiences for tourists. Thus, tourists cannot truly appreciate the importance of the tea industry. As a result, the activity's service quality and satisfaction level of tourists may not be so good to generate memorable tourism experiences, positive destination image, and loyalty. Planners and operators should be prepared to mitigate the negative impacts of tea tourism on tea-farming areas before promoting such tourism products.

As tea-farming regions usually have rich cultural assets and grand scenery, the concept and the practices of responsible travel and tourism can be a feasible alternative to maintain the economic and social benefits of tea tourism while preserving the heritage and environments of the destinations. When developing any tourism product, responsible tourism means there should be meaningful involvement of all stakeholders, including residents, business owners, tourists, and governments, to fulfil the UN Sustainable Development Goals to maximise benefits for everyone. Furthermore, all stakeholders should perform ethical and respectful behaviours in the destination with achieving economic, social, and environmental responsibility in mind (Chen *et al.* 2021b). Although the concept of responsible tourism has been applied to investigate hotel management and some niche tourism markets such as volunteer tourism, marine tourism, community-based tourism, or ecotourism, it has been rarely explored in the context of tea tourism. Developing the responsible framework and guidelines suitable for tea tourism is required to attract contemporary tourists with high environmental awareness. It is time to advance the destination management and marketing paradigm in tea tourism.

Information and communications technology in tea tourism

For planners and tour operators, how to make tea tourism more accessible for everyone is always a critical issue. Given most of the tea-growing areas are in remote areas, it is not easy for tourists to reach those places if there is no proper transportation. The internet, the innovation of information and communications technologies (ICTs), and other digital applications such as virtual reality can be used to provide tourism experiences even though tourists cannot be present at the sites. The contributors in this book have discussed the usage of social media to engage tea tourists. Still, there is a multitude of other ICTs that have not been investigated extensively in the tea-related literature. Before COVID-19, some of tea tourists' attractions used digital marketing tools such as websites, emails, Facebook, or Instagram to provide information and obtain feedback from customers, but the majority of them did not respond to the evolution and opportunities provided by ICTs to offer an augmented and supplemented experience for tea tourists. The exploration of innovative technology seems to be motivated by the social-distancing and lockdown policies during the COVID-19 pandemic. Among all applications, virtual reality is the most broadly used technology to allow tea tourists to visit the tea farms and learn the process of making tea. For example, Obubu Tea in Japan offered online tea tours to allow tourists to experience the tea fields in Wazuka, a tea-farming community in Kyoto Prefecture. The organiser sends tea wares and tea samples to the tourists before the virtual tour. During this 1.5-hour journey, the tourists can learn about tea farming, processing, tasting, and interact with local tea farmers (Kyoto Obubu Tea Farms 2021). The main purpose of such a tour is to increase the awareness of the brand and to simulate potential customers' interests. This experience may motivate them to pay a real visit to the destination after the COVID-19 pandemic.

Since smart tourism has become an important tourism product, the DMOs or owners of the tea tourism business should make substantial investments in the improvement of ICTs to remain relevant in this competing market. The usage of digital tools to enhance the tourism experiences during the pre-visit, on-site, and post-visit stages of tea tours should be carefully designed. To support the tea tourism destinations and businesses in adapting such trends and modifying the management strategy, more in-depth studies are needed to explore the challenges and benefits of this transition process. Furthermore, tea tourists' behaviours in the digital world should also be examined. For example, future studies may focus on the characteristics of electronic word of mouth or the immersive experiences in virtual tours or the metaverse, which are essential to understand modern tourists, especially Millennials and Generation Z. Compared to the extensive research in technological applications in the wine tourism context (Zamarreno Aramendia *et al.* 2021), a dearth of research is found regarding the exploration of smart tourism theories and practices in tea tourism. Nevertheless, the absence of such investigation should not be regarded as a weakness but rather a research opportunity to explore the unknown knowledge that can be used to benefit the practitioners.

Emerging markets and innovative products in tea tourism

Besides ICT, the global tea industry involves innovations in developing tea products and services that directly contribute to the tourism business. For example, tea entrepreneurs in Japan produce

Japanese black teas with their unique tastes and offer tea tourism experiences with the authenticity of the products. Innovation also applies creative ideas to tea products and services, such as tea-based cuisine with storytelling, making the tourism experience more valuable.

Tea tourism in tea-growing countries encourages linking tea-farming communities in rural areas with urban visitors by providing pastoral and urban tourism experiences. This linkage requires innovative marketing strategies relating to heritage, agriculture, and gastronomy trials, such as tea cafés and tea-processing factories. But this contribution has less attention in taking this advantage of developing the tea industry. Even in tea consumption countries, tea is part of gastronomy and tourism with different tea cultural heritage. However, drawing attention to taking those tea cultures to attract tourists in tea tourism is not much mentioned. Future research, therefore, should consider this notion with innovative issues to enhance a new form of tea tourism in tea-consuming destinations.

Accordingly, the emerging tea tourism market intends to connect tea-growing communities with urban associations and tea-consuming destinations through innovative products and creative ideas to differentiate the regions. This phenomenon has occurred in some tea-growing areas, for example, an effort to develop tea tour guides in Turkey and linking urban restaurants and tea cafes to rural tea gardens through tea trails in Japan. This trend emerges from the notion of branding a tea tourism destination, an essential tool for marketing tea tourism and tea industry communities.

Tea tourism and peace

As the editors prepare the final manuscript, it seems appropriate, if not essential, to add these remarks regarding tea tourism and peace. Tea tourism offers a unique opportunity to be employed as a means to foster peace and understanding. Tea is acknowledged to be a sign of hospitality, as well as a welcome gift, and is offered within many cultures and traditions as a means of fostering a calming and peaceful situation for meetings, discussions, and negotiations. In many cultures, tea is often linked with philosophy and a peaceful mind. When we want to stay away from worldly disputes and calm our thoughts down, drinking a cup of tea leads us into a world of quiet contemplation. The ancient Chinese poet Lo Tung said, "When I drink tea I am conscious of peace. The cool breath of heaven rises in my sleeves, and blows my cares away." The philosophy of drinking tea, or "teaism," should be practised more in modern society to sustain mindfulness in every moment of daily life.

Tourism is known to be instrumental in increasing understanding across cultures, as evidenced by the work of the International Institute for Peace Tourism, whose founder noted that tourism transcends government boundaries, bringing the peoples of the world closer together through an understanding of different cultures, environments, and heritage (D'Amore 1988). As the editors complete their work for this volume during the first few months of the Russia-Ukraine War (2022), it is more important than ever to note the great potential that tea tourism has to contribute to world peace. The war is also disrupting both the flows of tea and tourism to both countries. As global trade is impacted, tea producers no longer able to export to Russia (e.g., formerly 13% of tea exports went to India) may need to look to alternate revenue sources if the war continues (Shrivastava 2022). Domestic tea tourism in exporting countries could possibly address some of these losses.

Final remarks

This concluding chapter reflects on what has been done and asks what else should be done to advance our understanding and studying of tea tourism. Although this *Handbook* was planned to be as comprehensive in covering all relevant topics about tea tourism, some subjects and regions are inevitably missed in this volume. Thus, we highlight six areas that are required for further explorations. Furthermore, most of the chapters in this *Handbook* focused on tea-producing areas more than tea-consumption places. This limitation offers abundant opportunities for future research. Last, we hope the findings identified from these 32 chapters provide helpful solutions and examples for practitioners. As the world is expecting the recovery of tourism after the COVID-19 pandemic, all stakeholders involved in tea tourism should be well prepared to welcome tourists who seek memorable tea tourism experiences. The editors thank the contributors for their valuable insights and hope the *Handbook* stimulates broader and deeper discussions in this emerging research area.

References

Chen, L.-H., Hsu, C.S., Muñoza, K.E. and Aye, N., 2021b. The industry-academia gap in responsible tourism management: an automated content analysis. *Journal of Responsible Tourism Management*, 1(2), 21–42.

Chen, L.-H., Wang, M.-J., Morrison, A.M., Ting, H. and Yeap, J.A.L., 2021a. Guest editorial. *International Journal of Culture, Tourism and Hospitality Research*, 15(3), 285–289.

Chen, S.-H., Huang, J. and Tham, A., 2021. A systematic literature review of coffee and tea tourism. *International Journal of Culture, Tourism and Hospitality Research*, 15(3), 290–311.

D'Amore, L., 1988. Tourism-the world's peace industry. *Journal of Travel Research*, 27(1), 35–40.

Doctor, V., 2021. Save the brew: Tourism will help Darjeeling, but the world-renowned tea shouldn't be flushed down in the process. *The Economic Times*, 12 July 2020. Available from: https://economictimes.indiatimes.com/magazines/panache/save-the-brew-tourism-will-help-darjeeling-but-the-world-renowned-tea-shouldnt-be-flushed-down-in-the-process/articleshow/76918179.cms [Accessed 4 February 2022].

Fact MR, 2021. Ice tea market [online]. *Fact MR*. Available from: www.factmr.com/report/ice-tea-market [Accessed 25 March 2022].

Kyoto Obubu Tea Farms, 2021. Online tea tour [online]. Available from: https://obubutea.com/online-tea-tour/ [Accessed 5 February 2022].

Meccano, 2022. The place Taipei [online]. Available from: www.mecanoo.nl/Projects/project/241/The-Place-Taipei [Accessed 25 March 2022].

Mondal, S. and Samaddar, K., 2021. Exploring the current issues, challenges and opportunities in tea tourism: a morphological analysis. *International Journal of Culture, Tourism and Hospitality Research*, 15(3), 312–327.

Purswani, T., 2021. Survey: Millennials, Gen Z big fans of bubble tea [online]. Available from: https://foodinstitute.com/focus/survey-millennials-gen-z-big-fans-of-bubble-tea/ [Accessed 4 February 2022].

Shrivastava, R., 2022. Russia-Ukraine war impact: India eyes increase in wheat exports. *India Today*. Available from: www.indiatoday.in/world/russia-ukraine-war/story/russia-ukraine-war-impact-india-eyes-increase-wheat-exports-tea-metals-1926391-2022-03-17 [Accessed 26 March 2022].

Zamarreno Aramendia, G., Cruz Ruiz, E. and Hernando Nieto, C., 2021. Digitalization of the wine tourism experience: a literature review and practical applications. *Doxa Comunicacion*, 33, 257–283.

INDEX

Page numbers in italic indicate a figure, and page numbers in bold indicate a table on the corresponding page.

Steiner, C. 135
Steyn, L.L. 92
stimulation 24, 109, 166, 169, 171, 172
storytelling 176
structural equation modelling (SEM) 198, 199,
    200–201, *200*, **201**
supporting facilities 197, 198, 201, 202, 203
supporting industries 216, 220–221
Suranga Silva, D.A.C. 7
sustainability 141–144
Suvantola, J. 126
Su, X 307
Swarbrooke, J. 90
SWOT analysis 53–57, **53**
symbiosis 100
Systrom, K. 175

Taiwan 49, 165–174
Takano, R. 8, 277
Takeguchi, A. 5, 48
Tamil community 135–136
Tan, A.Ö. 108
Tang dynasty 15, 17, 18
Taoism 16–17, 18, 21
tapioca tea *see* pearl milk tea
Tasci, A.D. 165
tea: consumption 1, 61, 108, 113; culture 1–2, 16,
    62–63; heritage 1–2; history 2, **2**; importance
    of 1–2; overview 1; preparation 20, 61, 62–63,
    112–113; processing 20–21, 23, 33, 102;
    traditions 61–62
Tea and Tourism (Jolliffe) 68
tea art 165
tea bars 120
tea cafés 259–268; categories of 262, **262**, 263–264,
    266; events **265**; literature review 259–261;
    methodology 261; nature tourism **264**, *266*;
    overview 259; routes 264–266; Shizuoka
    260–267, *260*, **262**, **265**; *see also* tea rooms
tea culture: dimensions of 118; social science and
    82
tea cups/glasses *110*
tea dictionary 114
tea factory tourism 166–174
tea gardens 108, 112, 124–125; Dooars, India 301,
    301; Meghalaya, India 127–133; Shizuoka, Japan
    271
teaics 76–88; definition 76; development of 77;
    disciplines of 77–78, **78–82**, 82–84; history 76;
    as knowledge framework 77; literature review
    76–77; methodology 77; overview 76; seon 83,
    84; tea tourism and 84–87
tea-infused meals 121
tea lounges 120
tea plucking 121, 122
teapots 111
tea roads/routes: Ancient Tea Horse Road 2, 23–30;
    case study 25–29; Great Tea Road 24; literature

review 24; methodology 25; overview 23–24; tea
    cafés 264–266
tea rooms 157–164; *see also* tea cafés
tea routes *see* tea roads/routes
tea spaces *see* tea cafés; tea rooms
tea tasting 122
tea time 327
tea tourism: challenges 327–328; definition 21, 76,
    90, 100, 117, 119, 214, 237, 259; description 3;
    development 27, 44, 53, 94, 215; origins 15–22;
    perceived value of 196; research on 328–331;
    stages of 207; theories 4; types of 118; *see also*
    tourism
tea types 327; black 239, 277–286; Ceylon 217, 221;
    Chinese 99; Darjeeling 2, 40; green 224–233;
    Oolong 40; Panchen Tuo 33–37; pearl milk
    (bubble, boba, tapioca) 165, 168–172; Puer 23;
    Rooibos 92–93; Shan Tuyet 39; Snow Shan
    39–47; wild 39–47, 99
tea villages: Huay Nam Guen village 102–105;
    Kandapola Village House 140–141, *142*, *143*;
    Nakeli village 23–30
technology 330
Telfer, D.J. 89
Thailand 99–107, 247, 249, 251–252, 256
theories *198*; development of 4, 77; embodiment
    31–32, 36; orchestra model 165, 166, *167*;
    Porter's diamond model 214–223; practice 109;
    teaics **78–82**, 84; tea tourism **85–87**
Tibet 33, 34, 35
Tokyo 208, 209, 210, 211
toolkit 193
Top Tea Place 157, 161
tour guides: curriculum **294**; definition 288;
    development of 292–293; employment issues 149,
    150; Iran 287–296; qualifications for 288–289;
    training 151
tourism: definition 214; destination 70; domestic 51,
    68; indigenous 39–47; niche 90, 185, 186–187;
    orchestra model 165, 166, *167*; religious 31–38;
    rural 126–127; tea factory 166–174; types of 3–4,
    90, 126; *see also* tea tourism
touristic modernity 28
tourists: allocentric 207, 212, 259; behaviour 170;
    motivations 225–226, 259–260; psychocentric
    207, 259
Traditional Tea-grass Integrated System (Chagusaba)
    method 269, 271, **272**, *273*
traditions 61–62
training 151, 271
travel: Ancient Tea Horse Road 2; history 2–3;
    spiritual 31–38; *see also* tourism
Turkey 108–116; case studies 239–241; ethnographic
    features 110–112; folkloric features 112–114;
    gastronomy 237–246; methodology 109–110,
    238–239; overview 108, 237; poster *241*; role of
    tea 288; sociocultural values 114; socioecological
    perspective 109; tea dictionary 114; tea harvest

For Product Safety Concerns and Information please contact our EU
representative GPSR@taylorandfrancis.com
Taylor & Francis Verlag GmbH, Kaufingerstraße 24, 80331 München, Germany

www.ingramcontent.com/pod-product-compliance
Lightning Source LLC
Chambersburg PA
CBHW081047220326
41598CB00038B/7009